Water Quality in Catchment Ecosystems

THE INSTITUTION OF ENVIRONMENTAL SCIENCES SERIES

Food Additives
R. Taylor

Water Quality in Catchment Ecosystems
A. M. Gower

Environmental Pollution
H. M. Dix

Water Quality in Catchment Ecosystems

Edited by
A. M. Gower
School of Environmental Sciences
Plymouth Polytechnic
Devon

JOHN WILEY & SONS
Chichester · New York · Brisbane · Toronto

British Library Cataloguing in Publication Data:

Water quality in catchment ecosystems.—
(Institution of Environmental Sciences. Series).
1. Water quality
2. Aquatic ecology
I. Gower, A M
II. Series
628.1′61 TD370 79-42907

ISBN 0 471 27692 8

Photoset and printed in Malta by Interprint Limited

Contributors

R. C. Ballance — *Division of Environmental Health, World Health Organization, 1211 Geneva 27, Switzerland.*

M. S. Farrimond — *Severn–Trent Water Authority, Abelson House, 2297 Coventry Road, Sheldon, Birmingham B26 3PU, UK.*

J. K. R. Gasser — *Agricultural Research Council, 160 Great Portland Street, London W1N 6DT, UK.*

A. M. Gower — *School of Environmental Sciences, Plymouth Polytechnic, Plymouth PL4 8AA, UK.*

A. James — *Division of Public Health Engineering, Department of Civil Engineering, University of Newcastle-upon-Tyne, Newcastle-upon-Tyne NE1 7RU, UK.*

B. H. Olson — *Programme in Social Ecology, University of California, Irvine, California 92717, USA.*

T. O'Riordan — *School of Environmental Sciences, University of East Anglia, Norwich NR4 7TJ, UK.*

W. Pope — *Public Health and Water Engineering Section, Department of Civil Engineering, Imperial College of Science and Technology, London SW7 2AZ, UK.*

D. C. Renshaw — *Department of the Environment, 2 Marsham Street, London SW1P 3EB, UK.*

D. Sheldon — *Thames Water Authority, New River Head, Rosebery Avenue, London EC1R 4TP, UK.*

D. J. Storey — *Centre for Environmental Studies, 62 Chandos Place, London WC2N 4HH, UK.*

D. E. Walling — *Department of Geography, University of Exeter, Exeter EX4 4RJ, UK.*

L. B. Wood — *Thames Water Authority, New River Head, Rosebery Avenue, London EC1R 4TP, UK.*

Contents

Contents

Introduction to Series

In the last two decades the environment has figured increasingly in our everyday discussions. From a matter of superficial interest to all but an enlightened few, it has become not just the calling of professors and politicians, but the concern of parliaments and the responsibility of us all. This explosive increase of interest has been associated with an ever-widening accumulation of related environmental facts. This series seeks to bring together these facts, and allied ideas, so that they may be readily accessible in a number of identifiable volumes. The Institution of Environmental Sciences in promoting the series has, as an objective, the express desire to diffuse a balanced professional view on all matters relating to the environment.

The environment now figures prominently as a curriculum subject and Environmental Science is established as a school and university examination subject at all levels. The need for a comprehensive series of volumes which disseminate the wealth of information relating to this science has become of paramount importance. The series should be of value both to students of the environment and to professionals working in related fields.

PETER FARAGO
DAVID HUGHES-EVANS
JOHN F. POTTER

Publications Committee,
Institution of Environmental Sciences,
14 Princes Gate,
Hyde Park,
London, SW7 1PU.

Preface

Water quality is a broad topic which is currently attracting considerable interest, but the relevant aspects are seldom brought together under one heading; this book is designed to meet such a need.

A study of water quality is fundamental to understanding the water resource, as it gives insight into the benefits to be gained from water management and the consequences of its mismanagement. There is increasing public interest in water as a resource, and an increasing awareness of the need to protect water quality. No doubt this is due, in part, to the demand for higher standards of living and, also, to an increase in the time devoted to leisure activities. The study of water quality is also of enormous educational value, especially for environmental science courses which demand a more interaction-orientated approach—for the quality of water is dictated by a pattern of interactions, rather than by any factors operating in isolation.

The processes that contribute to water quality can best be understood if they are examined in the context of a catchment ecosystem. This consists of a river or lake together with its associated drainage area, and is an integrated complex of non-living (abiotic) and living (biotic, including man) components and their interactions. Recognition of the catchment as an interrelated system is particularly instructive as it contains many of the factors that are relevant to the control and management of water quality. The hydrological cycle, which may also be considered in the catchment context, links the factors influencing water quality; the characteristics acquired by water during its passage through the catchment are due to chemical, physical, and biological processes, and the final adjustment occurs within the water body (river, lake) itself. There are of course some factors, including certain atmospheric influences and many of man's social, economic, and political activities, which can act across catchment boundaries.

Man modifies water quality not only directly through discharges but also indirectly through such activities as adjusting flow regimes, impounding and abstracting. Clearly, man is a major biotic influence, but it is also important to realize that his activities in a catchment modify a system which is already dynamic and varying. Nevertheless, the impact of man emphasizes the need for careful management, but any manipulation needs to be planned with reference to the whole catchment ecosystem in which land and water development are carefully coordinated.

In this monograph, the chapters are arranged to facilitate continuity and to provide a logical progression of topics. Chapter 1 discusses the way in which background water quality is governed by catchment processes and characteristics and sets the scene for the rest of the book. Chapters 2, 3, and 4 deal with man's impact in catchments and the associated modifications of water quality. The ecological effects of some of these modifications are considered in Chapter 5, followed by the health implications of waters of varying quality characteristics (Chapter 6). Water quality control (Chapter 7) includes methods of treatment and monitoring together with relevant legislation—which leads into an assessment of legislation and the setting of water quality objectives in the UK and EEC in Chapter 8. Economic aspects are dealt with in Chapter 9, which examines the costs of pollution and assesses policies for achieving a desirable water quality. Important developments in mathematical modelling and their application to water quality management are described in Chapter 10, and the monograph concludes (Chapter 11) with a discussion of the relationship between water quality and planning.

It is hoped that the reader will be stimulated by the approach adopted in this monograph, i.e. that of focusing on the quality of an essential but vulnerable resource, and by the material provided by the contributing authors. Although much of what is written is up-to-date at the time of going to press, no doubt some of the information will eventually become dated, but it is hoped that lasting value will be gained from the insight given into many of the processes, principles, and issues associated with water quality. This should be a timely publication in view of the current interest in water, and may provide the incentive for further integration of water quality data.

Finally, I should like to express my gratitude to the authors for contributing to the monograph and for their subsequent patience during editing. In addition, I should like to thank those people whose help and advice have been of great value in the production of this monograph.

Anthony M. Gower

(Note that metric units are used throughout.)

CHAPTER 1

Water in the Catchment Ecosystem

D. E. Walling

'I think it is time that hydrologists admit that water has quality as well as quantity dimensions' (Ackermann, 1969). This statement was made during an International Hydrology Seminar held at Urbana, Illinois, in July 1969. The speaker was attempting to review recent trends in the science of hydrology and to highlight profitable directions for future work. Thus, while it was possible to refer to the considerable developments that had occurred in understanding the processes, volumes, and rates involved in the movement of water through the hydrological cycle (i.e. the quantity dimension), much less was known about the natural quality of that water, particularly the changes associated with the various hydrological processes operating within a drainage basin and the precise controls on both temporal and spatial variations. A considerable upsurge of activity has subsequently marked the 1970s, and the hydrologist's knowledge of this quality dimension has been much improved. To some extent this is due to his own work, but important contributions have also been made by workers in adjacent disciplines.

This chapter will attempt to provide a hydrological overview of the current understanding of the quality dimension of the hydrological processes operating within the catchment ecosystem. The treatment will be focused primarily on natural or background water quality, although it must be recognized that few areas will be entirely unaffected by human activity, and both these effects and the pathways by which diffuse source pollutants may move through the system to the stream will be briefly considered. Attention will be given first to the various processes involved in the movement of water through a drainage basin and the quality dimension of those processes; secondly, to the resultant spatial and temporal variation in stream water quality; thirdly, to the impact of human activity on the system; and, finally, to the interaction of diffuse source pollutants with the processes of water movement. For the purposes of this account, a restricted range of water quality parameters will be discussed, and emphasis will be placed on the chemical parameters of inorganic ion concentrations and dissolved oxygen content and on the physical properties of suspended sediment concentration, or turbidity, and temperature.

HYDROLOGICAL PROCESSES AND THE QUALITY DIMENSION

Figure 1.1(a) presents a simplified view of the hydrological processes which operate in a catchment ecosystem. Inputs of precipitation are distributed through a number of stores by a series of transfer processes and are output as channel flow, evapotranspiration, and deep leakage. Thus, in simple terms, water entering a catchment is intercepted by vegetation and subsequently evaporated, or reaches the ground surface as through-fall and stem-flow. In the absence of vegetation, the precipitation reaches the ground directly, although interception by surface litter may still occur. At the soil surface, water infiltrates into the soil or is retained in surface storage. The surface water may move downslope as surface run-off or be slowly evaporated. Moisture held within the soil is subject to surface evaporation and plant transpiration, to downward percolation to the water table, and to downslope movement as through-flow and inter-flow. Water held within the groundwater body is similarly subject to losses by plant transpiration and evaporation when the water table is close to the surface, to upward movement into the unsaturated soil by capillary action, to deep leakage to adjoining groundwater bodies, and to slow surface outflow into springs, seeps, and river channels as base-flow. The river channel receives variable contributions of surface run-off, through-flow and inter-flow, and base-flow which are routed downstream and together contribute to the time-variant outflow hydrograph of the drainage basin.

As water moves through this system of stores and transfer pathways, its quality is modified, and the quality of the water leaving the drainage basin will be considerably different from that of the moisture entering the catchment as precipitation. Figure 1.1(b) attempts to portray a number of the more important mechanisms that modify the water quality, and this representation can be viewed as significant in two respects. First, it indicates that water quality will change as moisture moves through the catchment system and that water characterizing different stages of the hydrological cycle will exhibit different properties. Secondly, it demonstrates that the quality of the water flowing in a river or stream will have evolved through a complex series of interactions with the soil, rock, and biota of the catchment ecosystem, and that it will be necessary to consider these interactions if spatial and temporal variations in stream quality are to be fully understood. Furthermore, it must be appreciated that the various processes controlling stream water quality may be in a delicate balance and that a slight modification to the catchment, such as a change in land use, could generate significant changes in water quality.

Precipitation Inputs

Elaboration of the various mechanisms contributing to stream water quality shown in Figure 1.1(b) can usefully commence by considering the chemical composition of precipitation inputs to the drainage basin. This water will not be chemically pure but will contain material in solution. Table 1.1 lists some typical concentrations which have been reported for rainfall over the UK (Stevenson, 1968). Further discussion of precipitation quality must, however, clarify the precise nature of data reported in the literature. In many cases, measurements are made using a simple open collector emptied at daily or weekly intervals, and it must be recognized that both wet precipitation and dry fall-out will be collected. Chemical analysis of the resultant moisture will therefore reflect both components. Whitehead and Feth (1964) have termed this measurement one of *bulk precipitation* quality and, although it is of great interest in nutrient and geochemical budget studies where estimates of total atmospheric input are required, it is of less relevance to a study of water quality where information on wet precipitation is of more significance. The values cited in Table 1.1 are, in fact, values of bulk precipitation chemistry, and a study undertaken by Cawse (1977) at Chilton, Oxfordshire, suggests that concentrations of Na^+, Ca^{2+}, and Cl^- in wet precipitation may be less than 50% of these figures.

Whereas dry fall-out is associated with fall-out due to gravity and primarily involves relatively large particles (>20 μm diameter), the chemical composition of wet precipitation is commonly ascribed to rain-out and wash-out mechanisms. The first involves the solution of aerosol particles which have acted as condensation nuclei for raindrops, whilst the second relates to the solution of atmospheric particles collected by the falling raindrops below the cloud level through impaction. The aerosols dissolved by the rain-out and wash-out processes may be terrestrial or marine in origin. In the former case, they are associated with dust carried into the atmosphere and are high in Ca^{2+}, NH_4^+, NO_3^-, NO_2^-, HCO_3^-, and SO_4^{2-}, whilst in the latter instance they are derived from the sea surface and will be rich in Na^+, Cl^-, Mg^{2+}, and K^+. In general, terrestrial aerosols predominate in non-humid continental interiors where precipitation quality will reflect their composition, whilst in coastal locations the precipitation chemistry will be dominated by material of oceanic origin. Some indication of the relative importance of oceanic sources can be obtained by assuming that the supply of Cl^- is dominated by this source, and calculating the oceanic proportion of other ion concentrations by using their ratios to Cl^- in seawater. However, the possibility of chemical fractionation of sea-salt particles during production and transportation necessarily introduces problems.

In spite of their limitations, the data presented in Table 1.1 demonstrate significant contrasts in the chemical quality of precipitation over the UK which may be related to the sources of atmospheric aerosols outlined above.

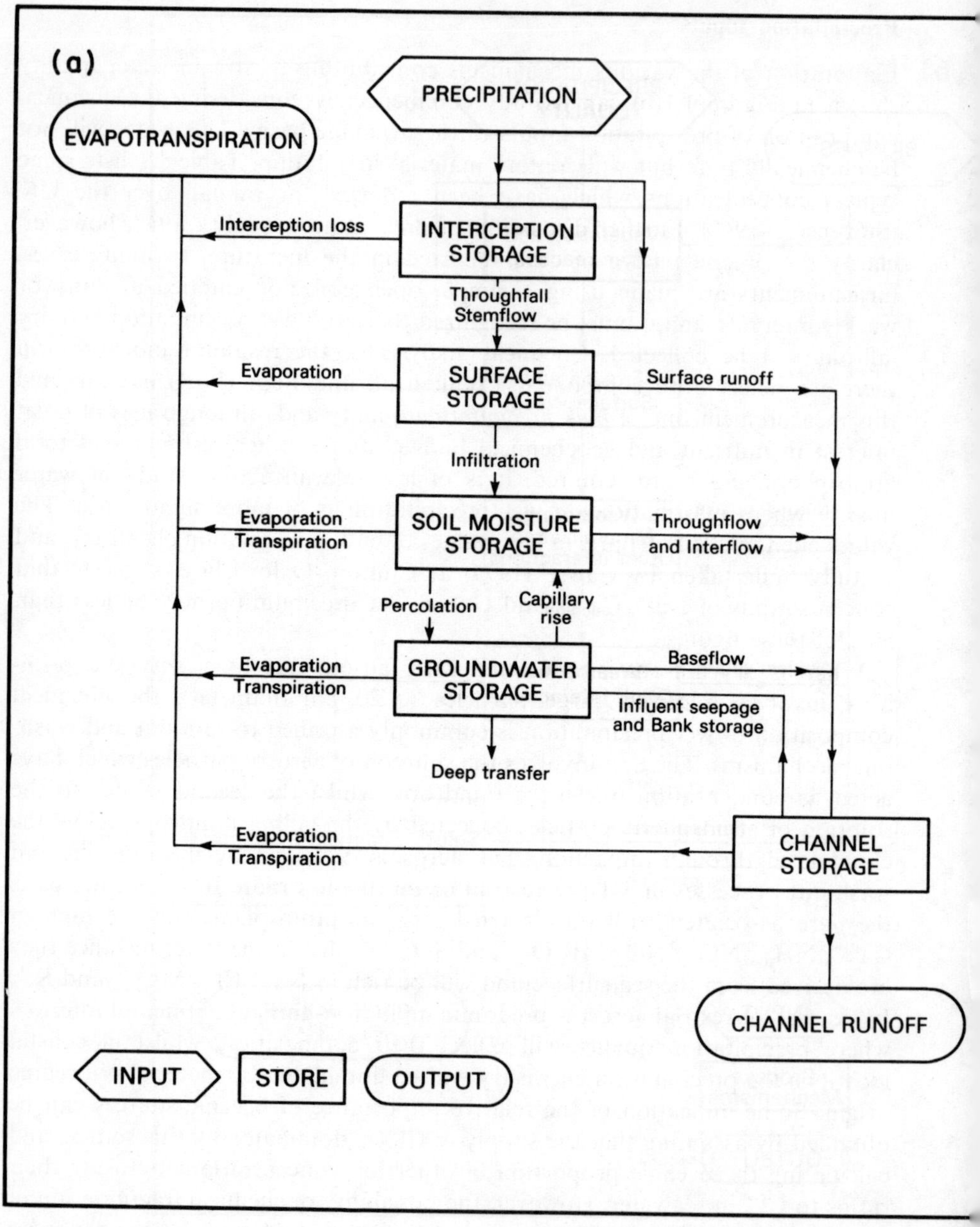

Figure 1.1. Quantity and quality processes. A simplified
in a catchment and (b) of the major mechanisms

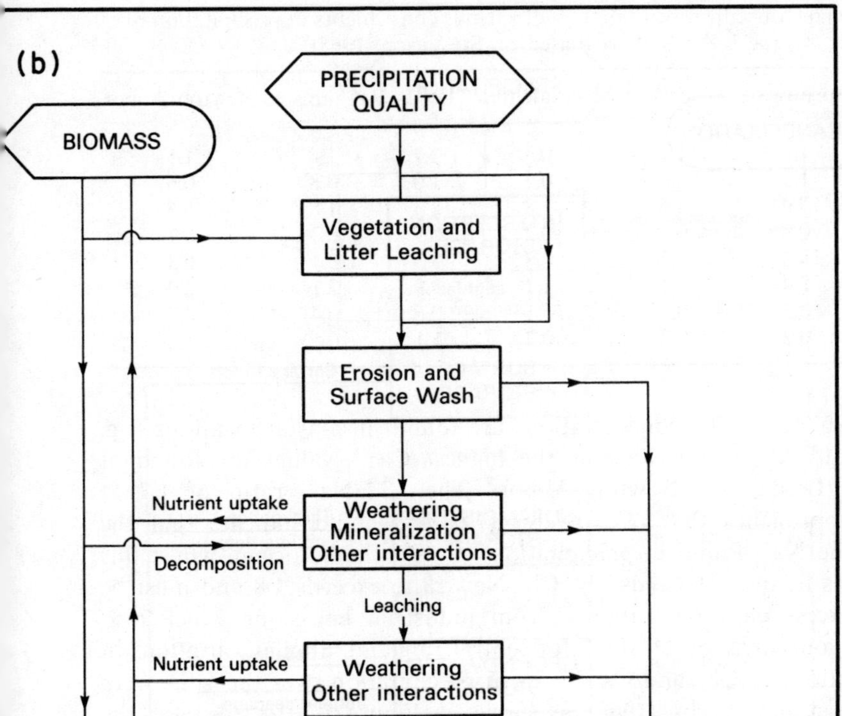

representation (a) of the hydrological processes operating influencing the quality of the water involved

Table 1.1. Median concentrations (mg l^{-1}) of various constituents in precipitation over the UK, 1959–64 (based on Stevenson, 1968)

	Stornoway	Aberdeen	Eskdalemuir	Leeds	Rothamsted	Newton Abbot
Calcium	1.1	1.1	0.3	2.7	1.4	1.5
Magnesium	1.9	0.6	0.2	1.0	0.3	0.6
Sodium	17.0	4.6	1.8	2.3	1.8	5.5
Potassium	0.9	0.4	0.2	0.5	0.3	0.7
Chloride	28.0	5.7	2.8	5.3	3.1	6.3
Sulphur	2.4	2.1	1.2	3.9	2.1	2.9
Nitrate N	0.2	0.3	0.1	0.6	0.4	0.7
Ammonia N	0.2	0.3	0.2	1.1	0.5	1.4

Thus, high Na^+ and Cl^- concentrations are found in coastal locations (e.g. Stornoway and Aberdeen), whereas the highest Ca^{2+} values are found at inland sites (Leeds and Newton Abbot). The Cl^-/Na^+ ratio of 1.7 at Stornoway approaches that of seawater (1.8) and could indicate that the majority of the Na^+ found in precipitation was of oceanic origin, whilst this ratio decreases inland. At Leeds, the Cl^-/Na^+ ratio exceeds 1.8 and must be related to 'excess chloride' supplied from industrial emissions which also influence the concentration of SO_4^{2-} for central England. Similar variations in the concentrations of Cl^- and Ca^{2+} ions in precipitation over the USA were found by Junge and Werby (1958), as shown in Figure 1.2. In this case, the collectors at the 67 measuring stations were only exposed during precipitation events and the concentrations effectively relate to wet precipitation.

Variations in the relative importance of marine and terrestrial aerosols over an area could also be expected to occur according to wind direction and air mass trajectories, and could produce variability in the chemical quality of precipitation through the year. Figure 1.3(a) depicts the annual variation of monthly mean Cl^- concentrations over eastern North Carolina and south-eastern Virginia, as reported by Gambell and Fisher (1966). Here, concentration values relate to bulk precipitation. Analysis of storm directions and patterns of air mass movement undertaken by these authors indicated that high concentrations, characteristic of most winter months, were associated with strong north-easterly flow across the area (i.e. off the sea). The low concentrations, typical of summer months, were in contrast linked to south-westerly air flow. The major source of Ca^{2+} in precipitation over this region is soil dust, and the pattern of annual variation in the monthly mean concentrations of this ion (Figure 1.3(b)) demonstrates a progression from low levels in winter to high values in summer, consonant with a dominance of continental air flow at the latter time of the year. The occurrence of dry soil conditions and greatest agricultural activity in the spring and summer months would further intensify this trend. The input of Na^+ in precipitation may be

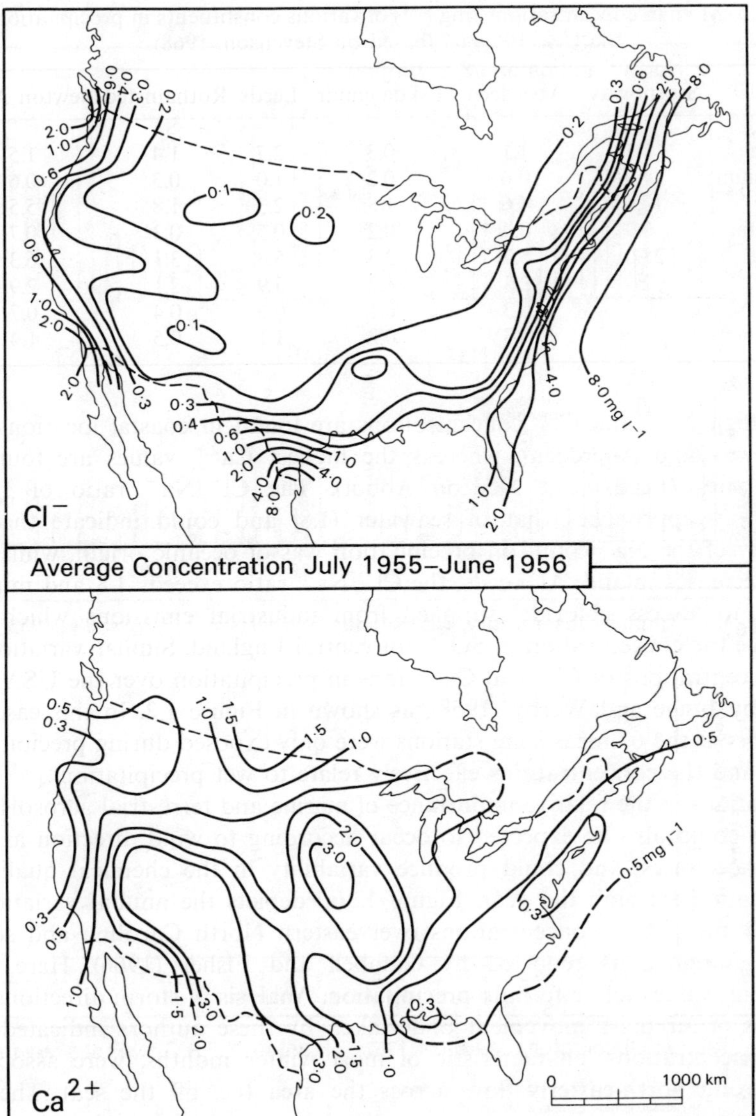

Figure 1.2. Spatial variation in precipitation quality. Average Cl^- and Ca^{2+} concentrations in precipitation over the USA, July 1955–June 1956, (based on Junge and Werby, 1958)

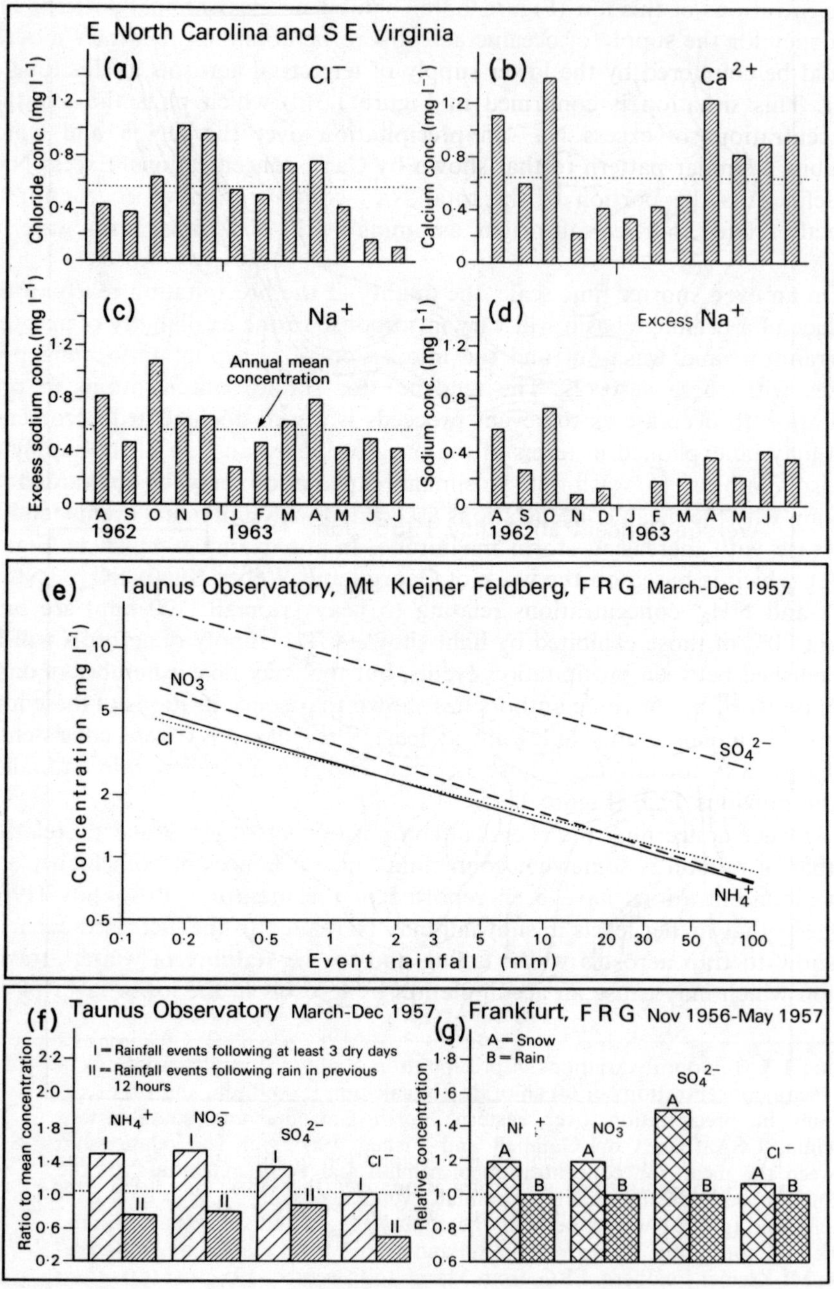
E North Carolina and S E Virginia
(a) Cl⁻
Chloride conc. (mg l⁻¹)
(b) Ca²⁺
Calcium conc. (mg l⁻¹)
(c) Na⁺
Excess sodium conc. (mg l⁻¹)
Annual mean concentration
(d) Excess Na⁺
Sodium conc. (mg l⁻¹)
A S O N D J F M A M J J
1962 1963
(e) Taunus Observatory, Mt Kleiner Feldberg, F R G March-Dec 1957
Concentration (mg l⁻¹)
NO3⁻
Cl⁻
SO4²⁻
NH4⁺
Event rainfall (mm)
(f) Taunus Observatory March-Dec 1957
I = Rainfall events following at least 3 dry days
II = Rainfall events following rain in previous 12 hours
Ratio to mean concentration
NH4⁺
NO3⁻
SO4²⁻
Cl⁻
(g) Frankfurt, F R G Nov 1956-May 1957
A = Snow
B = Rain
Relative concentration

associated with both oceanic and terrestrial sources, and the monthly mean concentrations of this ion (Figure 1.3(c)) exhibit no clear annual pattern. The tendency for the supply of oceanic aerosols to be dominant in winter months would be countered by the lower supply of terrestrial aerosols at this time of year. This situation is confirmed in Figure 1.3(d) which plots the monthly concentrations of excess Na^+ in precipitation over the region and which exhibits a similar pattern to that shown by Ca^{2+} concentrations. Excess Na^+ is defined as the portion of the total Na^+ concentration which cannot be directly related to a sea-salt origin, assuming a Cl^-/Na^+ ratio in sea water of 1.8.

On an even shorter time scale, the quality of the precipitation reaching the surface of a drainage basin will vary in response to the availability of aerosols for rain-out and wash-out and the interaction of precipitation-forming processes with these aerosols. The tendency for solute concentrations in precipitation to decrease as the event proceeds is a well documented occurrence and may be explained in terms of the progressive exhaustion of the supply of aerosols subject to wash-out. A similar explanation may be forwarded to explain why the mean concentrations associated with individual events tend to decrease with increasing storm magnitude. In the example shown in Figure 1.3(e), which is based on the work of Georgii and Weber (1960), SO_4^{2-}, NO_3^-, Cl^-, and NH_4^+ concentrations relating to heavy rainfall (100 mm) are only about 20% of those exhibited by light showers. The supply of aerosols will be replenished between precipitation events, but this may take a number of days, because work by the same authors has shown that concentrations of these ions found in rainfall events following at least three dry days were consistently higher by a factor of 2 than those associated with events where rain had fallen in the previous 12 h (Figure 1.3(f)).

Evidence concerning the chemical composition of snowfall and its relation to that of rainfall is somewhat conflicting, since instances of both higher and lower concentrations have been reported in the literature. Posokhov (1968) suggests that higher levels in snowfall may be related to the increased capacity of snow to trap aerosols whilst falling and to the stability of winter stratification which may cause an accumulation of aerosols in the lower layer of the

---

Figure 1.3. Temporal variations in precipitation quality. Monthly variations in the mean concentrations of (a) chloride, (b) calcium, (c) sodium, and (d) excess sodium in precipitation over eastern North Carolina and south-eastern Virginia, USA (based on Gambell and Fisher, 1966). (e) The relationship between the mean ion concentration of rainfall and event magnitude, (f) the distinction between the quality of rainfall events according to preceding conditions, and (g) the contrast between the quality of rain and snow (based on the work of Georgii and Weber (1960)). (e) and (f) are from the Taunus Observatory, Mt Kleiner Feldberg, FRG, from March to December 1957 and (g) is from Frankfurt, FRG, from November 1956 to May 1957

atmosphere. Reduced concentrations in snowfall may alternatively be ascribed to the blanketing of the ground surface by a snow cover and therefore to a reduction in the supply of terrestrial aerosols. These two mechanisms are clearly not mutually exclusive and in the comparison of the chemical quality of rain and snow presented in Figure 1.3(g) the higher concentrations found in snowfall may be accounted for by the former reasoning. The particularly marked increase in the SO_4^{2-} content of snowfall recorded in Frankfurt, which contrasts with the small increase in Cl^- content, must be related to atmospheric pollution.

It is clear that the chemical quality of the precipitation falling on a drainage basin will vary in both time and space in response to a number of atmospheric controls, and, just as it is wrong to view precipitation inputs as being composed of pure water, so it is misleading to think in terms of a constant chemical composition for the drainage basin input.

Interception by Vegetation

Significant changes in the chemical quality of precipitation may result form contact with vegetation during the interception process and Figure 1.4(a) indicates the potential extent of such changes by comparing the annual mean composition of incident precipitation over an oak woodland with that of through-fall under the trees, of through-fall under a ground cover of bracken (*Pteridium aquilinum*), and of stem-flow collected from the tree trunks. In all cases, except inorganic nitrogen, enrichment occurs as a result of contact with the vegetation. This enrichment results from both the leaching of metabolites and exudates from the plant tissue and the wash-off of atmospheric aerosols that have been deposited on the plant surfaces by dry fall-out and by impaction. In this study (Carlisle *et al.*, 1967), increases in the concentrations of K^+, Ca^{2+}, Mg^{2+}, and Na^+ were particularly marked for stem-flow, but the volumes of water involved are considerably less than those represented by through-fall. Changes in the pH of water at different levels within the forest canopy were also documented. Oak through-fall was generally slightly more acid than the incident precipitation which was in the pH range 4.1–4.6, but oak stem-flow (pH 3.5–3.9) demonstrated a consistent increase in acidity. These changes may be related to the adsorption of cations by the tree bark and to the leaching of organic acids from the vegetation. The decrease in inorganic N concentrations (Figure 1.4(a)) is associated with the uptake of this nutrient by plant leaves, by epiphytic lichens, and by the microflora on the surface of the leaves.

A review of other studies of through-fall and stem-flow quality reported in the literature indicates that the degree of modification of the quality of the incident precipitation varies from place to place and according to tree species. Greater enrichment under mature stands of conifers than from hardwood

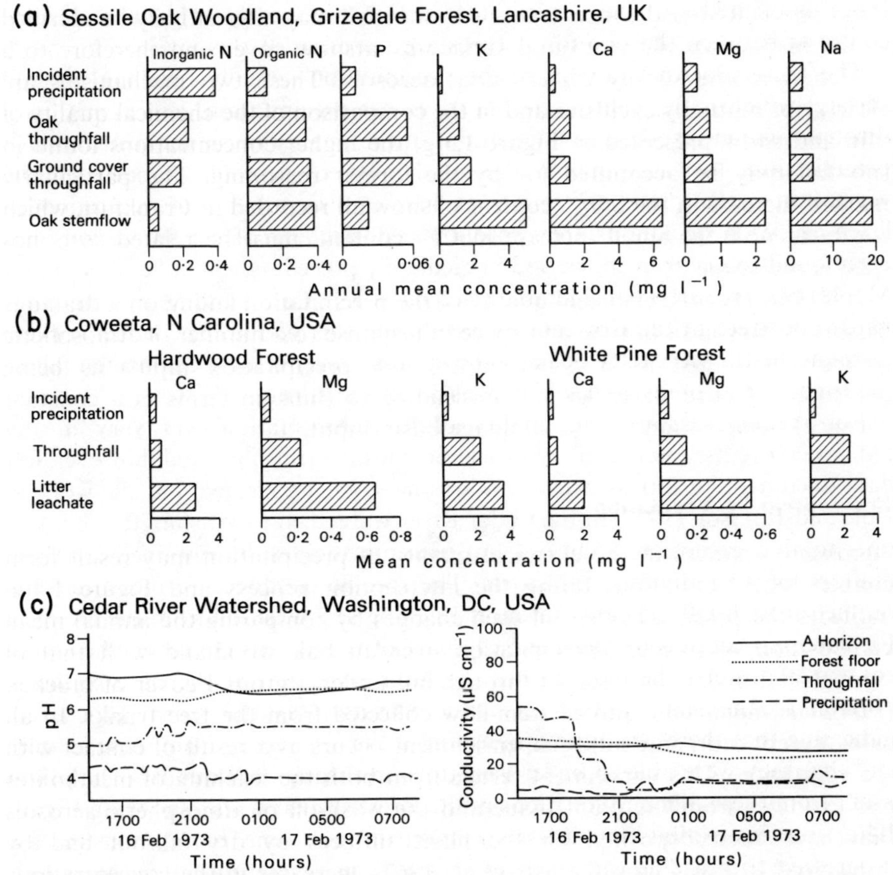

Figure 1.4. Interception and precipitation quality. Changes in precipitation quality at different levels within a vegetation canopy are shown in (a) and (b) which are based on measurements made by Carlisle *et al.* (1967) in sessile oak woodland in the Grizedale Forest, Cumbria, UK, and by Best and Monk (1975) in hardwood and white pine forests at Coweeta, North Carolina, USA. (c) The changes in net precipitation quality during a storm through the night of 16–17 February 1973 reported by Cole and Johnson (1977) from the Cedar River watershed, Washington, DC, USA

stands has been reported by Nihlgard (1970). However, the comparison of mean cation enrichment in through-fall under hardwoods and white pine at the Coweeta Hydrologic Laboratory provided by Figure 1.4(b) demonstrates little difference between the two forest types, although significant differences were reported for individual months. Figure 1.4(b) also includes information on the quality of precipitation after it had passed through the litter layer mantling the ground surface. Further increases in solute concentrations over those associated with vegetation interception are clearly evident, and some contrasts between the two forest types are apparent at this level.

The quality of the fraction of the precipitation actually reaching the soil surface, which is frequently referred to as net precipitation, will clearly vary through time in response to fluctuations in the quality of the incident or gross precipitation, to variations in the availability of aerosols trapped on the vegetation surface, and to variations in the susceptibility of vegetation to leaching. With deciduous trees, pronounced leaching of certain cations has been found to occur in the period immediately preceding leaf fall and Best and Monk (1975) report a considerable leaching of K^+, Ca^{2+}, and Mg^{2+} ions from hardwood trees in the Coweeta Forest during this period. The same workers distinguished a secondary peak in the cation concentration of through-fall at the time of bud break and leaf expansion in the spring.

For similar reasons, the quality of net precipitation will vary during individual events. In view of the measurement problems involved, however, few data documenting intra-storm variations have been collected, but the work of Cole and Johnson (1977) in the Cedar River watershed in Washington, USA, is one notable exception. Figure 1.4(c) presents the results obtained by these authors for an individual rainstorm. The specific conductance measurements on the through-fall indicate that most leaching and wash-off from the Douglas Fir canopy occurred during the early part of the storm. However, the solute content of water collected beneath the litter layer and at the base of the A horizon of the soil showed very little variation as the storm proceeded, indicating that there was little exhaustion of the ion supply from these sources and that they were able to buffer variations in through-fall quality. The pH of water from these different locations in the forest system is also shown in Figure 1.4(c) and this exhibits much less variation through time, but demonstrates a progressive increase as water moves through the vegetation, litter, and soil.

Water at the Surface

The soil surface is an important threshold in the hydrological system from the viewpoint of both quantity and quality. Water infiltrating into the surface will enter soil moisture storage and may subsequently percolate to the ground-water body (Figure 1.1(a)) and will be involved in a number of chemical reactions which exert an important influence on water quality (Figure 1.1(b)). Water remaining on the surface, however, has little opportunity to be involved in these reactions and will either evaporate or move downslope as surface runoff (Figure 1.1(a)). The residence time of subsurface water in the catchment may be considerable, whilst that of surface water will be relatively short.

A discussion of the quality characteristics of surface run-off must consider two main process mechanisms: first, the incorporation of soluble material at the surface and, secondly, the entrainment of sediment particles which may make the water muddy or turbid and produce high concentrations of suspended sediments. These sediment concentrations are significant both in terms of a physical water quality parameter and because ions may be adsorbed onto the surface of the sediment particles and may subsequently be released into solution. In the case of enrichment by solution of material at the surface, it might be expected that solute concentrations in surface run-off would exhibit marginal increases over those in the precipitation reaching the soil surface as a result of contact with dry fall-out deposited on the surface, with vegetation, and with evaporation residues produced either as a result of upward movement of soil moisture to the surface or the evaporation of water held in surface detention. However, the residence times associated with surface run-off will be extremely short and the extent of any enrichment will depend on the speed or kinetics of the solution reactions involved.

The entrainment of sediment particles will be clearly related to the processes of sheet and rill erosion which, in turn, depend on the energy associated with raindrop impact and with surface run-off and on the erodibility of the soil. Discussions of these erosion processes are traditionally concerned with the downslope movement of soil particles, but, in the water quality context, attention must necessarily focus on the transport of material in suspension and the sediment concentrations in the water involved. In addition, the sediment particle size characteristics may be of significance, particularly since adsorption mechanisms are preferentially associated with fine-grained material and colloids. The relationship between the magnitude of a particular fine fraction in the transported sediment and that in the soil may be expressed as an *enrichment ratio*, and this may vary both during a storm and according to the erosion processes involved. Changes in the particle size characteristics of the sediment transported in surface run-off may also result from deposition, where flow velocities are reduced by vegetation or decreased slope steepness.

Water in the Soil

Water infiltrating into the soil will be subject to a number of processes which will modify its quality from that of the moisture reaching the soil surface (Figure 1.1(b)). Discussion of these processes and their influence on soil water quality can conveniently commence with the idealized situation of an essentially inert soil, progress through a consideration of adsorption and exchange reactions, and finally review the significance of chemical weathering and biotic processes.

In the idealized case of an inert soil, the quality of the infiltrating water and the interplay of precipitation inputs and evaporation losses would be the dominant controls on soil water quality. The relative importance of various ions in solution would reflect the composition of the infiltrating water but their absolute concentrations would reflect the balance of input and evaporative loss. During dry periods, evaporative losses would concentrate the soil solution, whilst further infiltration would dilute the existing solution and cause a reduction in solute concentrations. A simple model of this nature will indeed often be applicable to constituents in the moisture of a real soil which have an atmospheric source and do not interact with the mineral and organic components of the soil. With extreme drying, concentrations within the soil solution could increase to the level where chemical precipitates are formed and, if certain constituents are precipitated before others, the composition of the soil water will change. The ease with which precipitates will form can be estimated theoretically by evaluating the *solubility products* of individual ion pairs (e.g. Dutt, 1962). Magnesium and calcium carbonates and calcium sulphate are examples of readily precipitated solutes. Subsequent dilution of the soil solution could reverse the process, although a delay may be involved and certain precipitates may not be readily soluble.

Interaction with the soil matrix can be introduced by considering the processes of adsorption and ion exchange. In the former process, ionic constituents will be removed from the soil solution and attached to soil particles, and in the latter the removal will be accompanied by release or exchange of replacement ions. Ion exchange mechanisms are primarily associated with cations and *cation exchange* is of considerable importance in influencing soil water quality. This process can be explained by the existence of negative electrical charges at the surface of individual soil particles, particularly colloids, which attract positively charged cations. Cations attached or adsorbed onto these negatively charged sites can enter into exchange reactions with those in the soil solution as follows:

$$\underset{\text{(solid)}}{\text{Soil} - A^+} + \underset{\text{(solution)}}{B^+} \rightleftharpoons \underset{\text{(solid)}}{\text{Soil} - B^+} + \underset{\text{(solution)}}{A^+}$$

The double arrow signifies the reversibility of the exchange. In soils, the most common exchangeable cations are Ca^{2+}, Mg^{2+}, H^+, K^+, Na^+, and NH_4^+, with Ca^{2+} generally being dominant. Ion exchange between the soil and the soil solution is a continuous process, and displacement of the equilibrium condition by the addition of foreign ions to the soil solution (or by simply altering its total concentration) will cause a new equilibrium to be established. The exchange processes are generally rapid and are governed by the relative quantities of individual ions in the solution and adsorbed phases and by variations in the strength of adsorption between different ions. A number of theoretical approaches are available to predict associated exchange reactions and equilibrium conditions (e.g. Bolt *et al.*, 1976).

Of particular importance to the soil solution is the influence of H^+ ions in cation exchange. H^+ ions supplied by rainfall with a low pH, or by leaching of organic acids from vegetation, may be removed from solution and replaced by exchangeable cations. The total amount of cations held exchangeably by a unit mass of soil is termed the *cation exchange capacity* of the soil. Colloids provide the major source of exchange sites and cation exchange capacity generally increases in fine-textured soils. Typical values for sandy, loam, and clay soils might be 2, 12, and 30 milliequivalents per 100 g of soil, respectively.

Anion exchange mechanisms involving adsorption of anions (e.g. PO_4^{3-}, SO_4^{2-}, and NO_3^-) onto positively charged sites operate in a similar manner to those of cation exchange. However, the adsorption capacity of the soil for anions is considerably smaller than that for cations and may be of little importance above pH 6–7. This mechanism is therefore of much less significance in regulating soil water quality.

In addition to adsorption associated with ion exchange, which can be viewed as reversible, there are also *specific adsorption* processes that are not reversible. Here the bonding is chemical rather than electrostatic in nature, and ions involved will be effectively removed from the soil solution. The adsorption of PO_4^{3-} ions by clay minerals and sesqui-oxides/hydroxides and the complexing of metal cations by organic matter ligands are examples of this mechanism. Precipitate formation may, in some instances, be closely related to adsorption of this type, since ions contained in infiltrating water could combine with other ions in the soil solution or associated with the soil matrix to produce an insoluble product.

Perhaps the most significant influence on soil water quality is the role of water as an agent of chemical weathering, which in turn releases additional material from the soil matrix into solution. Most work on chemical weathering has, however, been primarily concerned with the processes of breakdown and the residual products, rather than with its influence on the active solution. Of critical importance in this context is the *residence time* of the soil water, since this will determine whether there is sufficient time for a weathering reaction to proceed to the point where the solution is in equilibrium with that reaction. This will be related to the movement of moisture through the soil matrix, and to the interplay of saturated and unsaturated conditions.

It is beyond the scope of this account to discuss the various reactions associated with chemical weathering, but a general statement of their significance to soil water quality must point to dissolved silica, Na^+, K^+, Ca^{2+}, Mg^{2+}, and HCO_3^- as being important soluble products. Since silicate minerals make up 70% or more of the rocks in contact with the subsurface water of the hydrological cycle (Bricker and Garrels, 1967), the example of the dissolution of aluminium silicates may be introduced. In this case, the primary minerals are converted into hydrated secondary minerals with accompanying release of cations and silicic acid, for example:

$$\text{Cation Al-silicate} + \text{H(anion)} + H_2O \rightarrow \text{Al-silicate} + \text{Cation} + H_4SiO_4 + \text{Anion}$$

In most cases, dissociated H_2CO_3 (weak carbonic acid) is the H^+ donor in the hydration process, namely:

$$2KAlSi_3O_8(\text{K-feldspar}) + 2H_2CO_3 + 9H_2O \rightarrow$$
$$Al_2Si_2O_5(OH)_4(\text{kaolinite}) + 2K^+ + 4H_4SiO_4 + 2HCO_3^-$$

This acid derives partly from the solution of carbon dioxide from the atmosphere in precipitation and, more importantly, from the soil atmosphere which may have a CO_2 content up to a few hundred times greater as a result of plant respiration and bacterial activity. In this instance, therefore, the quality of water entering the soil may be markedly changed by weathering reactions and more particularly by the addition of cations, silicic acid, and the HCO_3^- anion. The two former constituents are directly related to weathering release, whilst the latter is essentially related to biotic activity in the soil. There are other sources of H^+ ions for weathering reactions, including atmospheric-derived sulphuric (H_2SO_4) and nitric (HNO_3) acids and the oxidation of sulphide minerals. H_2CO_3 will in fact be unimportant under acid soil conditions since it dissociates only very slightly at pH 5 and these other sources will then predominate.

Figure 1.5(a) outlines some of the major factors influencing weathering reactions and therefore the release of solutes to the soil solution. From these, three overriding controls may be isolated. First, mineral stability, which relates to the susceptibility of particular minerals to weathering within a certain weathering environment and therefore to the *thermodynamics* of the system; secondly, the rate of operation or the *kinetics* of the weathering reaction; and, thirdly, the residence time of the water involved, which, in association with reaction kinetics, will determine if that water attains equilibrium with a particular weathering reaction.

In addition to the essentially physical processes considered above, it is important to recognize the influence of biotic processes on soil water quality. The significance of plant respiration and bacterial degradation of organic material in conditioning the CO_2 content of the soil atmosphere, which in turn influences mineral weathering, cation release, and the production of the bicarbonate anion, has already been stressed. Furthermore, the influence of micro-organisms in catalysing oxidation and reduction processes, the production of water-soluble organic substances and their influence in complexing and mobilizing metal ions, and the uptake of nutrients by plants all exert an important influence on soil water quality. In the latter case, high rates of plant

---

Figure 1.5. Soil water quality. (a) Some of the major factors influencing weathering inputs to soil water quality. (b) The interaction between organic and inorganic nitrogen in the soil and (c) some results obtained by Tollenaar and Ryckborst (1975) at Castricum, northern Holland, who monitored the quantity and quality of drainwater from two large lysimeters

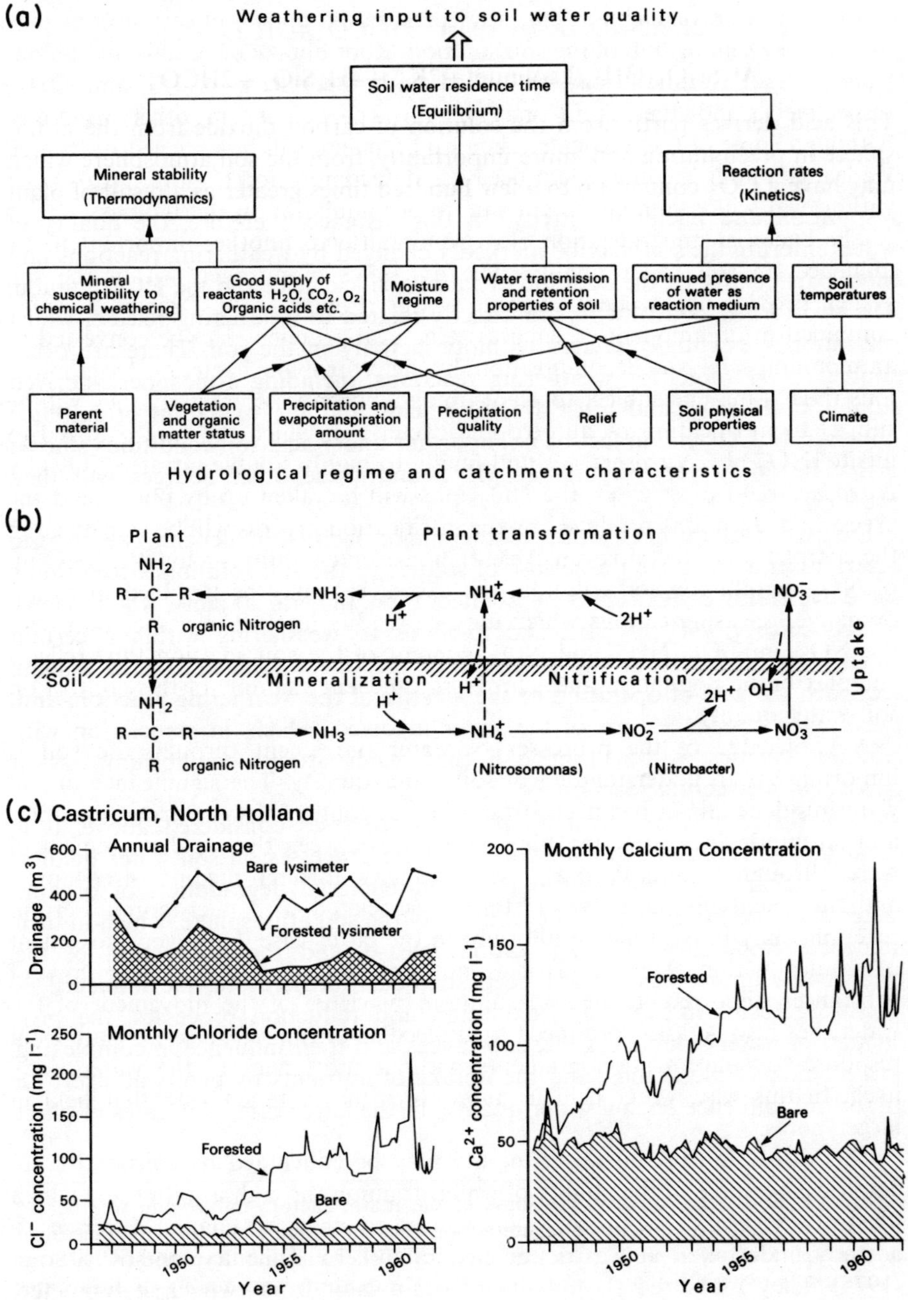
(a)
Weathering input to soil water quality
Soil water residence time
(Equilibrium)
Mineral stability
(Thermodynamics)
Reaction rates
(Kinetics)
Mineral susceptibility to chemical weathering
Good supply of reactants H_2O, CO_2, O_2 Organic acids etc.
Moisture regime
Water transmission and retention properties of soil
Continued presence of water as reaction medium
Soil temperatures
Parent material
Vegetation and organic matter status
Precipitation and evapotranspiration amount
Precipitation quality
Relief
Soil physical properties
Climate
Hydrological regime and catchment characteristics
(b)
Plant
Plant transformation
organic Nitrogen
Soil
Mineralization
Nitrification
Uptake
(Nitrosomonas)
(Nitrobacter)
(c) Castricum, North Holland
Annual Drainage
Bare lysimeter
Forested lysimeter
Drainage (m^3)
Monthly Chloride Concentration
Forested
Bare
Cl^- concentration ($mg\ l^{-1}$)
Year
Monthly Calcium Concentration
Forested
Bare
Ca^{2+} concentration ($mg\ l^{-1}$)
Year

uptake relative to availability may be the dominant control on the concentration of particular nutrients, and preferential uptake of certain ions may modify the composition of the soil solution. Root uptake of cations and anions from the soil solution may be linked with the release of H^+ and OH^-, respectively, and these ions may in turn participate in other reactions. Although some of the nutrients taken up by plants will be returned to the soil via leaf fall, litter leaching, plant decay, and decompostion (Figure 1.1(b)), a net withdrawal will commonly occur due to accumulation in the biomass.

The bacterial mineralization of organic matter is another important biotic influence on soil water quality. The dynamics of nitrogen within the soil system provide an excellent example of this (Figure 1.5(b). Organic nitrogen contained in organic matter, such as structural proteins, may be converted to ammonium salts via decomposition by heterotrophic bacteria. Ammonium ions (NH_4^+) may be subject to plant uptake, but more commonly they will be subject to nitrification by autotrophic bacteria in which they are oxidized to nitrite (NO_2^-) by *Nitrosomonas* and finally to highly soluble nitrate (NO_3^-) by *Nitrobacter*. In most cases, the NO_3 ions will be taken up by plants and the excess hydrogen ions produced by the nitrification process will be consumed in the completion of the nitrogen cycle. If, however, conditions allow the build-up of NO_3^- within the soil solution, the excess H^+ may be involved in cation exchange, releasing cations which may be readily leached in association with the NO_3^- ion. The NH_4^+ and NO_3^- content of the soil solution thus reflects complex biotic regulation and affords an excellent example of this control on soil water quality.

A knowledge of the processes of water movement through the soil is important to an understanding of soil water quality. The significance of soil water residence times has been stressed in the context of weathering reactions, and an important distinction may be drawn between the rapid movement of water through structural voids, the slower flow through medium-sized pores, and the essentially static water storage associated with micropores. Certain reactions may only attain equilibrium in the latter case. However, movement of solutes within the soil involves not only *convection*, the mass flow of substances dissolved in the soil solution by virtue of the movement of the water, but also *diffusion* produced by molecular or ionic diffusion of solutes in response to concentration gradients without movement of the water body itself. In this way, water held in micropores may interact with that held in larger pores.

This discussion of soil water quality may be concluded by considering an empirical example of the quality of water draining through a soil profile which illustrates several of the processes introduced so far in this chapter. It relates to a study undertaken at Castricum, in Holland, by Tollenaar and Ryckborst (1975). They measured the quality of drainwater from two large lysimeters filled with dune sand, one maintained in a bare condition and one supporting conifers. Since the dune sand contains very small quantities of soluble material, the results represent the response of a relatively simple soil system where adsorption, ion exchange, and weathering reactions would be at a minimum. Figure 1.5(c) presents data illustrating annual precipitation in the area and the annual drainwater volumes collected from both lysimeters. Drainage volumes from the vegetated lysimeter are less because of the increased evapotranspirational losses associated with the vegetation cover. Monthly measurements of NO_3^- concentration in the drainwater during the period 1946–62 showed that concentrations associated with the vegetated lysimeter were only about 10% of those measured for the bare lysimeter on account of plant uptake, and that the former values demonstrated a progressive decline over the period of recording as the trees developed. Figure 1.5(c) portrays monthly values of Cl^- and Ca^{2+} concentration for the two lysimeters and indicates that values for the vegetated lysimeter were higher and also showed a gradual increase through the period of study. In the case of Ca^{2+}, the increased levels were ascribed to the decreased drainwater volumes and with Cl^- both this mechanism and the interception of aerosols by the tree cover were important. In this example, differences in vegetation cover exert a profound influence on soil water quality, and it is clear that examination of water from soil systems containing material more conducive to soil/water interactions could evidence a rather more complex suite of factors controlling its quality.

Groundwater

In simple terms, the quality of water held within the groundwater reservoir (Figure 1.1) will reflect that of the soil water percolating to the water table and further water/rock interactions occurring within the groundwater body. In more detail, we may define the primary controls on the solute content of groundwater as the original chemical quality of the water entering the zone of saturation; the distribution, solubility, exchange capacity, and exchange selectivity of the minerals in the strata; the porosity and permeability of the rocks; and the flow path of the water. The thermodynamic and kinetic considerations introduced for soils are clearly also important here, although the long residence times that are generally involved mean that groundwater has a greater opportunity to achieve equilibrium with the surrounding rock. Trilinear diagrams developed from the work of Piper (1944) are frequently used to characterize the chemical quality of groundwater (Figure 1.6(a)) and these assume that this quality may be designated by three cation constituents, i.e. Ca^{2+}, $Na^+ + K^+$, and Mg^{2+}, and three anion constituents, i.e. HCO_3^-, SO_4^{2-}, and Cl^-. The proportion of each cation constituent, expressed as a percentage of the total equivalents per million (epm) is plotted as a single point in the left-hand triangular field and that of each anion in the right-hand triangular field. The central diamond-shaped field is used to show the overall chemical

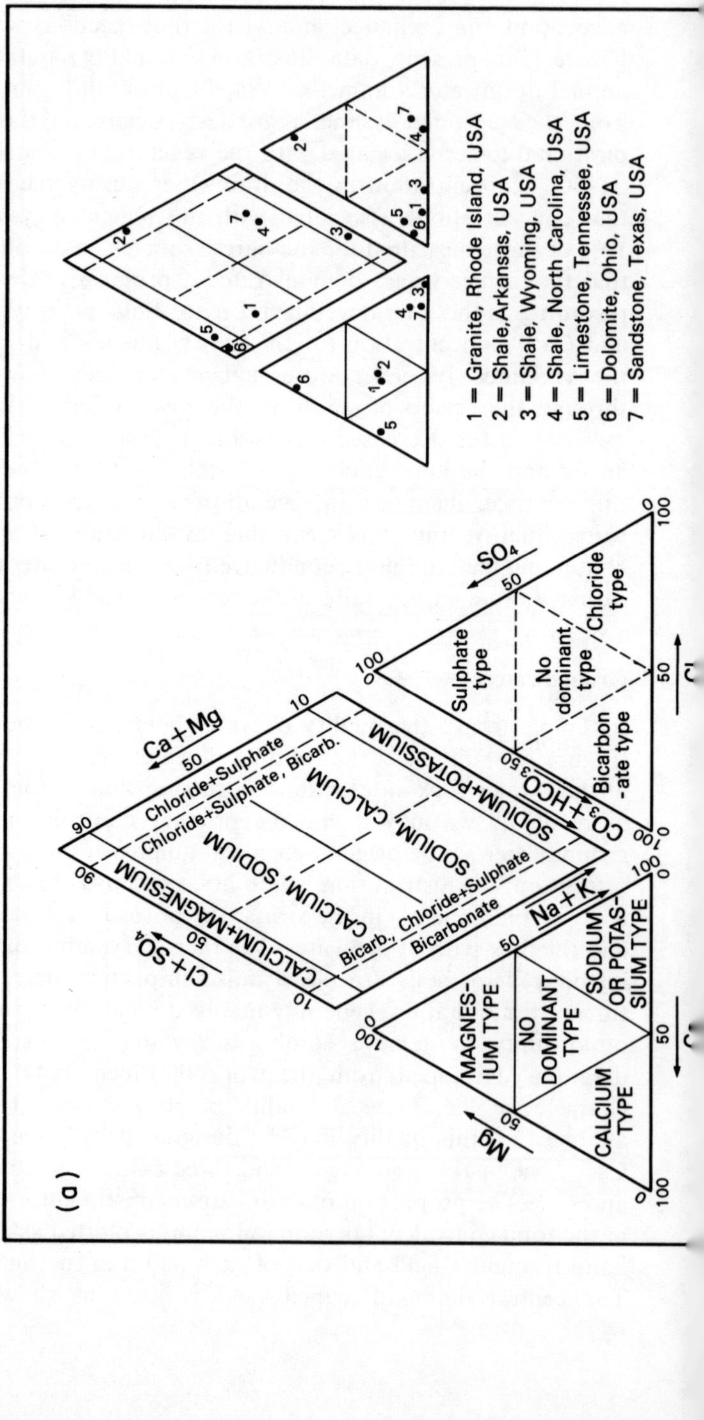
(a)
CALCIUM TYPE
MAGNESIUM TYPE
NO DOMINANT TYPE
SODIUM OR POTASSIUM TYPE
Mg
Na+K
Bicarbonate type
Sulphate type
No dominant type
Chloride type
SO_4
CO_3+HCO_3
Ca+Mg
$Cl+SO_4$
CALCIUM+MAGNESIUM
SODIUM+POTASSIUM
CALCIUM, SODIUM
SODIUM, CALCIUM
Chloride+Sulphate
Chloride+Sulphate, Bicarb.
Bicarb., Chloride+Sulphate
Bicarbonate
1 = Granite, Rhode Island, USA
2 = Shale, Arkansas, USA
3 = Shale, Wyoming, USA
4 = Shale, North Carolina, USA
5 = Limestone, Tennessee, USA
6 = Dolomite, Ohio, USA
7 = Sandstone, Texas, USA

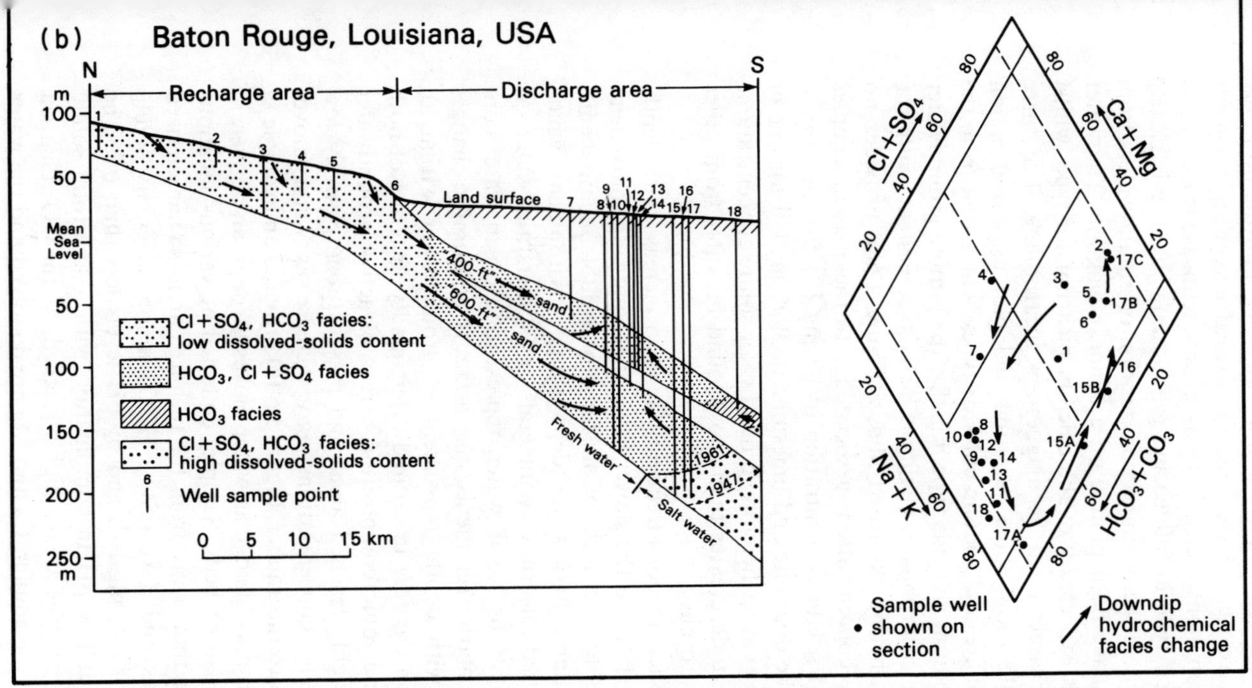

Figure 1.6. Groundwater quality. (a) The construction of a trilinear diagram and the chemical quality of a number of groundwater samples collected from different rock types (data from White *et al.*, 1963). (b) The changes in groundwater quality associated with groundwater movement through an aquifer at Baton Rouge, Louisiana, USA (based on Morgan and Winner, 1962)

character of the groundwater by a third single point which lies at the intersection of the rays projected from the cation and anion plotting positions. The fields shown in Figure 1.6(a) have been subdivided according to major water types, which are commonly referred to as hydrochemical facies.

The importance of rock type in influencing groundwater quality is clearly shown in Figure 1.6(a) where the chemical composition of groundwaters from seven contrasting strata have been plotted on a trilinear diagram. The water sample from limestone is expectedly of the calcium bicarbonate type, whilst that from dolomite is characterized as magnesium bicarbonate. The three water samples from shale show some variation and plot as mixed cation sulphate, sodium bicarbonate, and sodium chloride types, whilst that from the sandstone represents a sodium chloride type. Finally, the granite groundwater exhibits a mixed cation bicarbonate quality which is consistent with the products of feldspar weathering discussed above. Figure 1.6(a) provides no indication of the absolute concentrations associated with these seven samples which range from a total solute concentration of 82 mg l^{-1} for the granite through 579 mg l^{-1} for the dolomite and limestone to 3990 mg l^{-1} for one of the shale samples. In many cases, therefore, groundwater solute concentrations will be an order of magnitude greater than those found in water from other phases of the hydrological cycle.

Groundwater solute concentrations frequently tend to increase with depth, and this feature may be related to the slower rates of groundwater movement at depth and, therefore, to the lack of flushing of soluble components, particularly those attributable to a saline depositional environment, which may be retained as adsorbed ions on clay minerals or as interstitial water. At even greater depths, connate or 'fossil' water, representing ocean water contemporary with deposited marine sediments, may occur. In addition to changes in solute concentration with depth, movement of groundwater within an aquifer may also condition changes in groundwater quality in terms of both absolute concentration and ionic composition. Areas of active groundwater recharge are frequently highly leached areas and may exhibit relatively low solute concentrations whilst concentrations may increase as water moves downwards or outwards into the aquifer. Figure 1.6 (b) provides an example of how chemical composition may change as water moves through an aquifer, in this case two dipping sandstone beds. Samples collected at various locations within the aquifer indicate that water from the recharge area is characterized by a Na, Ca cation facies and a $Cl+SO_4$, HCO_3 anion facies, by solute contents ranging from 27 to 83 mg l^{-1}, and by pH levels less than 6.0. The presence of groundwater at low pH in combination with a low total solute content indicates that the recharge area is highly leached. As the CO_2 charged water moves downdip into the aquifer, Ca and Mg are dissolved and the water tends towards a Ca, Na facies (Sample 7, Figure 1.6(b)), but this action is countered by cation exchange, which replaces the Ca^{2+} and Mg^{2+} ions with

Na^+ ions. At the same time, the anion facies change downdip to an HCO_3, $Cl+SO_4$ facies and eventually to an HCO_3 facies (Sample 18). In the lower part of the aquifer, the pH rises to between 7.0 and 8.0 and the total solute content increases to 175–225 mg l^{-1}. At about 150 m below sea level, the fresh water—salt water interface associated with sea water intrusion is reached, and the cation and anion facies change to Na, Ca and $Cl+SO_4$, HCO_3, respectively. The change in facies produced over a period of time as this interface moved inland is illustrated by samples 15A and 15B, and 17A, 17B, and 17C, which represent sequential samples from wells 15 and 17.

The change from a hard calcium bicarbonate water to a soft alkaline sodium-rich type due to ion exchange is typical of groundwater movement through aquifers in many areas and has, for example, been well documented in the English chalk (Ineson and Downing, 1963). A more extended series of chemical changes has been proposed by Chebotarev (1955), who pointed to a general sequence involving the passage from bicarbonate facies at the outcrop through sulphate facies at intermediate depths to chloride waters at greater depths, the sequence being related to solubility equilibria as solute contents increase. Variations in the rates of facies change according to the permeability of the aquifer involved have been reported by a number of workers (e.g. Downing and Williams, 1969) and further demonstrate the importance of a knowledge of groundwater movement, as well as of rock lithology and mineralogy, in evaluating groundwater quality.

Run-off Processes

Water entering a stream channel as run-off is conventionally ascribed to three major sources or components, namely, surface run-off, through-flow and interflow, and base-flow or groundwater flow (Figure 1.1(a)), although in reality these form a continuum rather than being discrete sources. For example, surface run-off from one section of a slope may infiltrate lower down that slope and finally appear at the stream as through-flow.

Where surface run-off occurs, its quality might be expected to vary little from that of the precipitation that reaches the ground surface, but the processes of sediment entrainment, with attendant ion exchange activity and increased turbidity, and the possibility of solute pick-up, through contact with the soil surface and associated vegetation and litter, may be important. From studies of streams in the coastal ranges of California, Kennedy (1977) has argued that the dilution of solute concentrations in stream-flow during flood events is much less than would be expected from consideration of precipitation quality and the volumes of surface run-off involved, and that rapid release of solutes to surface run-off occurs.

The concept of residence time is of considerable value when assessing the water quality characteristics of various run-off components, particularly those associated with subsurface origins, since it may be argued that solute concentrations will be directly related to the time the water has remained in contact with the soil and rock of the drainage basin. Thus, the base-flow which feeds a stream during periods of dry weather might be expected to have a different chemical composition to that of through-flow and surface run-off associated with a storm event, and where only a few hours will have elapsed between precipitation input and the generation of the storm run-off hydrograph in the stream channel. However, this simple reasoning is complicated by the occurrence of translatory flow mechanisms (Hewlett and Hibbert, 1967). In this instance, the water moving to a stream during a storm event will be 'old water' displaced from the soil by the entry of 'new water' and may reflect a residence time several orders of magnitude greater than that suggested by the delay between precipitation input and run-off entering the stream. Isotope concentrations of precipitation and run-off have been used by Fritz *et al.* (1976) to apportion storm run-off hydrographs into water that existed in the basin before the storm and water that fell on the basin during the storm, and their work in the Wilson Creek Experimental Watershed in Manitoba, Canada, indicated that pre-storm water could represent about 60% of the stream-flow at peak discharge.

Studies of run-off dynamics can clearly shed light on variations in the quality of the water entering a stream, and there are also many examples in the literature where water quality parameters have been used to decipher run-off processes and to separate hydrographs into individual components (e.g. Gregory and Walling, 1973). Although the assumptions of some of these approaches may be questioned, they are, nevertheless, of interest in our context because they are based on clear evidence that run-off will vary in quality according to its origin. Figure 1.7 demonstrates how a knowledge of variations in the concentration of total dissolved solids and HCO_3^-, SO_4^{2-}, and Cl^- in the Pyalitsa River, USSR, has been used by Skakalskiy (1966) to apportion the annual run-off hydrograph into components of water of slope (i.e. surface), soil–ground, and groundwater origin. Equally, this interpretation assists in an understanding of the relationship between run-off quality and run-off source, because it demonstrates that waters of surface origin are associated with high Cl^- and SO_4^{2-} concentrations and low HCO_3^- concentrations.

The partial area and variable source area concepts of run-off generation (e.g. Dickinson and Whiteley, 1970; Nutter and Hewlett, 1971) have found increasing acceptance in recent years. These indicate that only a small, but possibly variable-sized, portion of a catchment may be hydrologically 'active' in the generation of storm run-off. Although the implications for water quality have yet to be fully investigated, it is clear that they are of considerable significance. For example, it will be the characteristics of these active areas, rather than

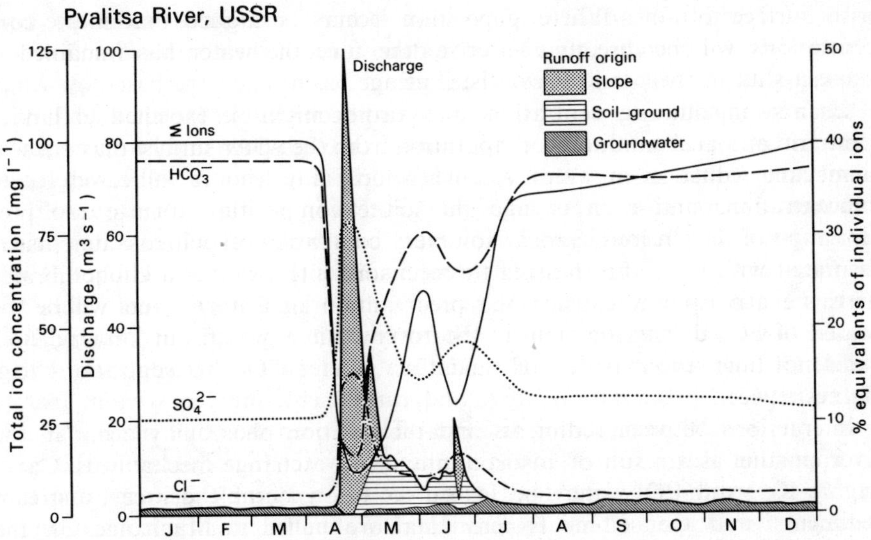

Figure 1.7. Run-off processes and water quality. This presents an attempt by Skakalskiy (1966) to apportion the annual run-off hydrograph of the Pyalitsa River, USSR, into a number of run-off components based on variations in ion concentrations

those of the entire catchment, that will control the quality of storm run-off, and where the source area expands under conditions of increasing catchment wetness it may incorporate zones with a contrasting response.

Channel Processes

Although the quality of the water leaving a drainage basin as stream-flow is essentially controlled by the run-off dynamics and their interaction with the numerous processes that may modify the quality of water as it moves through the catchment ecosystem, it is also important to take account of certain processes operating in the channel itself. These processes (Figure 1.1(b)) include a number which will modify the existing quality of the water and others which are responsible for the development of quality characteristics particular to channel flow, namely stream water temperatures and dissolved oxygen levels.

Channel processes are particularly significant in controlling suspended sediment concentrations in stream-flow through the processes of erosion and deposition. Bank erosion may occur during flood events as a result of increased water stage and velocity, and can provide a very important source of suspended sediment, which in many streams may outweigh the supply associated with surface run-off. Where deposition occurs, suspended sediment concentrations will decrease, and selective deposition of the coarsest fraction will cause a shift in the particle size distribution.

Changes in solute concentrations may also occur within the channel. Under semi-arid and arid conditions, evaporation from the water surface may cause a significant reduction in water volumes which may lead to increased solute concentrations and even changes in solute composition due to the precipitation of certain ions. Similar solubility controls apply where water that is saturated with CO_2 after passing through soil, enters a river and equilibrates with the atmosphere, causing the precipitation of $CaCO_3$; and where the release of CO_2 during the night by photosynthetic organisms in a river causes a diurnal fluctuation in the pH, and Ca^{2+} and HCO_3^- concentrations (e.g. Barnes, 1965).

Interactions between sediments and the solution phase may occur in the river channel as a result of adsorption and ion exchange mechanisms. Ca^{2+}, Mg^{2+}, K^+, and PO_4^{3-} may be sorbed to a considerable degree by stream sediments, and the sediments may therefore buffer small changes in the concentrations of these ions in stream water. Silica concentrations may also be buffered in a similar way by adsorption and desorption on the solid phase.

When studying the movement of water downstream in terms of its quality dimension, the physical processes involved in flood wave propagation and hydrograph routing must also be carefully considered; Heidel (1956) and Glover and Johnson (1974) have clearly demonstrated how the peak water discharge may move faster than the associated peak of sediment concentration or the minimum solute concentration. Similarly, within a large river network, the mixing and routing of the run-off contributions from various bributaries may produce a pattern of response which owes more to these network processes than to the processes occurring within the upstream drainage basins (cf Walling, 1978).

River water temperatures reflect the transfer of heat between the water flowing in a river channel and the overlying air and the channel margin. In a particular channel reach, the temperature will be conditioned by the temperature of the water flowing into that reach from upstream and by the various gains and losses of heat that occur. Figure 1.8(a) provides a diagrammatic representation of the energy budget of a reach and the various components of the energy budget are listed in Table 1.2. The major sources of heat may therefore be seen to be from solar radiation, from the air, from influent groundwater, and from the channel margin. Water temperatures will vary through time in response to changes in the energy budget. In most situations, air temperatures may be used as a surrogate for the complexities of the energy budget equation by assuming that river water temperatures constantly seek to achieve an equilibrium temperature represented by the adjacent air temperature. In practice, this equilibrium is rarely achieved and a phase lag is

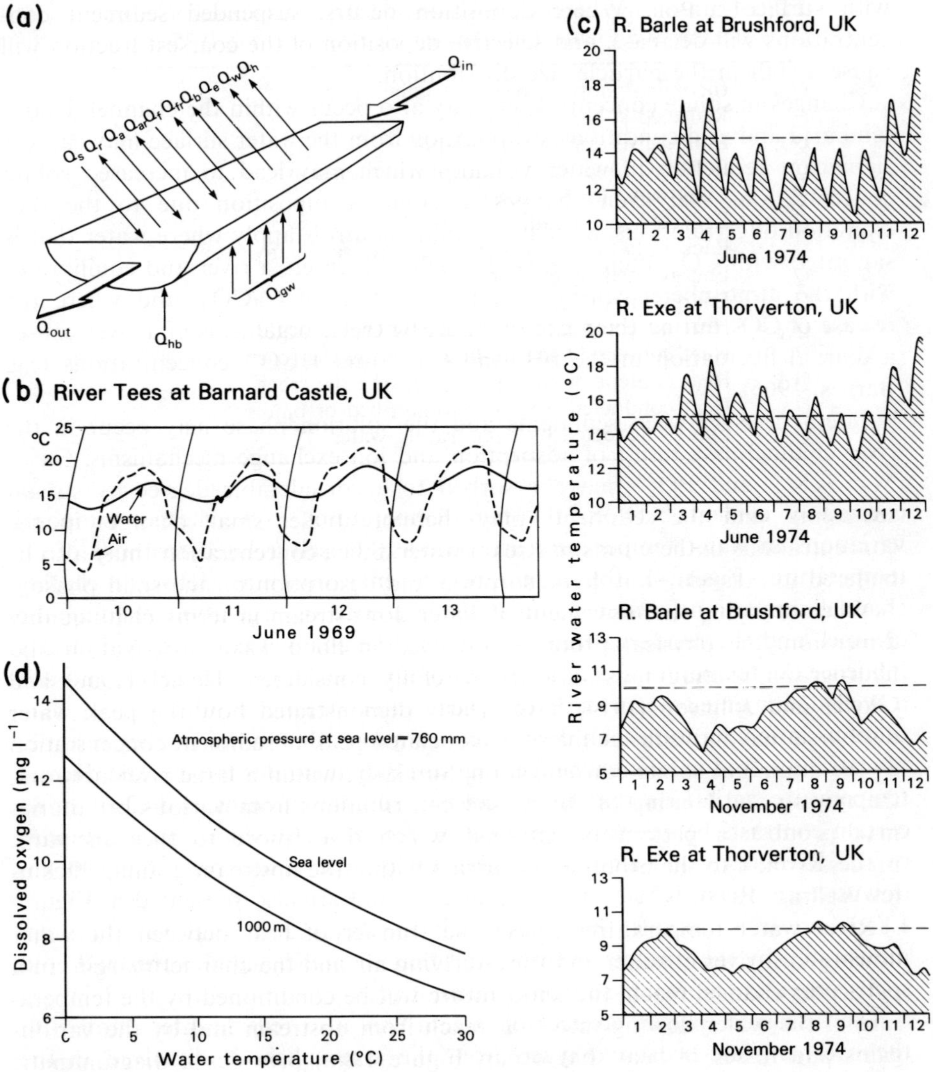

Figure 1.8. Channel processes and water quality. The major components of the energy budget of a stream reach shown in (a) are listed in Table 1.2. (b) Illustration of the close relationship between air temperature and water temperatures recorded in the River Tees at Barnard Castle, UK, during June 1969, and (c) illustrates portions of the water temperature records collected during summer and winter for two stations, on the River Exe in Devon, UK: at Thorverton, on the R. Exe and at Brushford, on the R. Barle; see map in Figure 1.12. (d) The relationship between the equilibrium solubility of oxygen and water temperature, and the influence of altitude on this relationship (assuming atmospheric pressure of 760 mm at sea level). ((a) is based on Stevens *et al.* (1975) and (b) on Smith (1975).)

Table 1.2. Principal energy components in the heat balance of small streams, see Figure 1.8(a) (based on Stevens *et al.*, 1975)

Q_s	Incoming short-wave solar radiation
Q_r	Reflected solar radiation
Q_a	Long-wave atmospheric radiation
Q_{ar}	Reflected long-wave atmospheric radiation
Q_f	Long-wave forest radiation
Q_{fr}	Reflected forest radiation
Q_b	Back radiation from the water surface (long-wave)
Q_e	Energy used by evaporation
Q_h	Energy lost by convection
Q_{gw}	Heat advected into the reach by groundwater
Q_{in}	Heat content of stream-flow entering the reach
Q_{out}	Heat content of stream-flow leaving the reach
Q_{hb}	Heat conducted from the stream bed or banks
Q_w	Energy advected by evaporating water

commonly evident between the two temperatures. Nevertheless, temporal variations in water temperature are generally closely related to those of air temperature (Figure 1.8(b). A number of environmental factors, including channel dimensions, water volume, the presence of vegetation shading the channel, and the residence time of water in the channel network, will clearly influence the temperature characteristics of a particular channel reach, and it is a well documented occurrence that headwater streams exhibit greater fluctuations in temperature than downstream stations, due to the smaller volumes of water involved in heat exchange. Figure 1.8(c) illustrates the marked diurnal temperature variations that may occur in summer months and also shows certain contrasts between the temperature record at Brushford on a tributary of the River Exe and that at Thorverton on the main river some 30 km downstream. Records relating to a winter period are also presented in Figure 1.8(c) and it is notable that these lack the marked diurnal variation demonstrated by the summer months, and that the incidence of warm and cold spells provides the major source of fluctuation.

The dissolved oxygen content of stream-flow also reflects interaction with the overlying air because oxygen from the atmosphere is dissolved in the water. Assuming equilibrium is achieved, the concentration of dissolved oxygen in a stream will be related to the water temperature which influences its solubility, and to the atmospheric pressure which reflects the partial pressure of the gas. The solubility of oxygen at 0°C and 760 mm atmospheric pressure is 14.6 mg l^{-1} and this decreases with increasing temperature (Figure 1.8(d)). An increase in altitude will also cause a reduction in the equilibrium concentration since the partial pressure will be reduced compared to that at sea level (Figure 1.8(d)). Several factors act to prevent equilibrium or saturation being achieved, and one of the most important is the production of oxygen within the water body by photosynthesis of algae and macrophytes, which may give rise to supersaturation during daylight, set against the continual consumption of oxygen by respiration and the oxidation of organic matter; significant diurnal variations in the concentration of dissolved oxygen may thus occur (see Chapter 5). The extent of any oxygen deficit will be closely related to the re-aeration properties of the stream, since equilibrium solubility conditions will be regained more rapidly in a shallow turbulent reach, with maximum contact between the water and the overlying air, than in a deep sluggish channel section.

Stream-flow Output

Table 1.3 presents some data collected by Best and Monk (1975) for a hardwood watershed at Coweeta, North Carolina, USA. These show the changes in cation concentrations associated with movement of water through the catchment ecosystem and relate the precipitation and stream-flow concentrations in terms of the net increase between the two. It is clear that the cation concentrations of water moving through the system may be higher than those characterizing the stream-flow output, and this feature may be related to the varied sources of water contributing to stream-flow and to the internal nutrient cycle which may release and take up nutrients *within* the hydrological system (Figure 1.1(b)). The relationship between precipitation and stream-flow concentrations and the role of the internal nutrient cycle may be further clarified by considering the annual nutrient flux, or the total amount of material involved, as well as the concentrations, and Table 1.3 also details the annual nutrient budget for the same watershed. In general terms, there is little overall difference between the flux of these nutrients into and out of the catchment, but in the case of K^+, Ca^{2+}, and Mg^{2+} the flux associated with through-fall and litter leaching, and therefore the internal nutrient cycle, is

Table 1.3. Annual average concentration and fluxes of selected cations contained in water passing through various levels of a hardwood forest watershed at Coweeta, North Carolina, USA (from Best and Monk, 1975)

Level	Concentration (mg l^{-1})				Flux (kg ha^{-1})			
	K^+	Na^+	Ca^{2+}	Mg^{2+}	K^+	Na^+	Ca^{2+}	Mg^{2+}
Rainfall (*R*)	0.28	0.37	0.25	0.05	4.7	6.3	4.2	0.8
Through-fall	2.21	0.62	0.59	0.23	30.5	8.6	8.1	3.1
Litter leachate	3.50	0.94	2.58	0.65	31.7	8.5	23.4	5.9
Soil percolate	1.22	1.11	1.11	0.35	2.1	1.9	1.9	0.6
Stream discharge (*S*)	0.47	0.87	0.37	0.28	4.2	7.8	5.7	2.5
Net change (*S*/*R*)	1.68	2.35	1.48	5.6				

considerably greater. Thus, while the annual flux of K^+ into the hardwood watershed is 4.7 kg ha^{-1} (kilograms per hectare), as much as 30.5 kg ha^{-1} reaches the forest floor in through-fall. However, most of this is taken up by the vegetation and only 4.2 kg ha^{-1} leaves the catchment in stream-flow. A useful indication of the role of vegetation uptake in regulating nutrient concentrations in stream-flow can frequently be obtained by comparing estimates of annual rates of biological uptake with the annual transport by the stream (e.g. Table 1.4). In this example, from the Hubbard Brook catchment (Johnson *et al*, 1969), the ions K^+, Mg^{2+}, and Ca^{2+}, and N can be seen to be in critical supply and their concentrations in stream-flow are likely to be very strongly influenced by biological uptake.

The close correspondence between input and output fluxes for the Coweeta watersheds shown in Table 1.3 suggests that there is little release of these nutrients by chemical weathering of the resistant metamorphic rocks and that an approximate equilibrium exists between input and output. On other rock types, weathering release may provide outputs of certain ions far in excess of input and the control exercised by vegetation uptake will be correspondingly less. Table 1.5 provides a few examples of other catchments where detailed nutrient budgets have been evaluated. Although nitrogen and phosphorus outputs are consistently lower than inputs, cation yields are generally greater, and, in the Walker Branch Watershed in Tennessee, Ca^{2+} output is an order of magnitude greater than the input because of weathering release from the limestone bedrock. When a detailed study of the input and output of solutes from a drainage basin is undertaken, it is frequently possible to draw up a budget which distinguishes atmospheric, weathering release, and biotic sources of the solutes. In most situations, Cl^- will be derived from atmospheric sources, Ca^{2+} and Mg^{2+} from weathering, and HCO_3^- and NO_3^- from biotic sources.

Table 1.4. Flux of nutrient elements through the Hubbard Brook watershed ecosystem (from Johnson *et al.*, 1969)

Element	Mean annual transport by stream, A (kg ha^{-1} yr^{-1})	Annual rate of biological uptake, B (kg ha^{-1} yr^{-1})	$A/B \times 100$ (%)
Na	6.1	1.6	380
S	12.4	4.0	310
Cl	3.6	1.6	220
Al	1.9	1.6	120
Ca	10.6	32.0	33
Mg	2.5	16.0	16
K	1.5	32.0	5
N	1.9	44.0	4

Table 1.5. Nutrient budgets for selected terrestrial ecosystems (kg ha^{-1} yr^{-1}) (based on a data compilation in Likens *et al.* (1977))

	Calcium		Magnesium		Sodium		Potassium		Nitrogen		Phosphorus	
	Input	Output	Input	Output	Input	Output	Input	Output	Input	Output	Input	Output
(a) Hubbard Brook, New Hampshire	2.2	13.9	0.6	3.3	1.6	7.5	0.9	2.4	20.7	4.0	0.04	0.02
(b) Long Island, New York	3.3	9.6	2.1	7.3	17.0	23.0	2.4	3.9	—	—	—	—
(c) Walker Branch, Tennessee	14.3	148.0	2.1	77.1	3.9	4.5	3.1	6.8	8.7	1.8	0.54	0.02
(d) Silverstream, New Zealand	6.0	26.0	9.0	13.0	50.0	62.0	5.0	13.0	2.2	1.8	0.2	0.03
(e) Tesuque, New Mexico	6.8	13.1	0.7	4.2	0.8	10.0	1.3	3.2	—	—	—	—
(f) El Verde, Puerto Rico	21.8	43.1	4.9	15.0	57.2	64.5	18.2	20.8	—	—	—	—

Geology:
(a) Granite and metamorphic shale.
(b) Sandstone.
(c) Limestone.
(d) Sandstone shales and metamorphic.
(e) Granite and metamorphic.
(f) Basalt, sandstone, and shale.

VARIATIONS IN STREAM-FLOW QUALITY

Temporal Variations

Figure 1.9(a) presents the frequency distributions or duration curves of water discharge and a number of water quality determinands measured in a stream draining a small catchment (44.5 km^2) in Illinois, USA. Although turbidity exhibits almost as much variation as stream-flow, which ranges over three orders of magnitude, the other parameters exhibit much less variability than water discharge. Contrasts between the five measures of chemical quality are apparent, and may be related to the operation of different controlling processes. For example, the considerable range in manganese concentrations may be related to the fact that this determinand reflects material both in solution and adsorbed onto sediment particles and will therefore be influenced by the range in turbidity. Furthermore, the concentrations of Ca^{2+} and Mg^{2+} which are reflected in the hardness values must respond differently to that of NO_3^- which exhibits a much greater degree of variation. However, simple statistical descriptions of this type fail to demonstrate that the time-variant nature of stream-flow quality reflects the interaction of the quantity dimension of the hydrological cycle (Figure 1.1(a)) with the various mechanisms influencing the quality of the resultant stream-flow (Figure 1.1(b)). In the case of river water temperatures and dissolved oxygen levels, these interactions are relatively easy to decipher and both parameters could be expected to show annual variations in addition to diurnal variations. With river water temperature, the pattern of annual variation will closely follow the annual march of air temperature, and that of dissolved oxygen will be very similar, although reversed, since it will reflect the influence of the annual variation in water temperature on the equilibrium solubility of oxygen. With other water quality parameters, such as solute and sediment concentrations, the interactions and resultant patterns will be more complex since catchment-wide processes related to all stages of the hydrological cycle will be involved.

In the search for general patterns of water quality response, many workers have shown that the magnitude of both suspended sediment concentrations (or turbidity) and solute concentrations may be closely related to the level of water discharge. The former relationship is in nearly every case positive, whilst the latter is generally inverse. Figure 1.9(b) shows how measurements of sediment and total solute concentration from the River Exe at Thorverton conform to this pattern. The degree of scatter apparent in the relationship indicates that many other factors besides the level of stream-flow influence these two quality parameters, but discharge is a useful surrogate for the major processes involved. With suspended sediment, high concentrations are linked to the incidence of both slope and channel erosion, which in turn are effectively restricted to storm events and high discharges. The inverse relationship

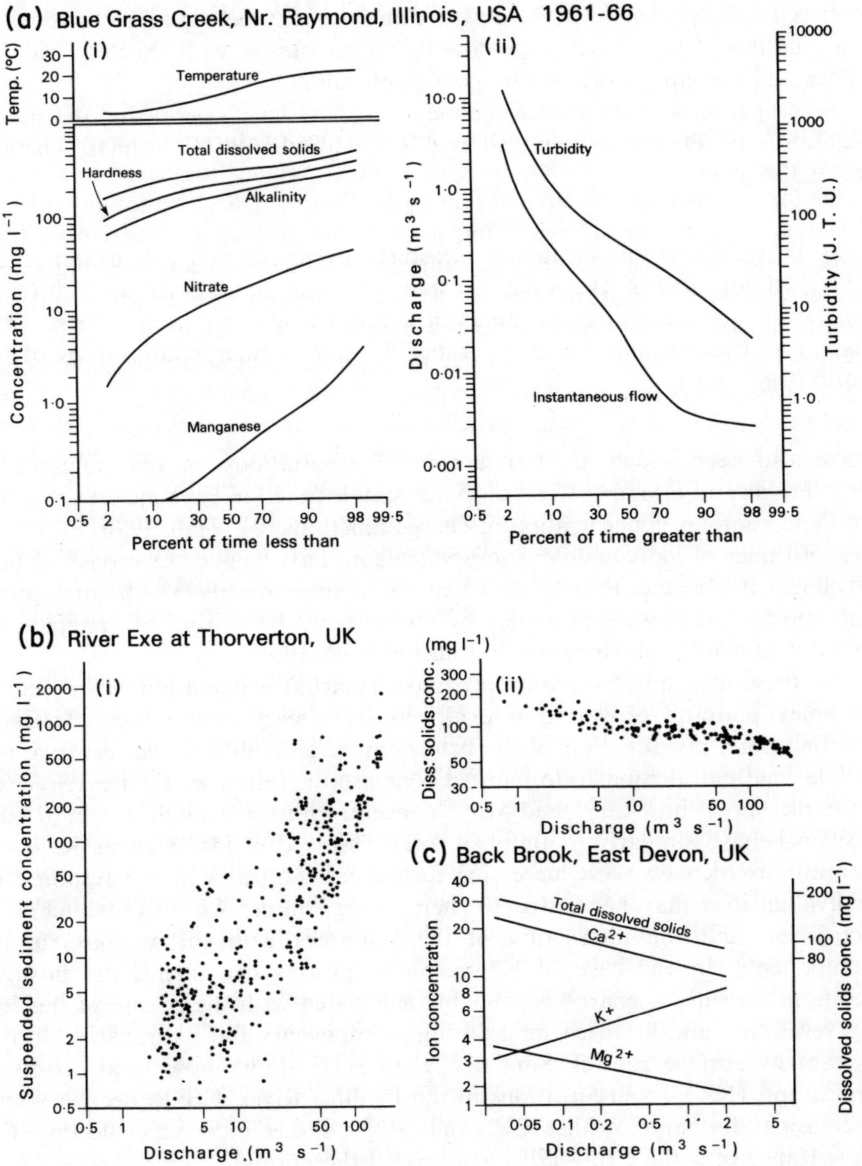

Figure 1.9. Stream-flow quality. (a) The frequency distributions of a number of water quality determinands reported by Harmeson and Larson (1969) for 1961–66 at Blue Grass Creek, Raymond, Illinois, USA. (b) The relationships between suspended sediment and dissolved solids concentrations and discharge for the River Exe at Thorverton, Devon, UK. (c) A demonstration of how individual ion concentrations may evidence contrasting relationships with discharge, for Back Brook, East Devon, UK

between solute concentrations and stream-flow may be explained in terms of the dilution of solute-rich base-flow by storm run-off with shorter residence times and therefore lower solute concentrations.

Several studies of suspended sediment response have shown that the simple relationship between concentration (C) and discharge (Q), which generally takes the form

$$C = aQ^b$$

may be seriously complicated by seasonal, hysteretic and exhaustion effects (e.g. Walling, 1974). However, in spite of these limitations, it is broadly applicable to rivers in many areas of the world and reported values of the exponent b commonly lie in the range 1.0–2.0. Similar relationships of the form

$$C = aQ^{-b}$$

have also been widely used to account for variations in total solute concentrations and attempts to develop this simple function have included the use of more complex concentration/discharge functions (e.g. Hall, 1970, 1971), the use of values of indivdual flow components in mass balance relationships (e.g. Walling, 1974), and the development of mixing models which attempt to incorporate basin-wide processes by considering the mixing of precipitation input with catchment storage (e.g. Johnson *et al.*, 1969).

All these attempts have only provided a partial explanation of the highly complex patterns of variation exhibited by stream solute concentrations. Furthermore, investigation of the behaviour of individual constituents of the solute load can demonstrate marked contrasts in response. Figure 1.9(c), for example, shows how Ca^{2+} and Mg^{2+} concentrations in a small stream in East Devon behave similarly to those of total dissolved solids, whereas K^+ concentrations increase with increasing discharge. Contrasts in the response of individual ions may be related to their major sources within the catchment ecosystem and the interaction of these sources with the various run-off components. In the case of K^+ response in this catchment, the positive relationship with discharge is probably associated with the release of this ion by vegetation and litter leaching during storm events and its transport to the stream by surface run-off. Similarly, Figure 1.7 shows how total dissolved solids and HCO_3^- concentrations in the Pyalitsa River, USSR, decrease with increasing discharge, whilst Cl^- and SO_4^{2-} levels increase, reflecting the importance of surface run-off as a source of these ions.

Figure 1.10(a) provides a further illustration of the complexity that may be demonstrated by the temporal behaviour of individual ions by considering data collected during a storm event on the Jackmoor Brook, a 9.8 km^2 catchment in Devon, UK. Whereas the specific conductance record indicates a general dilution in the total solute concentration during the event, the individual ions exhibit both positive and negative response to increasing

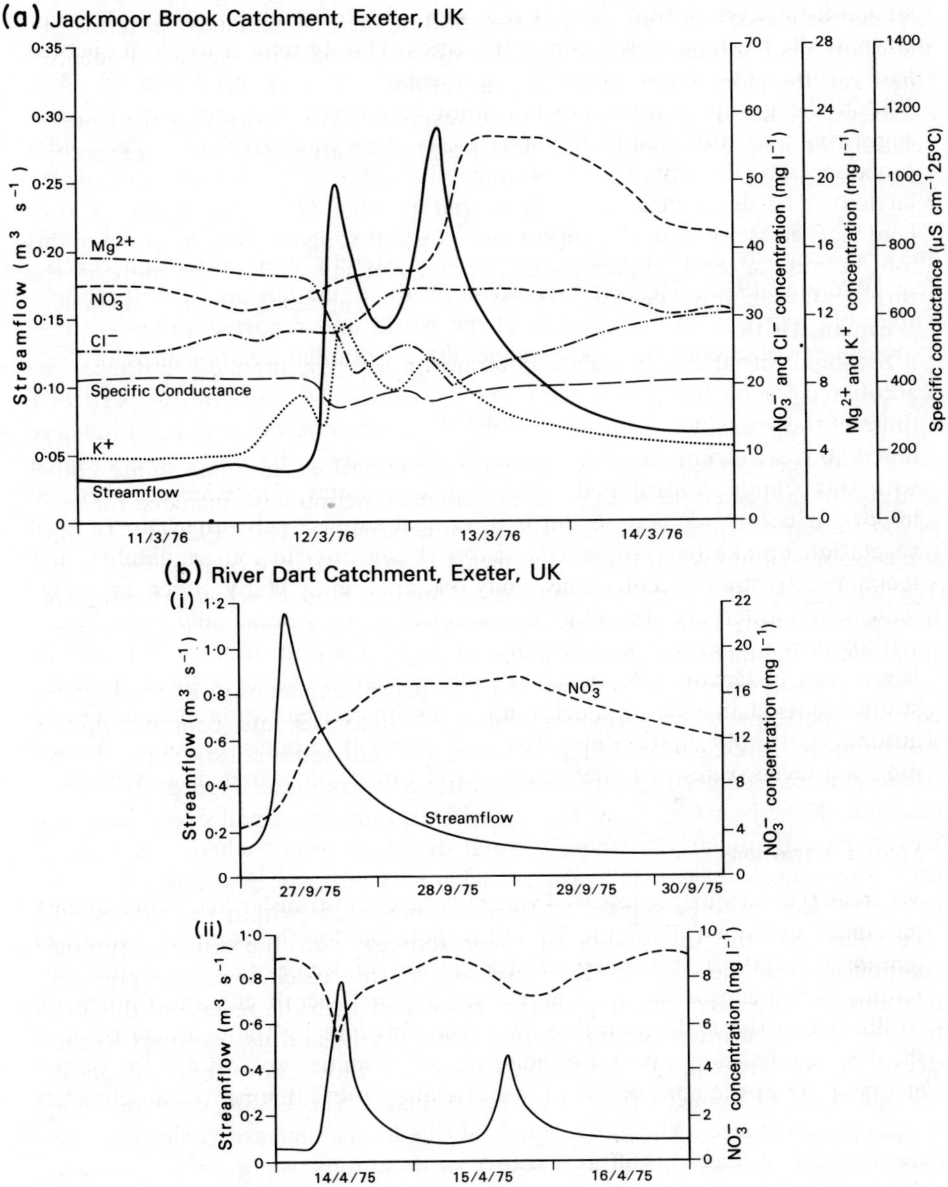

Figure 1.10. Temporal variations in stream-flow quality. (a) Variations in the behaviour of individual ion concentrations during a storm in March 1976 on the Jackmoor Brook catchment, Exeter, Devon, UK. (b) Seasonal contrasts in the behaviour of nitrate concentrations during storm events in 1975 on the River Dart catchment, Exeter, Devon, UK

stream-flow. Even within these two groups, marked contrasts are evident, for whereas the increase in K^+ concentrations is closely related to the individual hydrograph peaks, that of NO_3^- lags behind the peak discharge by 24 h. Within the group of ions showing dilution, different patterns are apparent, suggesting that, although their response is dominated by a dilution effect, other processes act differentially to produce the detailed variations in their behaviour. The delay between peak discharge and NO_3^- concentration shown here may be ascribed to the importance of soil through-flow as a source of this ion, because several studies have reported a delay of 24 h or more in the through-flow component of a storm run-off event reaching the stream (e.g. Weyman, 1970).

Seasonal variations in solute concentrations may, in many instances, be accounted for by the prevalence of high or low flow conditions at particular times of the year and by the relationship between concentration and discharge noted above. In some cases, however, the effect may go beyond that associated with this simple control and biotic regulation may, for example, exert an important control. When an ion is in critical supply with respect to rates of vegetation uptake (e.g. Table 1.4), seasonal contrasts in ion availability, and therefore stream concentrations, may be apparent. Many other processes influencing water quality could be expected to vary seasonally, and Figure 1.10(b) presents some measurements of NO_3^- concentrations taken on the River Dart in Devon, UK, which show how the behaviour of this ion during storm events changes from a delayed increase in concentration in summer and autumn to a rapid dilution in winter and spring in response to control by soil moisture levels, rates of mineralization and nitrification, and other factors.

Spatial Variations

Whereas the various processes of water storage and transfer and water quality modification shown in Figure 1.1 are responsible for the complex pattern of temporal variation shown by most measures of water quality, spatial variations in water quality are conditioned by the overall climatic framework within which the hydrological system functions and by the characteristics of the drainage basin involved. For convenience, climatic controls may be viewed as operating at the continental or world scale, while catchment characteristics

---

Figure 1.11. Spatial variations in stream-flow quality. (a) This illustrates Gibbs' (1970) attempt to portray the major factors influencing the solute composition of river water at the global level and (b) presents the relationships between maximum total solute concentration and mean annual run-off produced by Biesecker and Leifeste (1975) for streams draining US Benchmark catchments on different rock types in the USA. (c) The influence of altitude and age of forest stand on Cl^- and NO_3^- concentrations in streams draining the White Mountains, New Hampshire, USA (based on Vitousek (1977) and data in Leak and Martin (1975)

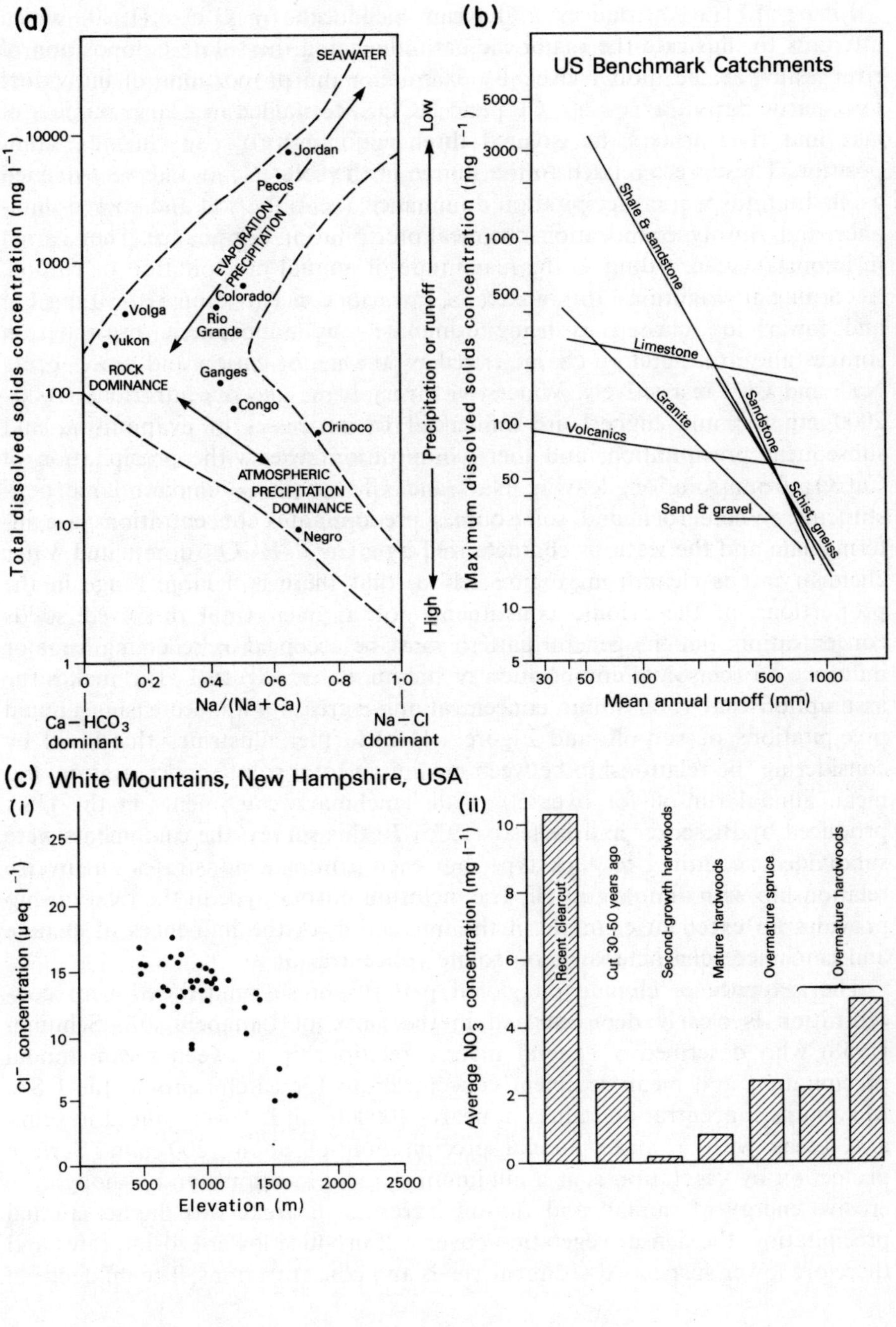
(a)
SEAWATER
Jordan
Pecos
EVAPORATION
PRECIPITATION
Colorado
Rio Grande
Volga
Yukon
ROCK DOMINANCE
Ganges
Congo
Orinoco
ATMOSPHERIC PRECIPITATION DOMINANCE
Negro
Low
Precipitation or runoff
High
Total dissolved solids concentration (mg l⁻¹)
10000
1000
100
10
1
0
0·2
0·4
0·6
0·8
1·0
Na/(Na+Ca)
Ca−HCO3 dominant
Na−Cl dominant
(b)
US Benchmark Catchments
Shale & sandstone
Limestone
Granite
Sandstone
Volcanics
Schist, gneiss
Sand & gravel
Maximum dissolved solids concentration (mg l⁻¹)
5000
3000
1000
500
100
50
10
5
30
50
100
500
1000
Mean annual runoff (mm)
(c) White Mountains, New Hampshire, USA
(i)
Cl⁻ concentration (µeq l⁻¹)
25
20
15
10
5
0
0
500
1000
1500
2000
2500
Elevation (m)
(ii)
Average NO3 concentration (mg l⁻¹)
10
8
6
4
2
0
Recent clearcut
Cut 30-50 years ago
Second-growth hardwoods
Mature hardwoods
Overmature spruce
Overmature hardwoods

become more significant at the local scale, although this division is clearly somewhat arbitrary.

Figure 1.11(a) introduces a diagram produced by Gibbs (1970) which attempts to illustrate the major factors influencing the solute composition of river water at the global level. By examining the proportions of individual ions, particularly Ca^{2+}, Na^{+}, Cl^{-}, and HCO_3^{-}, contained in a large number of lake and river waters, he isolated three major controls on chemical composition. These were related to the source of the solute load and were defined as, first, atmospheric precipitation dominance, secondly, soil and rock dominance, and, thirdly, evaporation–chemical precipitation dominance. They varied in importance according to the magnitude of annual precipitation or run-off. According to Figure 1.11(a), waters of low salt concentration (20–30 mg l^{-1} and lower) are classed as being dominated by atmospheric precipitation sources and are therefore characterized by the major cation and anion being Na^{+} and Cl^{-}, respectively. Waters with very high salt concentrations (1000–2000 mg l^{-1} and higher) are influenced by processes of evaporation and subsequent precipitation, and their composition reflects the precipitation of $CaCO_3$ from solution, leaving Na^{+} and Cl^{-} as the dominant ionic constituents. Where rock and soil sources predominate, concentrations are intermediate and the water is characterized by a Ca^{2+}–HCO_3^{-} dominated water chemistry. It is clear from Figure 1.11(a) that there is a large range in the proportions of these ionic constituents for a given total dissolved solids concentration, but the general pattern must be accepted as reflecting a major influence on the solute composition of stream water. Figure 1.11(a) makes the assumption that total solute concentrations increase with decreasing annual precipitations or run-off, and Figure 1.11(b) further illustrates this trend by considering the relationship between maximum total solute concentration and mean annual run-off for over 50 small benchmark catchments in the USA produced by Biesecker and Leifeste (1975). In this survey, the catchments were subdivided according to rock type, but each group demonstrates an inverse relationship with annual run-off. The inclusion of rock type in the relationship provides an excellent example of the interaction of the influences of climate and catchment characteristics on solute concentrations.

The influence of climate on global patterns of suspended sediment concentration is clearly demonstrated in the work of Langbein and Schumm (1958) who described a general inverse relationship between mean annual precipitation and mean sediment concentrations for catchments in the USA. Maximum concentration values of nearly 100000 mg l^{-1} were found in semi-arid areas with a mean annual precipitation of about 200 mm. Surface protection by vegetation is at a minimum in these locations, and although the erosive energy of rainfall and run-off increases in areas with higher annual precipitation, the denser vegetation cover will produce lower erosion rates and therefore lower suspended sediment yields and concentrations. The influence of climate on river water temperature is clearly illustrated in a map of median surface water temperatures within the USA produced by Blakey (1966). This shows a pronounced pattern of latitudinal zoning, with temperatures increasing southwards. Along longitude 100°W, values increase from less than 4°C in North Dakota, through 15°C in Kansas to 20°C or higher in southern Texas.

Because of the numerous mechanisms involved in regulating stream-flow quality, catchment characteristics will exert an extremely important influence on the precise nature of that quality, and small differences in these characteristics may produce very significant effects. Many studies have demonstrated the importance of rock type in controlling the solute content of stream-flow both in terms of total concentration and the levels of individual ions (e.g. Douglas, 1968; Webb and Walling, 1974). Similarly, there are numerous examples in the literature where the effects of individual catchment characteristics, such as relief, vegetation cover, and soil type, on sediment loads have been deduced (e.g. Schumm, 1954; Striffler, 1965; Wischmeier and Smith, 1965) and a brief reference may be made to the work of Brune (1951) in the mid-western United States which demonstrated that, even within relatively homogeneous physiographic areas, suspended sediment yields, and therefore average concentrations, could vary by two orders of magnitude or more.

Figure 1.11 (c) (i) introduces some results of a study of the spatial variations of solute levels in streams draining the White Mountains of New Hampshire, USA, given by Vitousek (1977). Both rock type and the forest vegetation cover are relatively uniform in this area and Vitousek was able to point to the importance of catchment altitude as a controlling factor. The levels of several ions including Cl^- decreased with increasing altitude (Figure 1.11 (c) (i)) and this may be explained by altitudinal variations in the water balance, with precipitation increasing and evapotranspiration decreasing at higher altitudes. Another study in this same area by Leak and Martin (1975) provides a useful example of the far-reaching effects of the vegetation cover in regulating solute levels. In this case, NO_3^- concentrations were measured during the spring–summer period in a variety of small forested watersheds, and the variations encountered were found to be closely related to the degree of disturbance of the watershed in recent years and the age of the forest stand (Figure 1.11(c) (ii)). The explanation proposed was that nitrogen uptake by forest vegetation was at its greatest in medium-aged stands and that streams draining such areas would exhibit the lowest NO_3^- concentrations. Higher concentrations would be associated with over-mature forest stands, areas subject to heavy cutting and disturbance, and young stands, where nitrogen uptake was less. A similar general relationship between nutrient output and changes in net ecosystem production through successional time has been suggested by Vitousek and Reiners (1975) for this area, although these authors proposed that nutrient losses would be at a minimum from successional ecosystems where productivity was greatest and that concentrations would increase in mature

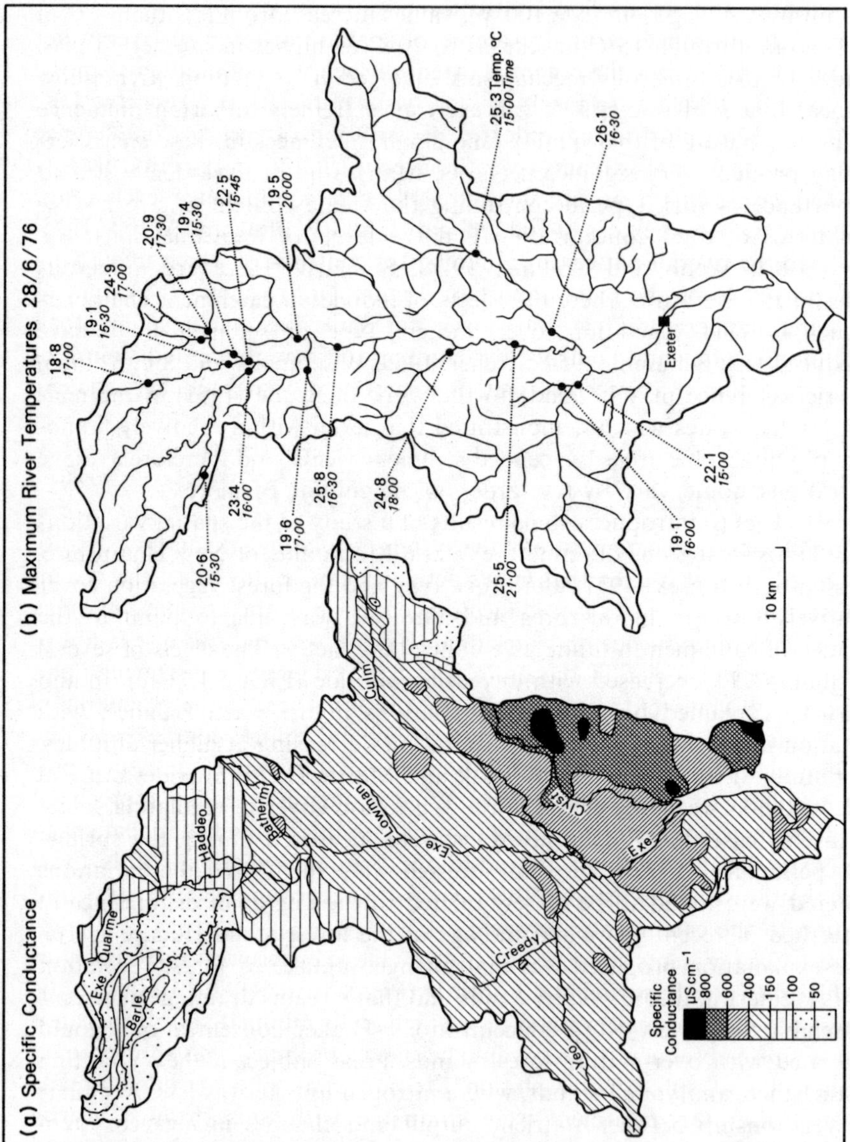

Figure 1.12. Local variation in stream-flow quality. Spatial variation in (a) the specific conductance levels measured in small streams within the Exe basin during a period of summer low flow and in (b) the magnitude and timing of maximum water temperatures recorded on a particular day (28 June 1976) are illustrated

(steady-state) ecosystems. They proposed that this control was applicable to all nutrients present in low levels relative to plant needs (e.g. Table 1.4) and it is clear that it could provide an important regulator of temporal as well as spatial variations in stream-flow nutrient concentrations.

Figure 1.12 introduces a final example of spatial variation in river water quality which emphasizes the potential extent of such variation rather than the precise controls. It relates to the 1460 km^2 basin of the River Exe in the UK, and Figure 1.12(a) illustrates the degree of variation of specific conductance levels measured in small streams over the area during a short period in July 1971. Even within this relatively small area, specific conductance levels range from 40 $\mu S\ cm^{-1}$ to in excess of 1000 $\mu S\ cm^{-1}$, and Walling and Webb (1975) have related the overall pattern to a number of controls, of which underlying geology proved to be the most important. Figure 1.12(b) shows the extent to which water temperatures can vary over this same basin in response to a number of environmental controls, by detailing the magnitude and timing of the maximum temperatures recorded at a number of sites on a particular summer day.

IMPLICATIONS OF MAN'S ACTIVITIES

If the complex and delicate nature of the system regulating stream-flow quality is recognized, then it is clear that man's activities can readily upset the system and causes changes. It is, in fact, difficult to conceive of the existence of areas where man has no influence on water quality processes, and this discussion of human impact will focus on the more substantial changes consequent upon human activity. Many of these activities will result in the *addition* of pollutants to the drainage basin system from both point and diffuse sources, and the succeeding chapters will demonstrate the considerable impact that such pollutants may have on river water quality. What is often less readily appreciated is that disturbance of the system may in itself condition significant changes in stream water quality without the addition of foreign material. It is these changes that must be discussed in this review of background water quality and its relation to the catchment ecosystem.

Figure 1.13(a) presents some results from a classic experiment undertaken at the Hubbard Brook Experimental Forest by Pierce *et al.*, (1970). In this study, the forest cover of a small watershed was cut with minimal soil disturbance and left *in situ*. Subsequently, an organic herbicide, bromacil, was sprayed over the basin to kill all sprouting vegetation. Measurements of water quality made before and after this treatment were compared with those from a nearby control catchment, and demonstrated that marked changes occurred as a result of the treatment (Figure 1.13(a)). NO_3^- concentrations showed most change, and exhibited an overall increase of about fifty-fold. Cation concentrations also increased, although not to the same extent as NO_3^-, but SO_4^{2-}

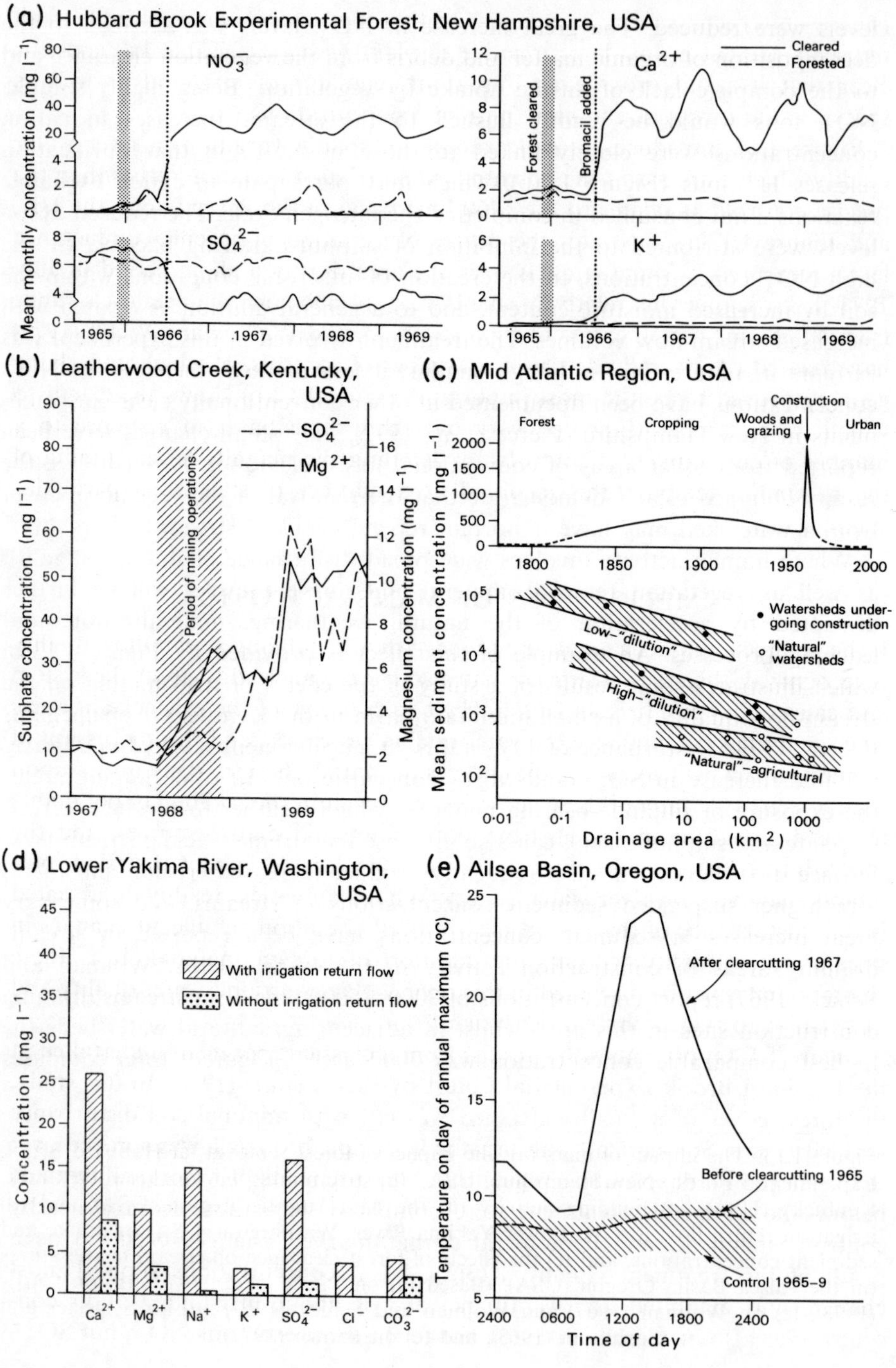
(a) Hubbard Brook Experimental Forest, New Hampshire, USA
Mean monthly concentration (mg l⁻¹)
NO3⁻
Ca2+
Cleared
Forested
Forest cleared
Bromacil added
SO4 2−
K+
1965
1966
1967
1968
1969
(b) Leatherwood Creek, Kentucky, USA
Sulphate concentration (mg l⁻¹)
Magnesium concentration (mg l⁻¹)
SO4 2−
Mg 2+
Period of mining operations
1967
1968
1969
(c) Mid Atlantic Region, USA
Forest
Cropping
Woods and grazing
Construction
Urban
Mean sediment concentration (mg l⁻¹)
1800
1850
1900
1950
2000
Watersheds under-going construction
"Natural" watersheds
Low-"dilution"
High-"dilution"
"Natural"-agricultural
Drainage area (km²)
(d) Lower Yakima River, Washington, USA
Concentration (mg l⁻¹)
With irrigation return flow
Without irrigation return flow
Ca2+
Mg2+
Na+
K+
SO4 2−
Cl−
CO3 2−
(e) Ailsea Basin, Oregon, USA
Temperature on day of annual maximum (°C)
After clearcutting 1967
Before clearcutting 1965
Control 1965–9
2400
0600
1200
1800
2400
Time of day

levels were reduced. The great increase in NO_3^- levels was ascribed to the decomposition of organic matter and debris from the vegetation clearance and to the complete lack of nitrate uptake by vegetation. Being highly soluble, NO_3^- ions would be readily flushed to the stream. Increases in cation concentrations were closely linked to those of NO_3^-, in that nitrification releases H^+ ions (Figure 1.5(b)) which may participate in cation exchange reactions if not consumed in completing the nitrogen cycle. The reduced SO_4^{2-} levels were attributed to the inhibition of sulphur-oxidizing bacteria by the high NO_3^- concentrations, to the creation of anaerobic conditions within the soil by increased moisture content, and to a general dilution associated with increased stream-flow volumes. The treatment involved in this experiment was extreme in nature, and although similar, if less marked, increases in NO_3^- concentrations have been documented in other conventionally clear-cut catchments in New Hampshire (Pierce *et al.*, 1972), only small changes have been reported from other areas of conventional forest clearance. Nevertheless, the results obtained clearly demonstrate that the removal of the vegetation cover from a watershed may have important repercussions in stream-flow quality.

Where human activity involves widespread disturbance of soil and regolith as well as vegetation removal, mineral and nutrient cycles will be further disrupted by acceleration of the natural weathering, mineralization, and leaching processes. An example of this effect is provided by Figure 1.13(b) which illustrates some results of a study of the effects of strip mining on the stream-flow quality of a small mountain basin in the Kentucky Appalachians (Curtis, 1973). Disturbance of 40% of this 70 ha catchment resulted in a nearly five-fold increase in SO_4^{2-} and Mg^{2+} concentrations. In some mining areas, the exposure of sulphide-bearing minerals, which weather to produce ferrous sulphate and sulphuric acid, can also give rise to extremely acidic stream-flow. Surface disturbance will also frequently produce increased erosion and therefore higher suspended sediment concentrations in streams, and some very great increases in sediment concentrations have been reported in streams draining areas of construction activity in the eastern USA. Wolman and Schick (1967) report concentrations of 3000–150000 mg^{-1} in streams draining construction sites in this area, whilst in adjacent agricultural watersheds the highest comparable concentration was 2000 mg^{-1}. Figure 1.13(c) compares

Figure 1.13. The impact of man. (a) The impact of forest removal (at Hubbard Brook Experimental Forest, New Hampshire, USA), (b) strip mining (at Leatherwood Creek, Kentucky, USA), (c) building activity (in the Mid-Atlantic Region, USA), and (d) irrigation return flow (on the Lower Yakima River, Washington, USA) on solute and sediment concentrations, and (e) the effects of forest clearance on stream temperatures (in the Ailsea Basin, Oregon, USA). (Based (a) on Pierce *et al.* (1970), (b) on Curtis (1973), (c) on Wolman (1967) and Wolman and Schick (1967), (d) on Sylvester and Seabloom (1963), and (e) on Brown (1971))

estimated discharge-weighted mean concentrations for natural watersheds and streams draining construction sites in the region of metropolitan Baltimore and Washington, USA, and also attempts to interpret the classic diagram produced by Wolman (1967) of the trend of sediment yields over the past 200 years in the Piedmont region of the eastern USA in terms of the discharge-weighted mean concentration for a typical river draining that area. Prior to the development of agriculture, concentrations were approximately 50 mg l^{-1} and these increased to 600 mg l^{-1} during the period of intensive farming. With the decline in agriculture during the period immediately preceding urban expansion, concentrations dropped to 300 mg l^{-1} but subsequently rose to 2000 mg l^{-1} during the period of construction activity. Finally, with the establishment of stable urban conditions, levels fell to less than 200 mg l^{-1}.

The development of irrigation can have important repercussions for water quality because of disruption of the natural water balance of the soil. Application of water for crop production in semi-arid areas results in increased evapotranspiration, and this can cause accumulation of soluble salts within the soil. In addition, increases in the level of the water table caused by water application may permit capillary rise of saline groundwater, again causing salt accumulation. Soil salinity problems of this type are one of the greatest problems facing the development of successful irrigation systems, and can only be overcome by careful regulation of groundwater levels and by leaching of accumulated salts from the soil. Some of the implications of increased salt concentrations are dealt with in Chapter 6. Where rivers draining irrigated areas receive return water highly charged in soluble salts, marked changes in water quality may result. Figure 1.13(d) illustrates this effect by comparing recorded water quality characteristics of the Lower Yakima River in Washington, western USA, during the irrigation season with those that would probably have existed in the absence of irrigation return flow. Sylvester and Seabloom (1963) estimate that Na^{+} concentrations, which are also influenced by ion exchange within the soil, show a fifty-fold increase.

The effects of human activity on river water temperatures may be introduced as another example of man's disruption of the natural system. Removal of the trees bordering a stream channel will completely change the energy budget depicted in Figure 1.8(a) and may radically alter the thermal regime of the stream. A number of studies of the magnitude of these changes have been carried out in Washington and Oregon, USA, because of the adverse effects on the fish population, and Figure 1.13(e) presents some results obtained for a small catchment in the Oregon Coast Range by Brown and Krygier (1970). In this figure, the diurnal temperature pattern on the day of the annual maximum in 1965, before logging took place, is compared with that for 1967 after the watershed had been logged. Comparison of these two patterns with that from a control watershed for the years 1965–69 indicate that the change in thermal regime is entirely the result of forest clearance and not of any contrast. in weather conditions. Maximum recorded temperatures were therefore increased by over 15°C as a result of forest clearance. Brown and Krygier (1970) related this increase predominantly to the effects of increased solar radiation receipt at the water surface rather than higher air temperatures, and other studies have shown that the detrimental effects of forest clearance can be largely overcome by leaving buffer strips of trees along the channel margin to maintain shaded conditions. The construction of impoundments along river courses may also cause marked changes in their thermal regime as well as in other water quality parameters, since the temperature of water released from a large water body possessing considerable thermal inertia will generally be rather different from that in a natural channel which will approach thermal equilibrium with the overlying air (e.g. Lavis and Smith, 1972).

The Movement of Pollutants Through the Catchment Ecosystem

A knowledge of the various processes involved in the movement of water through the catchment ecosystem and in the control of water quality is essential in evaluating the impact of pollutants on river water quality and recognizing potential hazards. It is the general field of non-point or diffuse source pollution that is of greatest significance in this context because, whereas point sources generally discharge directly into channels and their effects may be readily discerned in terms of increased stream loadings, non-point pollutants may reach the channel by a variety of routes and this movement may not be fully understood.

The effects of emissions of SO_2 from industrial zones in Western Europe and eastern North America in increasing acidity of precipitation over Scandinavia and the north-eastern United States have been well documented (e.g. Dochinger and Seliga, 1976). This acid rain may be viewed as a non-point pollutant, and there are many locations where it has been suggested that the increasing acidity of precipitation has been paralleled by a decrease in the pH of stream and lake waters. For example, Oden (1976) cites examples from Sweden which indicate that the pH of water in the River Klaralven has decreased from a mean pH of approximately 6.9 in 1965 to about 6.5 in 1974. Understanding of the precise impact of acid rain on stream-flow quality requires consideration of the pathways and residence times of water movement through the catchment ecosystem, and of the buffering capacity of the soil and rock of the catchment. For example, Oden (1976) shows that the first phase of snow-melt in some Swedish rivers may be associated with a rapid decrease in pH due to the rapid transfer of water to the rivers by surface run-off over frozen ground. Subsequent phases of the spring melt are, however, associated with increasing pH levels as the meltwater enters the soil and buffering takes place. The impact of acid rain on water quality must not be viewed only in the context of pH levels because the increased acidity of infiltrating water may cause increased leaching of cations from the soil as a result of cation exchange, and may induce accelerated weathering of silicate minerals and solution of heavy metals accumulated within the soil.

An important consideration in the fate of atmospheric contaminants, such as heavy metals and radionuclides deposited on the surface of a watershed in wet precipitation or dry fall-out, is whether they enter the hydrological system in solution or are adsorbed by surface soil particles. In the former case, they will be subject to processes of chemical exchange within the soil and other water/soil/plant interactions, but may eventually reach the stream in sub-surface run-off, whilst sediment-associated contaminants will only be moved to the stream by processes of surface erosion and sediment transport. Radioactive material is particularly susceptible to adsorption.

The distinction between solution and adsorption is also of great significance when assessing the fate of fertilizers and pesticides applied to agricultural land and their impact on river water quality. Nitrogenous fertilizers are generally associated with the NO_3^- ion, either directly or as a result of mineralization and nitrification, and this ion is readily leached through the catchment ecosystem. Phosphorus forms applied in fertilizer are, however, readily adsorbed by soil particles and their loss from the catchment may be primarily associated with erosion and transport of suspended sediment. Enrichment ratios discussed previously will be of considerable significance where fertilizer nutrients are preferentially adsorbed onto fine particles, and Miller (1977) has shown that the fraction of a soil composed of particles less than 2 μm in diameter may contain 2.5 times as much phosphorus as the soil itself. Similar contrasts exist in the dominant behaviour of pesticides applied to agricultural land (Table 1.6). The partial- and variable-source area concepts of run-off production also assume great significance in the context of the movement of

Table 1.6. Transport modes of selected agricultural herbicides, insecticides, and fungicides (based on a report by the US Department of Agriculture (1976))

Herbicides		Insecticides		Fungicides	
Atrazine	SW	Aldrin	S	Benomyl	S
Bromacil	W	Chlordane	S	Captan	S
Dalapon	W	DDT	S	Dinocap	S
Diquat	S	Dieldrin	S	Fenaminosulf	W
Paraquat	S	Lindane	S		
Simazine	S	Malathion	W		

S Chemicals that will move primarily in sediment-associated form.
W Chemicals that will move primarily in solution.
SW Chemicals that will move in both sediment-associated form and in solution.

fertilizer nutrients and contaminants from agricultural land because certain areas of the watershed may be much more susceptible to subsurface run-off and surface erosion than others, and judicious application with this hydrological phenomenon firmly in mind could greatly reduce their impact on stream water quality.

With subsurface disposal of wastes in landfill sites and septic tanks, the movement of polluting substances to stream courses will be governed by both physical and chemical factors. With the first, such variables as groundwater velocity, aquifer dispersivity, and hydraulic conductivity will be important, whilst, with the latter, ion exchange and adsorption mechanisms and biological activity must be taken into account. It is again essential to consider the total catchment ecosystem rather than the immediate zone of disposal if the impact on water quality is to be fully understood.

THE HYDROLOGICAL VIEWPOINT

This chapter has attempted to demonstrate how the natural or background water quality of a catchment is regulated by a large number of processes, and controlled by a variety of catchment characteristics. In this way, water quality reflects the overall structure of the catchment ecosystem and each individual drainage basin must be viewed as exhibiting a unique water quality response. The focus has been primarily hydrological rather than founded on a detailed discussion of the underlying chemistry, and it is hoped that some justification for this approach may be found in the words of an eminent water chemist, John D. Hem, who said:

> 'The hydrologist ordinarily needs no reminder that fresh water in the natural environment is in constant circulation and that his branch of science requires methods of study appropriate to dynamic systems. The chemist who is concerned with water quality, however, may tend to underemphasize the importance of the motion of water in understanding its chemistry. Certainly he most often sees only small segments of the system, held captive and immobile in water sample bottles' (Hem, 1971).

CHAPTER 2

Impact of Man in Catchments (i) Agriculture

J. K. R. Gasser

In the United Kingdom, most of the land surface is farmed, and agriculture therefore has a considerable influence on the quality of water leaving the soil and either entering a watercourse or percolating into deeper strata.

In considering the impact of agriculture, three main types of farming may be identified which have been shown to have very different effects on the quality of drainage water and surface run-off: (1) hill farming and low-intensity livestock farming, which includes rough grazing and amenity land such as moorlands for game shooting and walking and steep hills; (2) intensive animal production; and (3) arable farming. In general, there is more arable farming in the drier eastern parts of the country and more animal production in the west. However, there are centres of intensive pig and poultry production in the east and there are also large areas of mixed farming.

A number of ecosystems have been studied, including small complete catchments with a direct outfall to the sea when the quality of the water along the streams and rivers is important, and those systems containing a lake which acts as a buffer (the cognate problem of stream- and river-fed reservoirs has received much attention). The type of farming and its influence on water quality will interact within the ecosystem to give particular patterns of change and particular problems in water management.

The problems of recharge of groundwaters after pumping are complex. The rate of movement from the surface depends on factors such as fissuring and porosity of the water-bearing strata. The recharge around the point of pumping depends on the relative importance of older storage water and water recently arrived from the surface and the groundwater flow regime both above and below the water table. For example, the predominance of rapid fissure flow in the saturated zone of the Chalk probably accounts for the fairly even distribution of nitrate revealed in detailed surveys of groundwater in parts of east Yorkshire and north Lincolnshire. Differences in chemical composition of groundwaters are thought to be detectable in adjacent areas of differing land use only immediately after recharge (Foster, 1976).

THE COMPOSITION OF DRAINAGE AND BOREHOLE WATERS

Of the elements commonly found in water, agriculture can directly affect nitrogen, phosphorus, potassium, and calcium and, to a lesser extent, sodium, chlorine, magnesium, and sulphur. The burning of fossil fuels has a major effect on sulphur and nitrogen in the atmosphere and rainfall; individual industrial plants affect the area immediately around them. Table 2.1 gives Williams (1976) extensive results of analyses of rainwater for three sites, showing how comparatively uniform are the average values for many elements: the largest differences are the much larger phosphate at Rothamsted, and the increased sodium, magnesium, and chloride at Saxmundham, reflecting the proximity of this site to the sea. These values are useful as a baseline to measure the changes in composition on movement of water through the surface soil and also into deeper strata.

The effects of agriculture on nitrogen and phosphorus concentrations in water have caused concern and are termed 'pollution' when the magnitude of these effects exceeds predetermined values. Both elements are on occasion associated with algal blooms, and nitrate in excess of World Health Organiza-

Table 2.1. Average chemical composition of rainwater for 1969–73 at the Experimental Stations of Rothamsted in Hertfordshire, Woburn in Bedfordshire, and Saxmundham in Suffolk. (From Williams, 1976, in *Agriculture and Water Quality*, MAFF Technical Bulletin no. 32. Reproduced by permission of the Controller of Her Majesty's Stationery Office.)

	Rainwater composition ($mg\ l^{-1}$)		
	Rothamsted	Woburn	Saxmundham
Na	1.9	1.6	4.4
K	0.7	0.4	0.6
Ca	1.8	1.5	1.6
Mg	0.4	0.3	0.7
NH_4-N	1.7	1.2	1.8
NO_3-N	1.1	0.9	1.2
PO_4-P	0.13	0.03	0.02
Cl	5.9	4.6	8.8
SO_4-S	3.3	2.6	3.3

tion (WHO) recommended levels is considered undesirable for the reasons given in Chapter 6. The effects of agricultural practices on the other elements leave the concentrations within an acceptable range, and these will be considered first.

Salt is applied as a fertilizer for sugar-beet, and excess sodium and chloride will be leached into the drainage waters. However, the total area treated is small (only 200000 ha are grown in any one year) and the effects will be slight and intermittent compared with the total amounts in circulation. Similarly, little magnesium is applied to UK soils and the effects, for example on water supplies, will be small. Lime (calcium carbonate) is added to acid soils to bring the pH to within the range required for maximum crop production. However, large areas are naturally calcareous, notably the Chalk and limestone formations, and these are also important aquifers (Foster, 1976). Table 2.2 shows that drainage water from soils at Rothamsted, Woburn, and Broom's Barn all contained much calcium. The soils at Rothamsted are derived from Clay-with-Flint overlying Chalk. Woburn soils are more complex, being largely derived from the Lower Greensand with areas of lighter drift overlying Oxford Clay and some soils from re-sorted Oxford Clay. Broom's Barn clay loam or sandy loam soils are derived from glacial deposits overlying Chalk at variable depth. The deep waters from the Chalk at Rothamsted and at Broom's Barn also contain much calcium, whereas that from Woburn contains less (and much less than the land drainage). Liming the soil is normal agricultural practice and will increase the concentration of calcium in drainage waters. On soils with acid-permeable subsoils and deeper strata, movement of water through these

Table 2.2. Average composition of drainage water and deep borehole water at Experimental Stations of Rothamsted (Hertfordshire), Woburn (Bedfordshire), and Broom's Barn (Suffolk). (From Williams, 1976, in *Agriculture and Water Quality*, MAFF Technical Bulletin no. 32. Reproduced by permission of the Controller of Her Majesty's Stationery Office.)

	Drainage water composition ($mg\ l^{-1}$)				Borehole water composition ($mg\ l^{-1}$)			
	Rothamsted	Woburn	Broom's Barn	Mean	Rothamsted	Woburn	Broom's Barn	Mean
Na	14	11	18	14	7	10	12	10
K	16	5	3	8	1	3	4	3
Ca	113	140	184	146	113	33	119	88
Mg	4	9	4	6	2	4	4	3
NH_4–N	0.2	0.05	1.1	0.4	0.02	0.08	0.01	0.04
NO_3–N	10	13	18	14	6	4	7	6
PO_4–P	0.5	0.1	0.2	0.3	0.003	0.05	0.04	0.03
Cl	18	30	68	38	14	23	26	21
SO_4–S	61	55	48	55	3	10	8	7

layers will remove some of the calcium until a new equilibrium is reached. The potassium concentration of the drainage water is greater than that of the rainwater (see Table 2.1 and also Chapter 1), and the potassium concentration in the borehole water is sometimes slightly greater and sometimes very much less than in the drainage. All drainage and borehole waters contained little ammonium as this form of nitrogen is retained by the soil.

The values in Table 2.2 show greatly enhanced concentrations of nitrate in the drainage water, and it is this phenomenon which has caused much debate. In contrast, all the borehole waters contained much less nitrate. For recharge through fissured rocks, Foster (1976) suggested that a major factor in maintaining a satisfactorily low concentration of nitrate in the groundwater is the mixing of drainage water from arable areas containing 10–20 mg l^{-1} NO_3-N with that from grassland containing less than 4 mg l^{-1} NO_3-N. However, there is little information on changes in nitrate concentration during the slow passage of water through fine-grained aquifers such as some Chalk and Triassic sandstones. The effect of agriculture on the nitrate concentration in drainage water and its relationship to river systems, lakes, and deep water supplies is a major topic for discussion. The WHO has suggested an arbitrary limit of 11.3 mg l^{-1} NO_3-N (50 mg l^{-1} NO_3) as a desirable upper limit, and twice this amount as an acceptable limit. The major concern is that excessive nitrate can cause methaemoglobinaemia in infants (see Chapter 6), although less than 10 cases have been confirmed in the UK in 20 years. Also, it is the total burden of nitrate from food, beverages, and water that must be considered; for example, leafy green vegetables and root crops can contribute as much, or more, nitrate to the diet as drinking water (White, 1975).

Drainage water contains more phosphorus than rainwater, but inorganic phosphate is very immobile in soil, and losses by leaching into drainage are slight. Organic forms of phosphorus move more readily and, as will be discussed later, can cause pollution. Phosphate does not cause health problems for the consumer, as it is an essential element, for example, in bone formation.

EFFECTS OF AGRICULTURAL DRAINAGE ON THE COMPOSITION OF POND AND LAKE WATERS

Troake *et al.* (1976) found that drainage from Slapton Wood, which forms part of the total catchment of Slapton Ley, (the largest freshwater lake in south-west England and showing every evidence of being a eutrophic lake) had an average 6.6 mg l^{-1} NO_3-N. In 1963, the lake had levels of between 0.2 and 2.5 mg l^{-1} NO_3-N, whereas measurements made from 1970 onwards on the streams feeding the Ley showed that they contained much more nitrate. Similarly Williams (1976) found that water from the land drain feeding a small lake at Woburn contained about 20 mg $l^{-1}NO_3-N$ and 0.1 mg l^{-1} PO_4-P, whereas the lake water had relatively constant concentrations of 2 mg l^{-1} NO_3-N and 0.02 mg l^{-1} PO_4-P. The organisms in the lake use nitrogen and phosphorus, although an algal bloom rarely develops because the concentration of soluble P is limiting. However, during one bloom, the fractionation of the total phosphorus concentration of 0.08 mg l^{-1} into various forms showed that 63% was in the organisms forming the bloom, 25% was as soluble organic P and 12% as soluble inorganic P.

There is adequate evidence to show that drainage water from under grassland contains less NO_3-N than from under arable soil. A good example is shown by the results of Williams (1976) from the Saxmundham experiment. In the later stages of his investigation, drainage waters from an arable section and a grassland/lucerne section were collected and analysed separately. Average concentrations were 22 and 4 mg l^{-1} NO_3-N, respectively, and losses from December 1973 to February 1974 were 20 kg ha^{-1} and 4 kg ha^{-1} of NO_3-N, respectively. The flow rates of the drains were similar, and therefore the composition of the water entering a watercourse from these drains will depend on the proportions of arable and grassland soils. In the east of Scotland, Stewart *et al.* (1976) found that surface run-off and drainage from agricultural land (with 70% pasture or rough grazing and 30% soft-fruit or arable crops) to some shallow freshwater lakes contributed 8 mg l^{-1} N mainly as nitrate and 0.05 mg l^{-1} P. Lake waters receiving agricultural run-off as the main source of nutrients had 2 mg l^{-1} of soluble N and 0.06 mg l^{-1} P. There was thus a marked decrease in the concentration of nitrate but not of phosphorus. The standing crops of phytoplankton vary markedly from year to year, and eutrophic waters contain the largest concentrations together with, in summer, extensive growths of filamentous algae and aquatic macrophytes. They considered that sediment denitrification may be very important in removing NO_3-N from such freshwater lakes, and presented microbiological data to support the idea. Like the results from Slapton Ley (Troake *et al.*, 1976), the largest concentrations of nitrate were found in the waters of the lake during the winter months, and a smaller increase in mid-summer coinciding with a sharp maximum in the nitrate content of the stream feeding the lake. However, from February to May, NO_3-N in the waters of the lake fell from 6 mg l^{-1} to zero, while the incoming water contained from 7 to 12 mg l^{-1} NO_3-N. The large amounts of nitrate in the winter drainage may be the result of mineralization of soil N and crop residues, of animal excreta or of excess fertilizer dressings. Again, in the absence of soil values from the various types of land use, the source of nitrogen cannot be apportioned.

The ratio of N:P in the water is also important (Williams 1976), because with wide ratios, organisms using fixed nitrogen and phosphate will grow in preference to nitrogen fixers such as blue-green algae with less risk of a bloom.

Table 2.3. Rainfall, drainage, fertilizer N applied, and nitrate N in drainage at three contrasting sites. (From *Agriculture and Water Quality* (1976), MAFF Technical Bulletin no. 32. Reproduced by permission of the Controller of Her Majesty's Stationery Office.)

Site and reference	Area of catchment (ha)	Farming system	Fertilizer N applied (kg ha^{-1} N)	Year	Rainfall (mm)	Drainage (mm)	Nitrate N in drainage water: Concentration (mg l^{-1}) Min.	Max.	Mean	Total lost (kg ha^{-1})
Jealott's Hill, Berkshire (Hood, 1976)	5	Intensive grassland	250	1971–72	709	191	1.7	15.7	10.7	20
			250	1972–73	422	12	0.3	17.8	9.4	1
			250	1973–74	732	122	(0.1)*	(23.4)*	(8.2)*	(10)*
	5		750	1971–72	709	201	0.1	58.3	35.5	71
			750	1972–73	422	29	1.4	60.7	18.8	5
			750	1973–74	732	148	13.1	58.6	20.5	30
Slapton Wood, Devon (Troake *et al.*, 1976)	1159	Mixed Farming i.e. catchment consists of 72% grass 27% arable 1% woodland	To fertilized area / On total area of catchment							
			54 / 42	1971–72	832	333	4–8†		6.4	32
			59 / 45	1972–73	917	436	4–8†		6.4	28
			77 / 58	1973–74	1029	450	3–8†		6.9	31
Great House, Yorkshire (Webber and Wadsworth, 1976)	41	Intensive grassland	190 (+ additional 94 as FYM and slurry) each year				Monthly averages			
				1972	1442	744	3.3	7.4	5.5	41
				1973	1274	572	2.3	7.5	7.1	41

* Values estimated by author because of incomplete results.
† Approximate values; exact values not given by authors.

NITRATE FORMATION AND MOVEMENT IN AGRICULTURAL SYSTEMS

Grassland Ecosystems

Ecosystems affected by agriculture may vary greatly in size and complexity, and examples of a number have been investigated. Those likely to lead to results which can be readily interpreted include small closed catchments with surface run-off and field drains, and upland rivers flowing through lowland or into lakes. The movement of water into deep aquifers is of great importance but is very complicated. The intensity of farming also affects the total loss of nutrients into drainage water and the likelihood of episodic increases in nitrate concentration.

Three areas of grassland (or predominantly grassland) have been studied which had varying intensity of farming systems, in areas with very different rainfall. Some details are given in Table 2.3. Hood (1976) studied the effects of intensive grassland management on the composition of the drainage water from a single field of 10 ha; Troake *et al.* (1976) investigated drainage from a small closed catchment of 1159 ha, mainly rough grazing and pasture but with just over one-quarter of the area in arable cropping with comparatively little fertilizer N applied; Webber and Wadsworth (1976) monitored the drainage from an area of 41 ha of upland grass in Yorkshire, which received above average dressings of fertilizer N, farmyard manure, and slurry. The results in Table 2.3 show that the average concentrations of nitrate were largest in the area with least drainage and smallest with most drainage. Total losses of nitrogen were the reverse. Table 2.3 also shows that mean values and ranges of values varied more on the driest site and varied least at Slapton. The individual values given by Webber and Wadsworth (1976) for Yorkshire showed that nitrate concentrations in the drainage water were related to slurry or manure applications where the highest value was 26 mg l^{-1} NO_3-N, following heavy rainfall after a dry period during which pig slurry and farmyard manure had been applied to the catchment.

At all sites, maximum concentrations of nitrate were found in the winter period, particularly December–March, and, at Slapton and Great House, largest flows were associated with largest concentrations. Thus, maximum losses, both in terms of concentration and weight, occurred at times which were the furthest removed from those of fertilizer application. In contrast, Hood found that, at Jealott's Hill during the wet June of 1971, there was a marked loss of nitrate which would reasonably be expected to be partly or mainly from fertilizer applied earlier. The source of the nitrate leached in winter is less certain. It may come from unused fertilizer N, from mineralization of soil N, or from decomposition of animal excreta or slurries or farmyard manure. There is no way of apportioning these sources unless one or more of them are suitably labelled, for example with ^{15}N.

In Table 2.3, the range of values for concentration of nitrate from Jealott's Hill are similar to those from arable sites. For example, at Rothamsted, drainage waters from an arable field (Barnfield) contained from 3.8 to 78.3 mg l^{-1} NO_3-N with a mean value of 9.8 mg l^{-1} (Williams, 1976). Again, large concentrations were associated with large rates of flow, but, in contrast to the grassland sites, largest concentrations occurred in September with little in January–February, suggesting different sources of nitrogen were important on the arable and grassland soils.

At Jealott's Hill, losses following application of 250 kg ha^{-1} N may be compared with losses following the very large experimental applications of 750 kg ha^{-1} N. The results show that, although the extra fertilizer N increased the nitrate concentration of the drainage water to above the upper WHO limit of 22.6 mg l^{-1} NO_3-N on 71 out of 116 weeks (1971–74) (the mean annual concentrations were 35.5, 18.8, and 20.5 mg l^{-1}), the extra lost from the additional 500 kg ha^{-1} N as fertilizer was 51, 4, and 20 kg ha^{-1} N in 1971/2, 1972/3, and 1973/4, or 10, 1, and 4% of that applied, respectively.

Hood (1976) stresses that these values represent those in the drainage water leaving the field and not those reaching the river. Nutrient concentrations over the 12 km meandering course through farmland and including a sewage works showed a value of 7 mg l^{-1} in the river close to the stream inlet. This is comparable to the average values found by both Troake *et al.* (1976) and Webber and Wadsworth (1976) whose studies may therefore be considered to relate more to catchment drainage than to field drainage.

Losses of Nitrate from Arable Soils

The major loss of nitrate from arable soils occurs during the winter months following mineralization during rewetting of the soils in autumn. With recommended dressings of fertilizer N, cereals and sugar-beet leave little or no residues of mineral N in the soil (Widdowson, 1974). Potatoes receive considerable amounts of fertilizer N, and experimental evidence shows that potatoes leave residues benefiting following winter wheat (Widdowson *et al.*, 1967). Therefore, the total area with fertilizer N at risk at the end of a summer with at least average rainfall is not likely to be more than the 200000 ha growing potatoes out of the six million hectares of arable soils. In Holland, where the amount of fertilizer used is greater than in the UK, Kolenbrander (1972) has estimated that fertilizer N adds only 5% to the total load of NO_3-N in drainage. Fertilizer N applied in the spring is at risk of loss by surface washing or leaching by drainage until crops are established and growing freely. Such episodes have been recorded and represent a loss to the farmer, but pose little threat to water supplies because they occur at a time of maximum flow, and water extraction can usually be stopped for the duration of high NO_3-N concentrations. Direct drilling of crops is one agricultural practice leading to less nitrate formation in the soil and less loss to drainage water. Recent experiments have also shown that the wetting pattern of soils under direct drilling differs from that of cultivated soils. On direct-drilled heavy soils, the surface soil wets more slowly, and the deeper soil faster than cultivated soils (Letcombe Laboratory, 1978).

The movement of nitrate from the surface of soils by winter rainfall has been investigated (Wild and Babiker, 1976) in relation to soil physical properties. Movement within and to below the cultivated layer has also been measured (Burns, 1977). Models were constructed allowing prediction of movement—for example, on bare fallow plots given large amounts of nitrate (840 kg ha^{-1} N). The depth in the soil at which the maximum concentration of nitrate was found was correlated with the apparent depth of movement of the winter rainfall calculated as Q/θ, Q being the amount of effective rainfall (total rainfall − evaporation) and θ the volumetric water content of the soil. The regression equation was

$$y = 0.57x + 4.1$$

where y is the depth of the nitrate peak (cm) and x is Q/θ (cm); that is, the depth of movement of nitrate was about 0.6 of the apparent depth of movement of the rainfall which followed the application of the fertilizer. The recovery of nitrate in the top 50 cm of the soil profile at the end of winter also varied with Q/θ (Wild and Babiker, 1976).

Ley–Arable Farming Including Leguminous Leys

A shift from grassland to arable farming greatly increases the amount of NO_3-N in drainage water. Growing legumes also does, and particularly the killing or ploughing of clovery swards (Gasser, 1968; Low and Armitage, 1970).

Low and Armitage (1970) measured the amount of nitrogen in drainage water from cropped and uncropped lysimeters. The crops grown were white clover or grass (Meadow Fescue) without fertilizer N. Later the clover died and nitrogen released to the drainage water was measured.

These authors found that the amounts of nitrogen (NO_3-N, NH_4-N, and other N) leached annually from the lysimeters were equivalent to 30 kg ha^{-1} N for clover and 2.5 kg ha^{-1} N for grass. Drainage averaged 153 mm during the growing period, so that these weights correspond to average concentrations of 19.6 and 1.6 mg l^{-1} N for clover and grass, respectively. When the clover died, decomposition of the residues released 130 kg ha^{-1} N into the drainage water, and during this period there was 284 mm of drainage, so that the average concentration was 46 mg l^{-1} N.

One year grass, grass/clover, and clover leys were grown as preparation for winter wheat (Widdowson *et al.*, 1963), and after ploughing the leys in autumn and sowing the wheat, samples of soil were taken for measuring the mineral N (NH_4-N and NO_3-N) in the fresh soils and the increases in mineral N (Δ (mineral N)) on incubating the re-wetted air-dry soils. A second set of samples was taken in the spring. The values for mineral N in the soil and mineralizable N, given in Table 2.4, show that soil contained most mineral N and mineralizable N in the autumn after clover, and swards containing clover, but that these differences had almost disappeared by the following spring. Comparing these values with the yields of winter wheat without fertilizer N shows that all estimates undervalued the nitrogen-supplying power of clover, which may be taken to indicate the rapid decomposition of the plant residues and movement of the nitrate below the cultivated layer (Gasser, 1968). Wheat is known to be able to recover nitrate from the subsoil on this site (Gasser, 1962). These results accord well with those from the lysimeter studies (Low and Armitage, 1970) and indicate the large amounts of mineral N released very quickly after ploughing clover or leys containing clover.

Draining of Soils and Nitrate Movement and Loss

Draining of soils also affects the amount of nitrate entering drainage and deep waters, because the downward movement of water and the aeration of the subsoil increases and denitrification decreases. For example, in work in which nitrate movement down the profile was followed in well drained heavy and light soils, fertilizer N applied to the surface could be found as it moved down the profile (Cunningham and Cooke, 1958; Gasser, 1959), whereas in a soil

Table 2.4. Measurements of available N in soil in autumn after ploughing one-year leys and sowing winter wheat, and in the following spring; and the yields of winter wheat without fresh fertilizer N. Measurements were the ammonium N + nitrate N (mineral N) in the fresh soil and the increase on incubating the re-wetted air-dry soil (Δ (mineral N)). From Gasser, 1968, *J. Agric. Sci., Camb.* Reproduced by permission of Cambridge University Press.)

	Fertilizer N to ley (kg ha^{-1} N)	Soil measurement (mg kg^{-1} N)				Yield of grain without fertilizer N (tonnes ha^{-1} at 15% moisture)
		Mineral N		Δ (mineral N)		
		Autumn 1961	Spring 1962	Autumn 1961	Spring 1962	
	0	9	8	21	29	3.99
Rye grass	26	6	8	25	30	4.32
	251	7	9	28	32	4.32
Rye grass/	126	8	9	30	32	4.55
clover	0	10	9	32	31	4.93
Clover	0	12	9	34	32	6.06

showing gleying, nitrate disappearance was attributed to denitrification in the gleyed horizon (Shaw, 1962). Draining such heavy soils with a reducing zone will increase the amount of water moving through the profile and therefore the possibility of transporting nitrate and will decrease the amount likely to be denitrified. Work on the effects of drainage conditions on crop growth and the nitrogen economy of soils is being done both in the field and in large undisturbed cores of soil in lysimeters (Letcombe Laboratory, 1978) which should enable the effects of drainage on losses by denitrification to be quantified.

Increases in Mineralizable N from Additions of Farmyard Manure

The addition of farmyard manure (FYM) also increases NO_3-N in drainage water. Cooke (1976) described the earlier results of Lawes and Gilbert (1874) and Lawes *et al.* (1882) on the effects of additions of FYM and fertilizers on drainage water from Broadbalk field at Rothamsted, on which winter wheat has been grown each year since 1843. In the winter period, drainage water from the plot receiving FYM contained more nitrate than from plots where no additional nitrogen was applied or where fertilizer N was only applied in the spring. Soil measurements made in the 1930s and 1950s confirmed these earlier findings (Gasser, 1962). The mineralizable N in soil receiving FYM was much greater than in soils without it, although applying fertilizer N in spring as ammonium sulphate also tended to increase the mineralizable N. On average of eight samples taken throughout a year, the amounts of mineralizable N in the soil were 19 mg kg^{-1} with FYM, 10 mg kg^{-1} with the treatment described as 'minerals only' (i.e. fertilizers supplying P, K, Na, Mg) or with nitrogen as sodium nitrate, and 12 mg kg^{-1} with minerals and ammonium sulphate. Also in the autumn, the profile to 690 mm contained the equivalent of 109 kg ha^{-1} N with FYM and 81 kg ha^{-1} N with fertilizer N compared with 57 kg ha^{-1} N in soil without FYM or fertilizer N.

Because of large seasonal variations, the total weight of nitrogen entering the drainage water can change greatly, but mineralization of 25 kg ha^{-1} N and its subsequent nitrification would provide 10 mg l^{-1} NO_3-N to 250 mm of drainage water. Therefore, if all nitrogen in the unfertilized soil was leached by 250 mm of drainage over the winter, this would give an average concentration of 23 mg l^{-1} NO_3-N in the water. However, total leaching does not take place.

PHOSPHORUS MOVEMENT IN SOILS

Inorganic phosphate is retained by most agricultural soils and the evidence points to very little fertilizer P moving into drainage water or deep aquifers.

Fertilizer P only appears in drainage water from heavy dressings applied to sandy soils containing very little clay, so that the amount of phosphorus in drainage water is controlled more by the nature of the soil than by the amount of fertilizer P applied, but surface run-off containing soil can add much phosphorus to the watercourses. However, the main source of phosphorus in surface waters is from sources other than agriculture, particularly sewage effluent. For example, most of the phosphorus entering Loch Leven is derived from the effluent from a woollen mill, sewage effluent provides a fifth and agriculture, directly, less than a sixth (Holden, 1976). Organic P moves more readily in soil, and, in areas of intensive cattle or dairy production, the concentration of phosphorus in the drainage water is much increased. For example, Table 2.5 gives some results of Smith (1976) for drainage water from the hill land and lowland farms into the River Main in County Antrim, showing that increasing stocking rate from hill farming with a stocking intensity of 0.62 dairy cow equivalent (DCE) per hectare to lowland farms with a stocking intensity of 0.86 DCE ha^{-1} increased the total soluble orthophosphate, but there was little further increase with more intensive stocking.

Table 2.5. Mean concentrations of nutrients (mg l^{-1}) in rainwater and land drainage from hill land and three areas of lowland with increasing intensities of livestock farming expressed as dairy cow equivalents (DCE) per hectare. Rainwater was analysed only for total P and nitrate N, values for orthophosphate P and total N were not available. For drainage water, samples were filtered before analysing for soluble components. (From Smith, 1976, in *Agriculture and Water Quality*, MAFF Technical Bulletin no. 32. Reproduced by permission of the Controller of Her Majesty's Stationery Office.)

Element	Rainwater	Drainage water			
		Hill land	Lowland		
		0.62*	0.86*	1.66*	3.98*
Soluble PO_4-P	nd†	0.02	0.04	0.07	0.05
Total P	0.06	0.12	0.13	0.17	0.17
NO_3-N	0.4	0.7	5.5	6.1	7.2
Total N	nd	1.5	6.4	7.1	8.2
Cl	nd	16.7	19.1	19.7	22.5
Na	4.3	16.7	16.9	20.0	19.6
K	1.0	0.9	1.1	1.9	1.1
Ca	2.0	17.6	32.2	35.9	36.8
Mg	0.6	9.7	17.4	15.5	17.8

* Intensity of livestock farming (DCE ha^{-1}).
† nd = not determined.

However, the other forms of phosphorus remained almost constant from the hill land and lowland areas. The problems of intensive animal husbandry are discussed later.

INTENSIVE LIVESTOCK FARMING AND WATER QUALITY

Results in Table 2.5 show how increasing the intensity of livestock farming from 0.62 DCE ha^{-1} to 3.98 DCE ha^{-1} progressively increased the inorganic N in the drainage water, but the change from hill to lowland farming increased the soluble orthophosphate irrespective of the intensity of lowland farming (Smith, 1976). The intensification of livestock farming has been taking place over many years and is frequently quoted as a factor increasing the problem of animal slurries as potential pollutants. The nature and scale of the problem are related to the amount of time that animals are housed. Sheep spend almost all their time in the field and no problems have been reported with them. Lowland beef and dairy cattle are normally housed for part of the year, and usually the area required for grazing is sufficient to accept the slurry accumulated during the housing period, so that few problems of slurry disposal have been reported with cattle (Richardson, 1976). In marked contrast, pigs and poultry are increasingly housed in intensive units, and pose the most frequent and most serious pollution problems. Although we are not directly concerned with smell nuisance, this topic is one of great importance, and much effort is being devoted to ways of decreasing smells from intensively housed pigs and poultry and during the storage and spreading of pig slurries. The number of cattle, pigs, and poultry kept on each holding has progressively increased, whereas the number of holdings has fallen (Foot, 1976). These trends are well illustrated in Table 2.6 by the data for pigs. During the period 1960–74 the total number of pigs increased from 2.46×10^6 7.71×10^6, so that the average size of holding increased from 22 to 189 head of pigs. Richardson (1976) calculated the approximate numbers of animals required to provide

Table 2.6. Numbers of pig herds by size of herd in England and Wales, 1960–74 (from Foot, 1976)

	Herd size			
Year	Over 5000	1000–4999	Under 1000	Total
1960	130		110534	110664
1970	11	714	59128	59853
1974	48	1249	39491	40788

most of the nutrient requirements of various crops. The values in Table 2.7 show that there is not likely to be a problem with dairy cattle, because the stocking rate of cattle grazing for six months of each year is unlikely to exceed the capability of the land to accept the slurry. However, the number of pigs and poultry that can be kept under intensive conditions can easily demand more land than is available for slurry disposal. Apart from these considerations of gross production, many livestock are housed in winter when crop growth is least and soil leaching is most intensive, and access to the land is often restricted by wet weather, particularly on heavy soils. Storage is therefore necessary, and some way of determining when and how much slurry may be applied is required.

Development of a Land-Spreading Model

Slurry has two main attributes, its volume and nutrient content. In order to achieve the maximum disposal of slurry on land, although the amount of nutrients applied may be ignored, the risk of pollution should be minimized. The efficient use of slurry as exemplified in Table 2.7 demands that sufficient slurry is applied to the soil at the correct time to meet the nutrient needs of the crop; this is normally less than that allowed by considering the maximum volume the land can accept. Grass allows much greater access and flexibility than other crops. For calculating the maximum amounts that can safely be applied to the soil (known as hydraulic loading of the land), the evapotranspiration is compared with the rainfall to give a soil moisture deficit, allowing slurry to be applied to remove this deficit (O'Callaghan *et al.*, 1971). O'Callaghan *et al.* calculated these loadings for different regions of the UK using published figures for rainfall and calculating the potential evapotranspiration. However, low hydraulic conductivity of the soil in the immediate vicinity of the roots may prevent the potential evapotranspiration being attained, which overestimates the deficit.

Table 2.7. Approximate numbers of animals supplying the nutrient requirements of 1 ha of various crops. (From Richardson, 1976, in *Agriculture and Water Quality*, MAFF Technical Bulletin no. 32. Reproduced by permission of the Controller of Her Majesty's Stationery Office.)

Crop	Dairy cattle housed for 6 months	Pigs housed for 12 months	Poultry housed for 12 months
Cut grass	6.2	100	700
Cereals	1.2	20	220
Potatoes	4.2	50	350

The soil moisture deficit approach was applied to four contrasting cropping patterns. The ability to spread slurry, represented by 'go/no go' matrices, determined in which weeks slurry may be spread. The four crops chosen as examples were grassland, grassland that is late in drying (for example, water meadows), cereals, and sacrifice areas assumed to be cropped. On a sacrifice area, access is permitted each week when there is an appropriate soil moisture deficit; for grassland, cyclical patterns of spreading and conservation have to be established; for cereals, least disposal is possible, because less slurry can be applied and only during limited periods.

Figure 2.1 gives a simplified flow diagram of a land-spreading programme indicating the relevant steps in calculating the amount of slurry that may be applied. For each region, values for the potential evapotranspiration during the summer are determined from the national data. The decrease from the potential evapotranspiration is read from a matrix, followed by the rainfall through the summer months for the last 15 years. For a particular farm under consideration, data required are the total area, the number and area of the fields, the cropping pattern on these fields, and the elevation of the farm together with the average elevation of the county.

The programme was applied to three widely separated places with differing rainfall and insolation. They were Durham City, Vange Reservoir in Essex, and Swyddflynon, Dyfed (previously Cardiganshire), with average maximum soil moisture deficits through the year of 36, 173, and 8 mm, respectively, representing the great range experience in England and Wales. In each case, cereals provided least opportunity for spreading, grassland more, and sacrifice about three times as much as cereals. As indicated by the values for soil moisture deficit, Essex allows most to be spread and Dyfed least.

Field Experiments Testing Land Spreading

The management implications under real farming conditions of a hydrological limit on the amount of slurry which could be applied to a grass crop in the growing season was demonstrated in a two-year field trial at two farm sites (O'Callaghan and Pollock, 1976). Sufficient slurry was spread on grass plots during the growing season to hold the soil near the field's slurry capacity and the drainage water was monitored for changes in suspended solids, biochemical oxygen demand BOD_5, nitrate and ammonium nitrogen, phosphorus, and potassium, which showed that some slurry leaked to the drains after an application, especially when rain fell within a day after the slurry was applied.

Drain flow only occurred at one site where the drainage water had BOD_5 values up to 800 mg l^{-1} when 6.25 mm of slurry was applied and 1000–3000 mg l^{-1} when 12.5 mm was applied. There were correspondingly large increases in the concentrations of N, P, and K. Most measurements returned to their

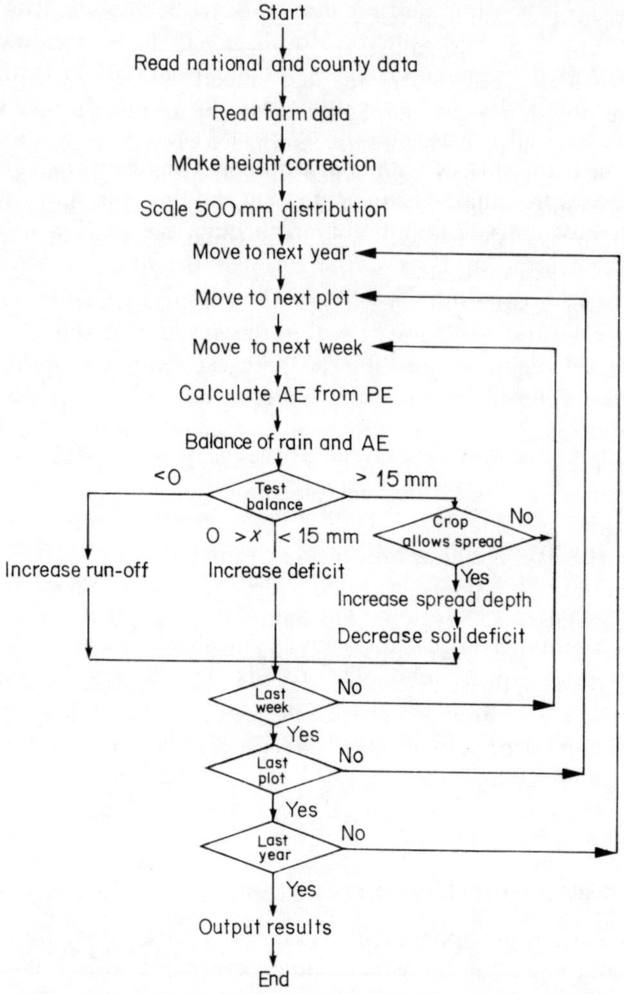

Figure 2.1. Simplified flow diagram for land-spreading model (AE, actual or calculated evapotranspiration; PE, potential evapotranspiration). (From O'Callaghan *et al.* 1971, *J. Agric. Eng. Res.* Reproduced by permission of The British Society for Research in Agricultural Engineering.)

background levels within 36 h. Although the maximum concentrations are unacceptably high for continuous discharge, they are of short duration and the total BOD and other constituents of the slurry found in the drainage water accounted for less than one-fiftieth of that applied. Not only the soil type and the amount of slurry applied affect the movement of slurry into drainage water; the depth of the drain and the nature of the backfill are also important. This is illustrated by some results of McAllister (1976) who found that water from two contrasted drainage systems contained markedly differing amounts of slurry constituents after its application. One site had fairly deep drains (1.25 m) which were backfilled with soil; the other had shallow drains (0.5 m) which were filled to near the surface with coarse stones. The results in Table 2.8 show that there were approximately similar flows at the time the slurry was applied. The analyses show considerable pollution through the shallow drains for two hours after application and very little through the deep drains. The values of chemical oxygen demand (COD) and phosphorus content of water from the shallow drains before slurry was applied suggest that solid material was entering the drains, and in the final period, 48 h after treatment, heavy rain again increased the drainage and the values for COD and phosphorus.

These results show how quickly constituents of slurry can enter drainage water, and also that slurry rapidly reaches equilibrium with the soil so that there is no further loss. From these findings, the conclusion can be reached that, immediately after application, slurry does not enter the soil uniformly but that a proportion follows an easier and preferred path to considerable depths in the soil. At the present time, we can only conjecture that these paths represent discontinuities in the soil, such as voids between structural units, old root channels and worm holes. The results also demonstrate that simple models of land spreading need to be tested under real conditions to identify important parameters not included in the model. This is true for all effluents spread on land and not just animal slurries.

Table 2.8. Composition of drainage water from sites with deep, D, (1.25 m) and shallow, S, (0.5 m) drains before and after the application of slurry. (From McAllister, 1976, in *Agriculture and Water Quality*, MAFF Technical Bulletin no. 32. Reproduced by permission of the Controller of Her Majesty's Stationery Office.)

Hours after spreading	Flow rate ($l\ h^{-1}$)		BOD ($mg\ l^{-1}$)		COD ($mg\ l^{-1}$)		N ($mg\ l^{-1}$)		P ($mg\ l^{-1}$)	
	D	S	D	S	D	S	D	S	D	S
0	955	1022	3	4	14	128	18	6	0.06	1.27
0.5	964	1076	10	530	21	947	18	91	0.28	9.54
1	976	1106	5	215	15	427	19	36	0.20	3.69
2	970	1031	3	34	18	179	17	12	0.13	1.17
4	942	1104	2	7	18	90	8	6	0.07	0.37
8	888	570	2	7	14	46	10	7	0.06	0.45
24	1042	783	neg.	3	16	27	10	5	0.05	0.16
48	1042	4200	neg.	3	12	69	7	5	0.04	0.65

In the above experiments, the slurry was uniformly distributed on the surface of the land in moderate quantities. In the field, cattle void both urine and dung in discrete patches containing relatively large amounts of nutrients. Urine contains much nitrogen and potassium and dung contains nitrogen and most of the excreted phosphorus. These patches act as sources of point pollution and it has already been demonstrated that on intensifying cattle production, more nitrogen and phosphorus are found in the drainage water. In a study of the effects of spreading heavy dressings of cattle slurry on grassland, Burford *et al.* (1976) measured the effects on yields of herbage, and on the composition of the soil atmosphere and drainage waters. One plot received no slurry, the intermediate rates were 150, 300, and 450 tonnes ha^{-1} and the maximum rate applied was 550 tonnes ha^{-1} of wet material containing 156 g l^{-1} of dry material. The dry material contained 2.15% N, 1.02% P, and 2.03% K, so that the largest amount of slurry supplied 1845 kg ha^{-1} N. This may be compared with the values of 450 kg ha^{-1} N in a urine patch and 850 kg ha^{-1} N in a dung patch (Petersen *et al.*, 1956). The largest rate used by Burford *et al.* is slightly greater than the combined voiding of urine and dung at one point, but is useful in providing information on the processes that may occur. The best early growth of grass was with 150 tonnes ha^{-1} and best late growth was with 300 tonnes ha^{-1} of slurry applied in spring. Growth was depressed by larger amounts.

Methane was evolved in the soil for several months after applying the slurry, indicating the presence of intense reducing activity at shallow depth. Nitrate initially present in the soil and slurry was lost by denitrification, and nitrate accumulation was inhibited for some time, either by poor aeration itself or by products resulting from restricted aeration. Aeration gradually improved with time from the slurry surface downwards, presumably allowing nitrification to proceed and the nitrate formed moved to the deeper soil layers. However, a zone of poorer aeration and intense reducing activity below the zone of nitrate accumulation resulted in denitrification. Nitrous oxide was found in both these zones after rainfall, confirming that denitrification was occurring and thus lessening the amount of nitrate leached during this time. These processes are illustrated diagramatically in Figure 2.2 where the dotted area represents the zone preventing downwards movement of nitrate, and the shaded area represents the zone of intermittent or slow gaseous nitrogen loss depending on soil moisture. Losses of nitrogen as N_2O cannot be calculated accurately from observed concentrations in the soil atmosphere because of the difficulty in establishing transport rates between the source of production in the soil and the soil surface (Burford and Stefanson, 1973). In addition to the need for assessments of the magnitude of N_2O production, there is also a need for measurement of losses as N_2, which may be the significant or dominant product in denitrification. Knowledge of the ratio of N_2O:N_2 in gases evolved from the soil may assist in interpretation of field data.

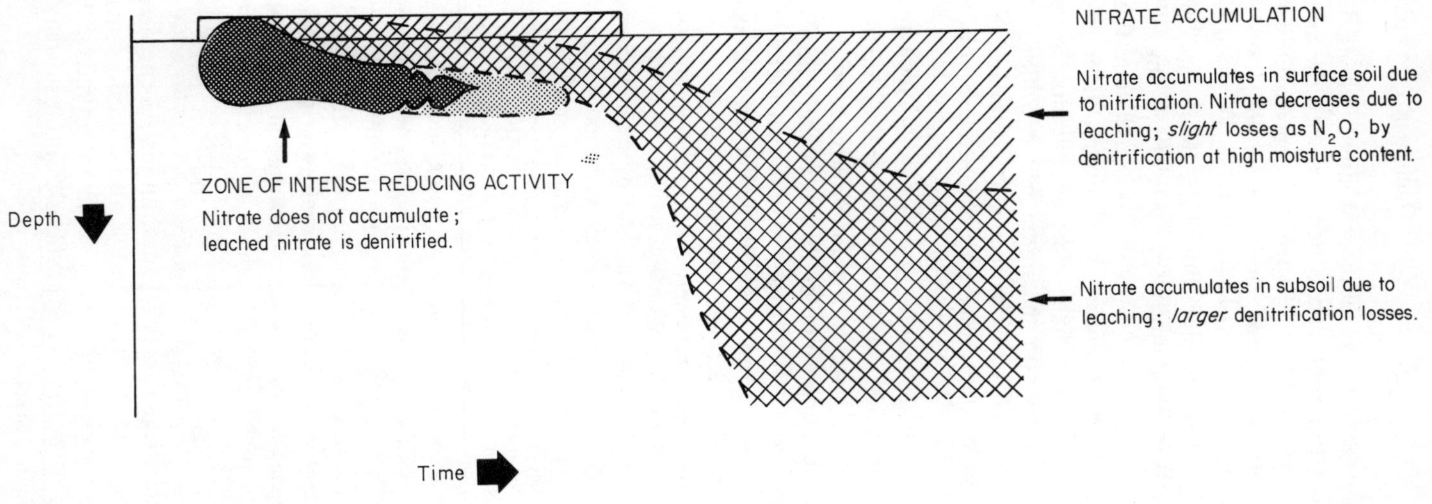

Figure 2.2. Diagram illustrating the pattern of nitrate accumulation and loss following a heavy slurry application. (From Burford *et al.*, 1976, in *Agriculture and Water Quality*, MAFF Technical Bulletin no. 32. Reproduced by permission of the Controller of Her Majesty's Stationery Office.)

Analyses of the drainage waters eight months after applying the cattle slurry showed that the concentration of nitrate increased with the amount of slurry applied (Burford *et al.*, 1976). Table 2.9 gives the concentrations in December 1972 for the four plots with slurry and one without, together with the calculated values for the loss of nitrogen. These data show that, with 150 tonnes ha^{-1} of slurry, on two occasions out of five the nitrate concentration of drainage exceeded the WHO recommended value of 11.3 mg l^{-1} NO_3–N and with 550 tonnes ha^{-1} of slurry, values were two to five times this value. Table 2.9 also gives the weight of nitrate in the drainage water and shows that maximum flow was at an intermediate value of concentration so that maximum losses do not coincide with maximum concentrations. Although the figures for losses from the different treatments are comparable, they are not absolute because the drains intercepted an unknown proportion of the drainage water.

Qualitative models, such as that of Burford *et al.* (1976), are useful in providing information on likely pathways for the transformations and movements of nutrients in slurry. However, the quantitative aspects of the effects of slurry applications on the composition of drainage water and the reduction of nitrate to nitrogen and oxides of nitrogen require considerably more knowledge of the various processes before a reliable model can be constructed. Both the biological and chemical aspects are receiving attention in North America and Europe, so that further advances may be expected in constructing models and their quantitative application.

Table 2.9. Nitrate N in drainage water in December from plots receiving slurry in March. (From Burford *et al.*, 1976, in *Agriculture and Water Quality*, MAFF Technical Bulletin no. 32. Reproduced by permission of the Controller of Her Majesty's Stationery Office.)

	NO_3-N concentration (mg l^{-1} N)						NO_3-N removed (kg ha^{-1} N)				
	Slurry application rate (tonnes ha^{-1})						Slurry application rate (tonnes ha^{-1})				
Date	0	150	300	450	550		0	150	300	450	550
	Weight of N applied (kg ha^{-1} N)						Weight of N applied (kg ha^{-1} N)				
	0	503	1006	1509	1845		0	503	1006	1509	1845
6 Dec.	4	7	10	20	27		0	0.2	0.3*	0.4	1.4*
7 Dec.	7	9	15	26	43		0.5	0.7	2.1*	8.5	8.3
8 Dec.	7	12	14	18	56		0.2	0.5	0.2	0.3	2.4
10 Dec.	6	10	12	16	59		0.1	0.5	0.1	0.2	0.5*
12 Dec.	7	17	23	31	61		0.1	0.6	0	0.3	0.1
Mean	6	11	15	22	49	Total	0.9	2.5	2.7	9.7	12.7

*Assumed values; incomplete flow measurement.

PESTICIDES AND WATER QUALITY

The use of chemicals for control of weeds, pests, and diseases has increased enormously from 1950 onwards. The term 'pesticides' usually includes insecticides, acaricides, nematicides, molluscicides, herbicides, fungicides, soil fumigants, and growth regulators. Before 1939, a few natural insecticides such as derris, pyrethrum, and nicotine were used, together with largely inorganic compounds such as lead arsenate, sodium arsenite, copper in various forms, sulphur, and mercury. Winter washes such as tar oil were common. In 1950 there were 15 chemicals in the Approvals List of the Ministry of Agriculture, Fisheries and Food, used in nearly 300 formulations. By 1975, the number of chemicals had increased to 200 and the number of formulations to just over 800; today there are in excess of 600 chemicals and 1500 formulations. Most of these statistics are taken from the introductory paper by Sly (1977) to a symposium of the Linnean Society on 'Ecological Effects of Pesticides', which discussed many aspects of the problems raised. So far as water quality is concerned, a number of issues need to be defined and briefly discussed, particularly the movement of pesticides and their decomposition products into water supplies, the persistence of chemicals in the environment, and the importance of these to ecology, including man. Probably the most persistent pesticides are the organochlorine compounds such as DDT and HCH (formerly called BHC). Traces are to be found almost throughout the world and can be identified because modern analytical methods can measure concentrations so small that their biological significance is uncertain. Today we are aware of the effects of passage through food chains in increasing the concentration of organochlorine compounds in the fatty tissues, and, for this reason, their use is banned in many countries. On the positive side, these untoward effects intensified efforts to find substitutes which would be at least as effective as the organochlorines, but would decompose more quickly, either photochemically or by biodegradation. One outstanding example is the development of the pyrethroid insecticides which are based on the natural pyrethrins found in pyrethrum flowers. One such synthetic compound, bioresmethrin is not only more active than natural pyrethrin 1, but is also even safer to mammals. On the criterion of relative acute toxicity, it is the safest insecticide known and is virtually harmless to mammals (Graham–Bryce, 1977).

Graham–Bryce also discussed the problems of the design of pesticides with appropriate mobility and stability, and he writes:

> 'The use of pesticides with the appropriate stability and mobility for each crop protection problem would do much to minimise damage to non-target organisms. The numerous compounds now in widespread use encompass a wide range of properties, in many cases allowing a reasonable match with the theoretical requirements. There

> remain, however, many situations where the appropriate combination of toxicity, persistence and mobility is not available and there is scope for considerable improvement.
>
> As with physiological selectivity, improved knowledge of degradative processes and of fundamental principles determining behaviour in the environment is providing an increasingly comprehensive basis on which to design toxicant molecules with the desired stability and mobility....
>
> ... An important, but often unrecognised implication of the dependence of stability and physicochemical behaviour on molecular structure is that the compounds which have the greatest intrinsic toxicity in any given chemical series may well differ from those which are most effective in the field. There are thus sound arguments for devoting more attention to the physicochemical properties of potential pesticides at an early stage in their development, not only from environmental considerations, but also because this would help to ensure selection of the most effective products. While such suggestions might be widely accepted in principle, it would require considerable resolution deliberately to set aside the most active compounds in favour of less active ones at an exploratory phase of development.' (Graham–Bryce, 1977, in Perring and Mellanby (eds), *Ecological Effects of Pesticides.* Reproduced by permission of he Linnean Society of London.)

These findings were also echoed for herbicides by Fryer (1977) during his presentation to the same symposium. Intensification of agriculture requires the use of herbicides; they are widely used in arable farming, and will be required increasingly for the improved utilization of grassland. Most of the recently developed herbicides also have a low mammalian toxicity, and the repeated use at the recommended rates has not led to the build-up of phytotoxic residues over a period of 15 years.

There are also a number of other developments taking place which minimize the impact of herbicides on the environment, including water. One of the disadvantages of the normal spray application of herbicides is the large volume of water which has to be supplied, transported, and moved across the field. This amounts to 300 l ha^{-1}. By using a spinning disc to produce droplets of better defined size range and other properties, the volume of liquid can be as little as 15–30 l ha^{-1} and the weight of chemical applied is also much less. There is much interest in this controlled droplet application (CDA) of pesticides.

The concept of integrated weed control involves modifying cultural practices to exploit characteristics of the growth cycle, and this, together with the use of some chemicals, allows more effective control of the weed. This principle has been applied with considerable success to the control of couch grass (*Agropyron repens*) and wild oats (*Avena spp*) in intensive cereal systems (Fryer, 1977). For example, in order to control couch grass with minimum use of herbicides, the farmer must grow good crops of cereals, harvest and clear the field quickly, cultivate immediately and kill any shoots of couch grass before they develop more than three leaves by further cultivation or by a low dose of paraquat.

From time to time, dramatic episodes are caused by the unintentional entry of pesticides into waterways. These are usually caused by carelessness, ignorance or failure to observe the properly prescribed regulations for handling the chemicals. As such, they are really industrial incidents and should not be taken as indications of dangers from the use of chemicals. These episodes do not reflect the inherent safety of the chemicals properly used. The use of pesticides in crop production, protection, and storage is essential if maximum yields of crops are to be attained and the subsequent losses by insect and disease attack are to be minimized.

CONCLUSIONS

Modern intensive agriculture affects water quality in catchment ecosystems. These changes are inevitable and, in order to feed ourselves, we have to accept them; their magnitude needs to be more accurately quantified and the impact on water quality determined. The major implications are concerned with the effects on the nitrate and phosphate concentrations of drainage and deep water supplies. Several farming practices tend to increase the amount of nitrate entering groundwater and watercourses—the shift from grassland to arable agriculture, growing legumes, intensifying animal production—and since in all these systems more nitrogen is in circulation, the use of fertilizer N will increase the leakage. The direct contribution of fertilizers to nitrate in water supplies is slight on present evidence, and, when properly used, fertilizers pose little risk. Organic manures and crop residues, particularly legumes, provide more nitrate during the winter months, which may be lost by leaching or denitrification. Much more work is needed to quantify the contributions from the various sources of nitrate to the drainage and deep water supplies. In addition, much better models of water and nitrate movement, and changes in nitrate concentration during passage through systems, are needed to combine the agricultural, biological, and hydrological information.

Finally, this author considers that the proper use of pesticides applied to soils and crops poses little threat to water quality in ecosystems. Unpredictable effects occur from time to time, and constant vigilance is necessary. Their direct application to watercourses does have profound effects and methods are being devised to minimize these.

CHAPTER 3

Impact of Man in Catchments (ii) Roads and Urbanization

W. Pope

INTRODUCTION

Roads and urbanization are not new phenomena—man's constructional activities have been inducing environmental change for as long a period of time as his agricultural techniques and domestic and industrial wastes. The significance of construction processes in contributing to environmental deterioration in recent times is now being increasingly recognized. A number of factors are responsible for this, including the present-day scale of construction processes, the position of a process within a catchment ecosystem, and the mode of use of the finished product.

Although small-scale urban developments which form part of a catchment ecosystem may be assimilated into stable environments with relative ease, this occurs less frequently with larger developments, especially where an urban area not only occupies the whole of any one catchment but may also occupy several adjacent catchments. For example, Greater London, UK, covers 12 catchments, wholly or partially, with some 7.45 million people and 1580 square kilometres of industrially and residentially developed land. These catchments ultimately drain into the River Thames, a far larger and tidal catchment—yet the impact of surface water run-off on Thames water quality is clearly demonstrable.

The London sewerage system has evolved from the Middle Ages when the natural streams provided the necessary drainage and many of these channels have now become combined sewers, with the original outlets remaining as stormwater overflows, of which there are some 200 discharging into the tidal Thames. A storm on the 16–17 August 1977 contributed high oxygen demanding loads to the river from such overflows and sewage works. The storm commenced in east London and as it moved west with the rising tide, the same body of water in the Thames received successive discharges of storm water resulting in an increased oxygen demanding load and the eventual death of substantial numbers of young smelt (*Osmerus eperlanus*), a species intolerant of dissolved oxygen concentrations below 30% saturation.

In the past, urban development has often been at a site of plentiful water, facilitating dilution of effluents. It is often feasible for small developments to release surface water into a large aquatic ecosystem for natural cycling without deleterious effects—for example, those on seaboards and mature rivers. However, present-day growth is forcing both the congestion of established sites and the expansion of development in less desirable situations, thus generating problems, for example, of water re-use where an abstraction point is downstream from a discharge of surface water effluent.

With the improvements in water quality resulting from treatment of point sources, it is only recently that urban run-off has come to the fore as a potentially undertreated source of pollutants. Research work has shown that this run-off may be the source of some subtly deleterious and insidious compounds. However, what generates this water quality degradation? Precipitation as overland flow is the mechanism of removal of certain compounds from the impervious surfaces of the urban area. When this urban run-off—the releasing of excess urban surface water into the nearest collection channel, man-made or natural—reaches a monitored river, it is noted that the quality of the latter may be lowered.

However, roads and the urban areas they both serve and generate cannot be considered 'apart from' but as a 'part of' the natural environment. Where a whole catchment is urbanized, then that urban area may become the 'natural environment', since with present social and technological trends, there is little evidence to suggest a foreseeable return to a pre-urbanized society. The natural environment for any organism is that which it creates and in which it survives—man creates urban society. The man-made environment we know as 'urban' affects both the quality of water and its movement in the fresh water part of the hydrological cycle. Some changes—notably techniques associated with water re-use and dirty water clean-up—are desirable or necessary for the high material welfare achievements of the modern urban area. Without certain modifications—for example, channelling of effluents away from sources of potable supply, and pathogen removal from drinking water—environmental health, as reflected in the quality of the water resources and consequent public health, would be so lowered as to prevent man assuming the material and social welfare which he now enjoys. However, it is a deterioration in environmental health as reflected in either public health or water clean-up economics/technology which gives attention to a problem, and it is in this respect that more subtle effects of urbanization and industrialization are now of interest.

To be weighed against the general environmental debits recorded in this chapter must be the pre-eminent contribution made by construction and transportation industries to the economies and welfare of developed countries.

In 1974 91% of passenger kilometres in the UK were travelled by road (cf. 8% by rail, with only 3% of passenger journeys). Roads carried 65% of freight tonne kilometres, and 88% of the total weight alone. Transport costs take 11% of total private expenditure (not counting transport for which we pay indirectly in the cost of other produce in the 'consumer society') and accounts for around 10% of the gross domestic product of the UK.

THE NATURE OF DRAINAGE SYSTEMS

Overland flow is usually detrimental to man's activities—and, in excess, can be both dangerous and destructive. The basic necessity for drainage networks is thus two-fold. Water on the road surface (which, by design, has a low porosity in order to increase usable life), during high-intensity rainfall, must be removed to prevent local flooding and inconvenience and ingress to the subsurface layers with consequent damage to the hard core, and also to maintain a relatively safe travelling environment. Although the impervious nature of roads, pavements, car parks, drives, and shopping areas aids communications and improves those environments under heavy use, it also increases the amount of precipitation which becomes overland flow. Hence, urban areas tend to find themselves with flood problems exacerbated by their own development, which they must solve at their own expense. Storms outside the design period of a drainage system may cause that system to function unpredictably, as the water will follow the path of least resistance. This will not be the surcharged sewer, and water will flow overland, down streets, and through property. Major events can inflict great damage and cause considerable inconvenience in urban areas. Two techniques are available for controlling the water once it has reached the ground; the water can be removed as efficiently as possible from the area, or it may be stored, and discharged slowly during and after the storm.

Mechanisms of Stormwater Control

Urban areas increase the flood hydrograph peak by up to six times, the effect decreasing with decreasing flood frequency, and it would appear that increasing the capacity of urban depression and conduit storage may offer the most cost effective solutions to this problem, as discussed by Whipple (1977). However, few urban areas have yet to implement such measures on any scale, and flood alleviation must still be dealt with by the drainage system which creates it.

Draining by water removal is satisfactory for small areas, but unfortunately, in large areas, the channel sizes required to cope with the amounts of water released become increasingly cumbersome. Storage, however, not only reduces the volume and velocity (with associated reduction of scour and erosion) of stormwater passing through the system per unit time, but also aids settlement of solids and hence improves water quality. As with rapid removal, the scale of construction is a problem for large developments—storage basins may not be economically favoured because of the cost penalty incurred on valuable urban land. Similarly prohibitive can be the cost of providing a service that is infrequently used—for example, the opportunity costs on the margin of a design to cope with the once in 100 years storm may be unjustifiable. If designing for the once in 5 years storm, then storage may be a feasible approach for a medium-sized urban area—if land is cheap—since around 1% of the urban surface area is needed to store the water generated. Also, free discharge, and hence storage, must be above the level of the local surface water, or the expense and maintenance of pumping systems must be borne if below-level storage is used. The competition for land in the urban area generally does not permit the full development of this technique.

There has been a compromise reached, therefore, in the construction of drainage systems, where some facets of both techniques are almost universally used. Local parameters dictate the actual combinations of mechanisms employed. The complexity of an urban drainage system is further increased by the incorporation of a network for sewage disposal—both domestic and industrial (see also Chapter 4). The sewage disposal system and stormwater disposal system are both high investment enterprises—and it will not be surprising to find that many areas have compromised with the use of 'combined' systems, especially older developments. More recent developments may have 'separate' systems depending largely on local economic constraints. In a 'partially separate' system (a method evolved in the 1930s in an attempt to save money and drain length), road drains may be connected into the sewer, but drainage from other impervious areas may be separate or combined, depending on physical location.

Within the drainage system, there must be a mechanism of alleviating the surcharging created by an event outside the design period; such mechanisms are the 'storm overflows' referred to earlier. If the system is separate, then this may be less important in water quality, but where there is a dry weather flow of waste water diluted by stormwater, the overflows will also contain a percentage of untreated sewage. The flow rate at which such overflows occur is clearly important in the determination of their effect—in the UK, up to three times dry weather flow is taken at the sewage works for full treatment, three to six times is settled on a short retention time, and anything over six is released without treatment. Overloading causes many discharges to take place below these levels. Although there are 10000–12000 storm overflows in England and Wales, the number of cases where separate systems can be economically justified on present criteria is limited. The established nature of the combined network, and the impracticability of dispensing with overflows altogether, is forcing more use of basic treatment such as settlement. A system such as that in Figure 3.1 may well aid in reducing the pollution from a combined system, and is yet another permutation of the 'storage' or 'removal' problem. This addition to an already complicated network works on the principle that there is a 'first flush' of pollutants in stormwater and that this first flush can be retained, as in Figure 3.2. However, the theory has its limitations. The first flush may not be an accurately described phenomenon to date, and designing on its absolute importance could be a mistake. Also, the overflow will still contain a diluted dry weather flow of foul sewage, and the practical difficulties of siting and designing tanks will still have to be solved at each location. Sewerage systems and stormwater overflows are also dealt with in the next chapter.

The Modified Hydrological Cycle

An ever-increasing proportion of the world population is living in urban areas—63% of the population in the USA and 80% of that in the UK is already sited in 9% and 8.5%, respectively, of the land space. This means that the average population densities of 22 and 608 persons per square kilometre, respectively, are somewhat higher in the urban areas. Such people live in a

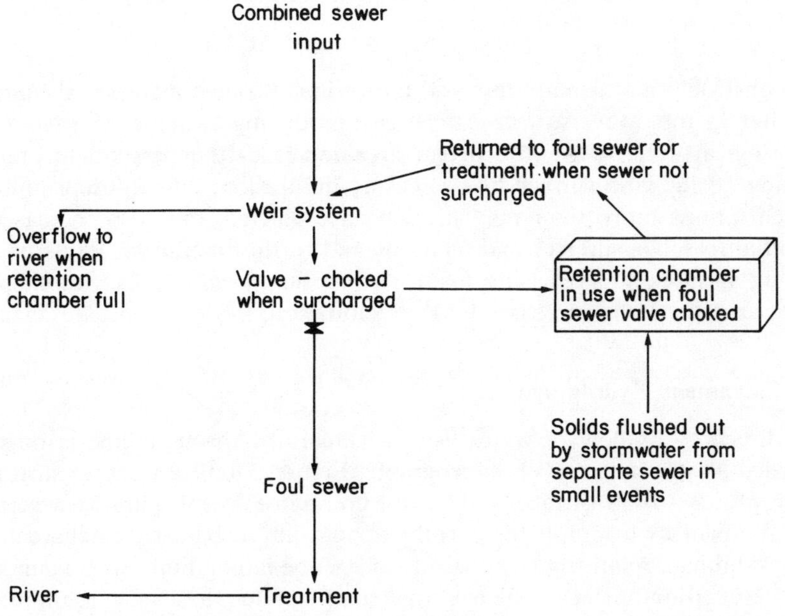

Figure 3.1. Schematic diagram for the addition of a partial retention tank system to a combined sewerage network—attempted removal from overflow of dirtiest water which is returned to foul sewer for treatment

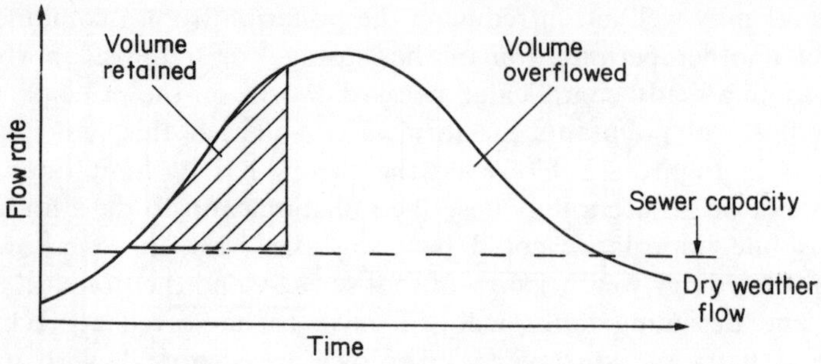

Figure 3.2. Theoretical removal of 'first flush' water from storm-water overflow, by selective retention, for later treatment (from Hedley and Lockley, 1978; reproduced by permission of the Institute of Water Pollution Control)

modified environment in hydrological terms. Figure 3.3 summarizes the effects of the urban area on the hydrological cycle, on the nature of the drainage system and on its points of influence. This leads into the next section which amplifies the nature and magnitude of hydrological changes brought about by the urban area and its drainage system.

HYDROLOGICAL IMPACTS

The construction of urban areas and impervious surfaces increases the amount of water in the above-surface catchment, modifying two major facets of the hydrological cycle. First, the urban area increases the percentage of water running off the land surface by decreasing infiltration, thus diluting pollutant concentrations more than may have been predicted; this also increases the potential to scour surfaces and conduits and to raise pollutant levels by both eroding surfaces and releasing material from temporary sinks. Secondly, the urban area generally increases local precipitation.

The Catchment Hydrograph

The discharge or flow of water from a catchment through time is normally recorded in the form of a hydrograph. Figure 3.4 illustrates a storm hydrograph for a road surface, and can be compared with Figure 3.5 which is a typical storm hydrograph for a rural or partially urbanized catchment. The latter exhibits a relatively steep rising and gentle falling limb, with some base-flow (i.e. groundwater), whereas the road surface has no base-flow and exceedingly sharp rising and falling limbs. The road surface catchment output is more responsive to small input (precipitation) changes, whereas the rural catchment tends to be damped in its response. The scouring capacity of the

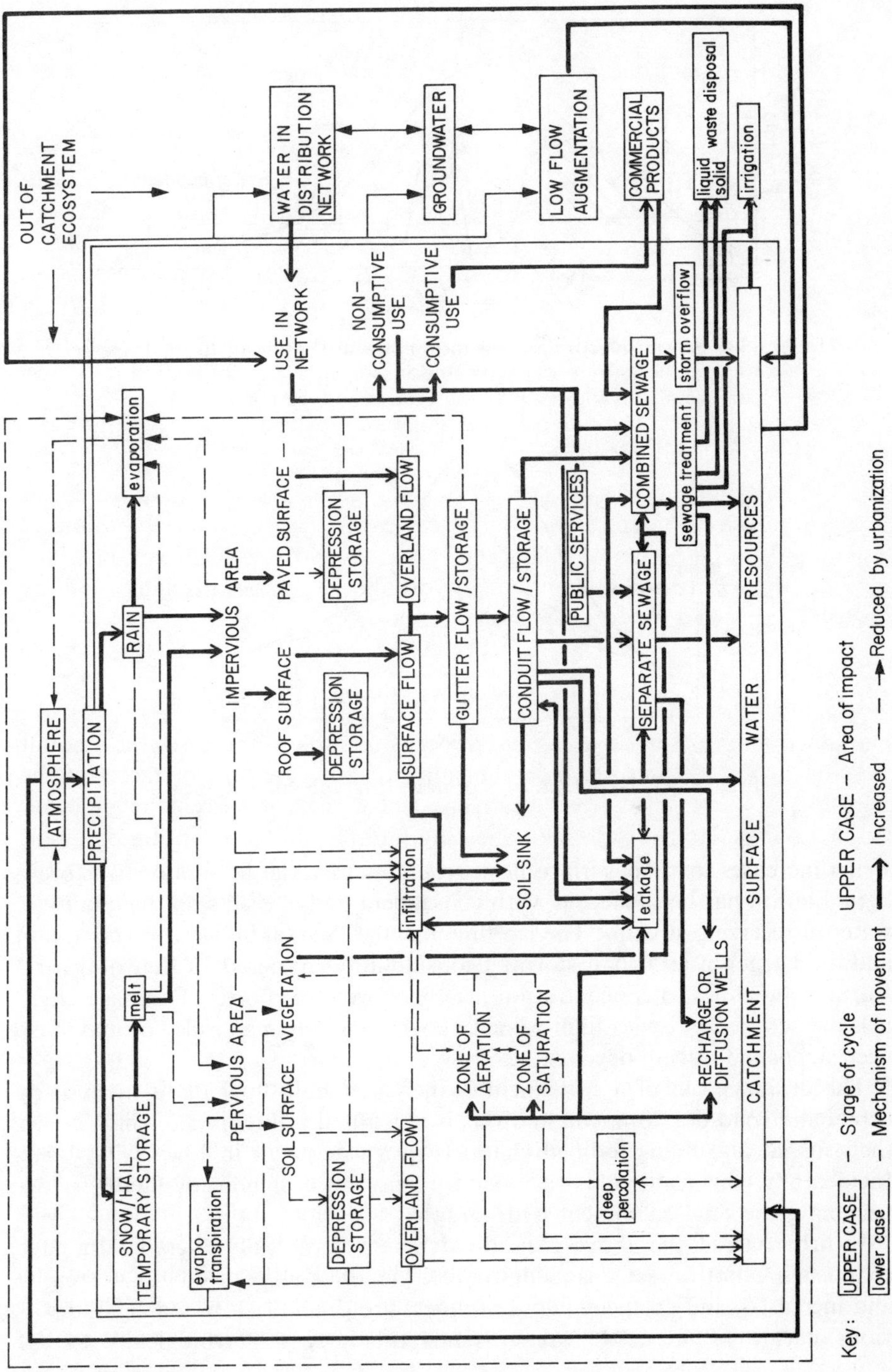

Figure 3.3. Water in the urban environment

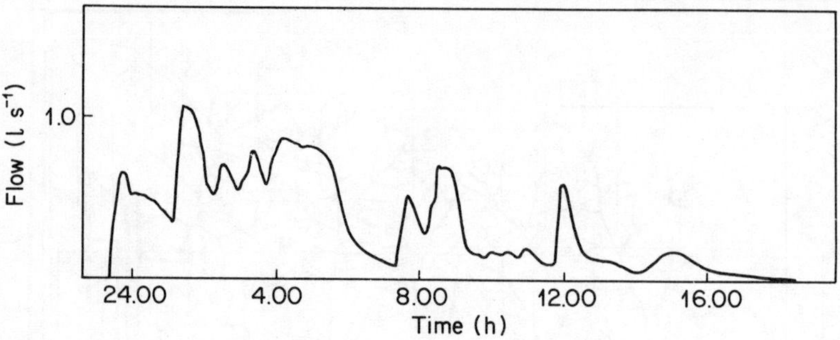

Figure 3.4. Storm hydrograph for motorway surface drain (from Pope *et al.*, 1978b)

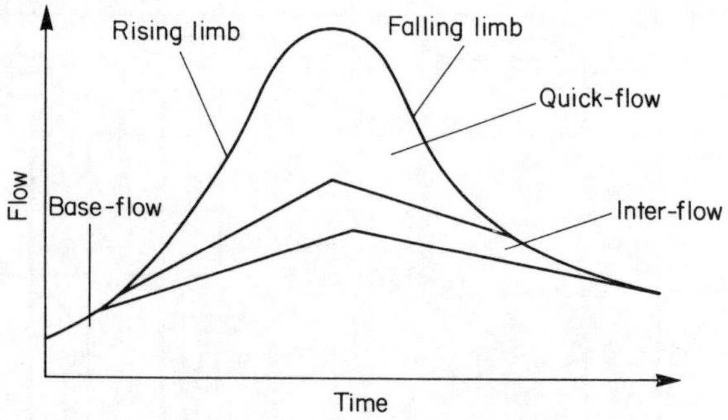

Figure 3.5. Hypothetical storm hydrograph for a typical rural stream

sharp increases in road surface flow generates material for removal before a high dilution has been effected, with consequent 'pulse' effects on the quality of water in receiving streams. The lag time (i.e. the interval between precipitation and discharge) of an urban storm is considerably shorter than that of a rural storm—and times to concentration are therefore also shorter. Drainage channels are scoured out with high-velocity flow, removing material that may have been deposited during dry weather flow.

The drainage net of a catchment is increased (and flow paths altered) by urban and road development, and this is illustrated in Figure 3.6 which shows the original and man-modified channels in a catchment in Iowa City, USA. Incidentally, the resultant street pattern may be a significant factor in determining the final catchment hydrograph.

In urban areas, the above-ground catchment is radically altered—drainage channels are both created and destroyed, angles of slope are both decreased and increased, and coefficients of roughness are generally lowered. The impervious nature of an urban area extends the impact hydrologically to the

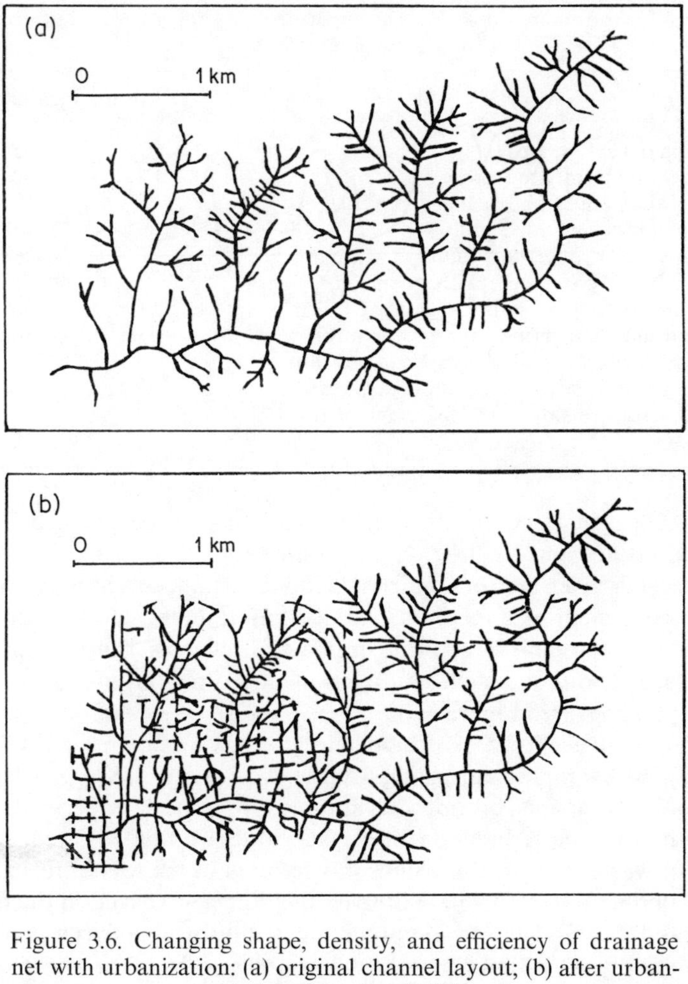

Figure 3.6. Changing shape, density, and efficiency of drainage net with urbanization: (a) original channel layout; (b) after urbanization (from Graf, 1977)

subsurface catchment by reducing infiltration, which together with precipitation governs the total volume of run-off in any one event. Hence, there is a direct relationship between land use (percentage impermeable cover, roofs, streets, and other paved surfaces) and run-off. Kuichling (1909) produced a widely accepted interpretation of this phenomenon which he termed 'impermeability factors' for various surfaces (Table 3.1). Such decreases in permeability tend to increase flood peaks not only directly through increasing run-off but indirectly by lowering the infiltration to groundwater and hence base-flows. The loss in infiltration is also demonstrated in evapotranspiration

Table 3.1. Approximations of the impermeability of various surfaces (from Kuichling, 1909)

Type of surface	Impermeability (%)
Watertight roof surfaces	70–95
Asphalt paving in good order	85–90
Stone, brick, and wooden block pavements:	
with tightly cemented joints	75–85
with open or uncertain joints	50–70
Inferior block pavements with open joints	40–50
Macadam roads and paths	25–60
Gravel roads and paths	15–30
Unpaved surfaces, railway yards, vacant lots	10–30
Parks, gardens, lawns, meadows—depending on the surface slope and character of the subsoil	5–25

changes—Gleason (1952) has estimated that this can be reduced by 30% on residential estates and by 85% on street surfaces.

The effect of a given area of urban land is not constant, and, in addition, both the spatial arrangement of impervious surfaces and the degree to which these are connected to the streams by artificial drainage affect urban run-off and its quality. Also, natural variations such as direction of storm movement can influence hydrograph shape (see Figure 3.7).

The cumulative effects of urbanization on river regime are well demonstrated by an example from Long Island, New York, USA (see Figure 3.8), where the 'conversion of open land areas to paved impervious surfaces, concurrent with the redirection of surface water run-off to streams, and the disposal of wastewater to the oceans has resulted in the loss of recharge to the groundwater reservoir'. Table 3.2 gives the pertinent basin characteristics of three areas (note that the topographic areas do not necessarily coincide with groundwater areas—of importance in considering subsurface transfers between basins). The extent and type of urban development varies widely among basins (Table 3.3), the most intensely urbanized being East Meadow. Figure 3.9 shows the decreases in the ratio of base-flow to total stream-flow caused by the storm sewer systems discharging directly to watercourses and the sanitary sewer network, instead of to soil sinks and recharge basins. The radical changes in stream-flow at East Meadow seem to correlate with sewer construction in the basin during three periods:

(1) recharge through sinks, cess-pools and septic tanks;
(2) construction of the sewer network; and
(3) full use of the network.

During the entire period of the study (1944–74), Carlls River had a slowly increasing proportion of stormwater diverted to its channel, and this is reflected in the slow fall in ratio (Figure 3.9), while the Connetquot River basin

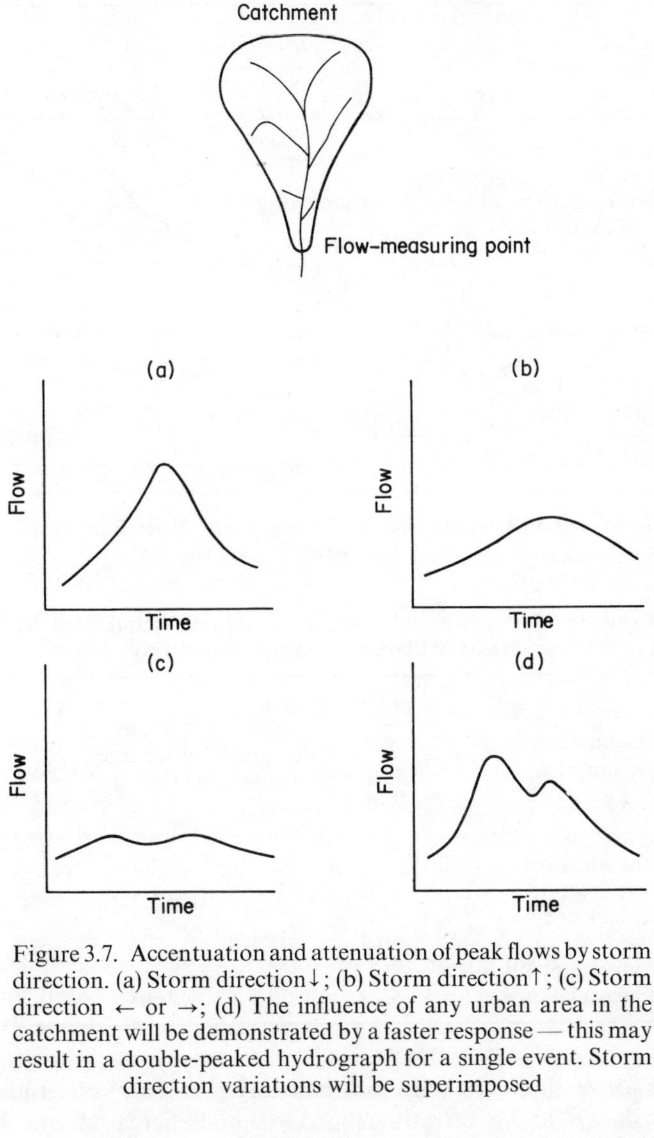

Figure 3.7. Accentuation and attenuation of peak flows by storm direction. (a) Storm direction ↓; (b) Storm direction ↑; (c) Storm direction ← or →; (d) The influence of any urban area in the catchment will be demonstrated by a faster response — this may result in a double-peaked hydrograph for a single event. Storm direction variations will be superimposed

exhibited a constant 96% contribution from base-flow, the catchment being forested, preserving its natural undisturbed conditions.

Precipitation

Urban areas have specific microclimates which may be able, under marginal conditions, to induce excess precipitation, the most significant contributing factor being the thermal characteristics of urban aerosols. Precipitation and

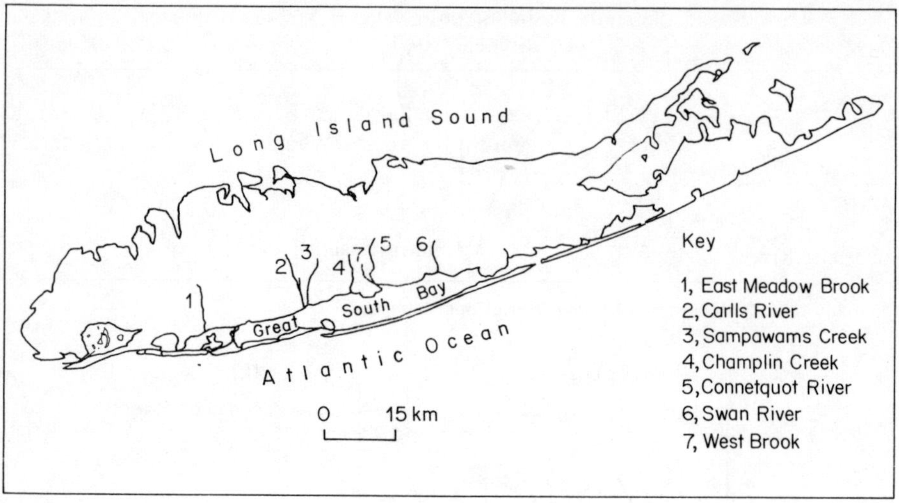

Figure 3.8. Drainage on Long Island, New York, USA. (from Pluhowski and Spinello, 1978)

Table 3.2. Selected basin characteristics, Long Island, New York (from Pluhowski and Spinello, 1978)

Catchment stream	Drainage area: Surface (km^2)	Drainage area: Ground-water (km^2)	Mean discharge ($l\ s^{-1}$)	Mean base-flow ($l\ s^{-1}$)
East Meadow at Freeport	80.0	31.0	416	340
Carlls River at Babylon	90.5	45.5	733	677
Connetquot River near Oakdale	62.0	70.0	1068	1028

the occurrence of thunderstorms and hailstorms at La Porte, Indiana, USA, have been shown to be directly related to, and increased by, the upwind Chicago complex (Changnon, 1968). Similarly, the urban production of heat, moisture, condensation, and freezing nuclei is probably responsible for Munich, West Germany, showing 11% more days of light rain, and Nurnberg 14% more thunderstorms than their rural surroundings (Barry and Chorley, 1968).

In general (partly as a consequence of the 'urban heat island effect'), European and North American cities apparently increase precipitation by 5–

Table 3.3. Land use in the study basins of Long Island, New York (from Pluhowski and Spinello, 1978)

Catchment stream	Percentage cover					
	Forest	Agriculture	Open	Water	Residential	Other*
East Meadow	5	1	22	—	58	14
Carlls River	7	10	25	1	54	3
Connetquot River	27	—	30	2	40	1

* Principally industrial, commercial, and institutional land use.

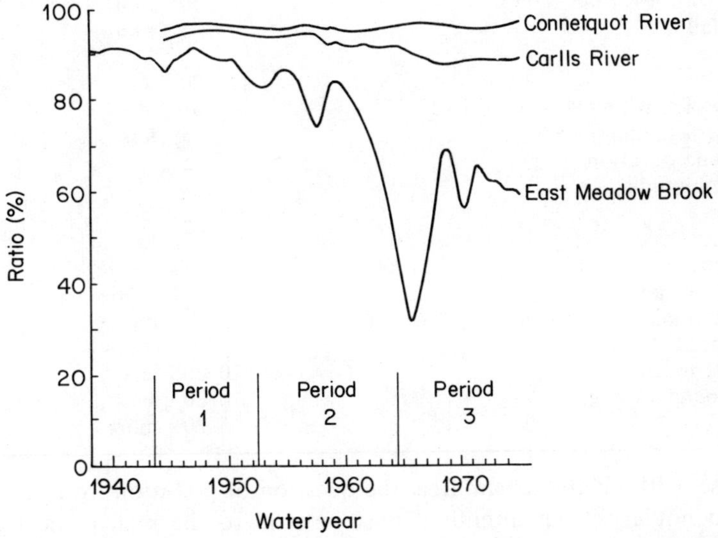

Figure 3.9. Ratio of base-flow to total stream-flow, for selected basins on Long Island, New York, USA (from Pluhowski and Spinello, 1978)

10%, and rain days by 6–7%. Table 3.4 summarizes the impact of urbanization on the atmospheric phase of the hydrological cycle, the precipitation increases, in turn affecting such factors as humidity, radiation, and fog.

It is apparent that the hydrological cycle is extensively modified by urbanization, and the consequent effects on water quality cannot be overlooked. Some factors associated with overland flow tend to increase total pollutant yield, while others tend to reduce pollutant concentration. Hydrological data for an undeveloped catchment cannot, therefore, be reliably used in the predictive sense for water quality modelling when that catchment is developed.

Table 3.4. Approximate microclimatological modifications as a result of urbanization (from Landsberg, 1970)

Element	Comparison with rural environment
Contaminants	
condensation nuclei and particles	10 times more
gaseous emissions	5–25 times more
Cloudiness	
cover	5–10% more
fog, winter	100% more
fog, summer	30% more
Precipitation	
totals	5–10% more
days with less than 5 mm	10% more
snowfall	5% more
Radiation	
global	15–20% less
ultraviolet, winter	30% less
ultraviolet, summer	5% less
sunshine duration	5–15% less
Relative humidity	
winter	2% less
summer	8% less
Temperature	
annual mean	½–1°C more
winter minima (average)	1–2°C more
Wind speed	
annual mean	20–30% less
extreme gusts	10–20% less
calms	5–20% more

Of course, without the urban area, the question of pollution from urban run-off could not arise, but attention must be paid to the hydrological changes concurrent with construction of roads and buildings.

IMPACT ON WATER QUALITY—CONURBATIONS

Water quality impacts of urban run-off must be seen in the context of an ecosystem. If effluents were released into sinks with an infinite capacity, then there would be less cause for concern—however, they are invariably released into systems which have some limit to their capacity for assimilation and recovery. This is increasingly important to human health and water quality, as we exploit water resources from lowland rivers, many of which will be of poorer quality than existing sources of supply. The aims of pollution reduction—or indeed elimination—will be thwarted if non-point sources continue uncontrolled; see, for example, US Public Law PL92-500 (see also Chapter 9) which, rightly or wrongly, aims at the elimination of all pollutants into receiving water by 1985. Such parallel aims in other countries cannot be fulfilled while urban run-off remains untreated. Table 3.5 presents a summary of the available information on the quality of urban run-off, which is compared to the quality of the initial input (precipitation) and the often confused 'combined sewage'. Clearly, there are inputs of many components to the surface waters of a catchment, and it is convenient to deal with these separately.

Atmospheric Inputs

As soon as water is condensed in the atmosphere, it is contaminated, and this is well demonstrated in terms of microclimatological effects. Run-off quality reflects to some degree the quality of precipitation, and rainwater in and around an urban area can be shown to have pollutant concentration gradients—for example, sulphur at Uppsala, Sweden, as shown in Figure 3.10. Sulphur is largely emitted from sulphur-containing fuels; also released into the urban aerosol are particulates, and free and associated organic and metal species. The electrical conductivity of water is a good indicator of dissolved load, and Burt and Day (1977) have demonstrated the conductivity of rainwater to be significantly higher downwind of the industrialized complex of Avonmouth, UK, during single storms (Figure 3.11).

Some material falls to the ground without the stripping or condensation mechanism of precipitation; this is normally wind transported 'dustfall', and its importance, scale, and significance as a contributing factor to urban run-off is poorly documented. However, Barkdoll *et al.* (1977) have estimated that there can be contributions to the chemical oxygen demand (COD), Hg^{+}, Cl^{-}, and PO_4^{3-} load from dustfall to the run-off of a primarily residential catchment. One of the more important effects of dustfall over impervious surfaces is to increase the percentage of small particulates which have a high surface-area-to-weight ratio and may radically influence the transport of pollutants. The smaller particulates are more difficult to remove in sediment traps, and are more readily stripped from these and other sinks as a result of heavy rainfall. They are very inefficiently removed from road and pavements by conventional street cleaning practices, and tend, therefore, to be concentrated relative to the larger sizes; conversely, unless attached to a larger particle for later release, they are easily resuspended in light wind conditions (less than 8 km h^{-1}), such as may be generated by traffic. This tends to redistribute them over the catchment.

Sediments

The total sediment yield of a catchment is governed by the total flow of water and the rate of flow of water. Randall *et al.* (1977) describe a storm in the urbanized Lower Bull Run catchment of the Occoquan watershed, Virginia,

Table 3.5. Summary of the range of parameters and concentrations reported in urban run-off—comparison with precipitation, strictly road surface water, and combined sewer overflow (see footnote for sources)

Parameter	Reported concentration range (mg l^{-1})*			
	Precipitation	Road/street run-off	Urban run-off	Combined sewer overflow
COD	2.5–322	300	5–3100	93–2636
BOD	1.1	25–165	1–700	15–685
Total solids	18–24	474–1070	400–15322	150–2300
Volatile total solids	—	37–86	12–1600	—
Suspended solids	2–13	11–5500	2–11300	20–1700
Volatile suspended solids	6–16	100–1500	12–1268	113
Settleable solids	—	—	0.5–5400	—
Total dissolved solids	—	66–3050	9–574	—
Volatile dissolved solids	—	1630	160	—
Conductance (μmho cm^{-1})	8–395	10000	5.5–20000	—
Turbidity (JTU)	4–7	—	3–70	—
Colour (Pt-Co units)	5–10	—	5–160	—
Total organic carbon	1–18	5.3–49	14–120	—
Total inorganic carbon	0–2.8	—	1–17	—
Oils/hydrocarbons	—	28–400	0–110	—
Phenols	—	—	0–10	—
Total nitrogen N	0.5–9.9	0.18–4.0	1.1–6.2	4.0–63.3
Organic N	0.1–0.32	0.18–3.23	0.1–16	1.5–33.1
Inorganic N	0.69	—	1.0	—
Ammonia N	0.01–0.4	1–2	0.1–14.0	0.1–12.5
Nitrate N	0.02–5.0	0.31–2.62	0.1–2.5	—
Nitrite N	0–0.1	—	0–1.5	—
Total phosphorus	0.001–0.35	0.3–0.7	0.09–4.4	1.0–26.5
Hydrolysable phosphorus	0.8–0.24	—	0.1–10	—
Aldrin	—	—	'trace'	—
Dieldrin	0.003†	6.8×10^{-6}‡	'trace'	—
p, p′-DDD	—	18.9×10^{-6}‡	—	—
p, p′-DDT	—	17.2×10^{-6}‡	—	—
Heptachlor	0.04†	—	'trace'	—
Lindane	—	—	'trace'	—
PCB	—	311×10^{-6}‡	—	—
Bromide	—	—	5	—
Chloride	0.1–1.1	4–70000	2–25000	—
Cadmium	0.013–0.056	0.002–0.01	0.006–0.045	—
Chromium	0.023–0.08	0.018–1.0	0.01–27.0	—
Copper	0.06–0.48	0.007–2.5	0.041–0.45	—
Iron	0–3.05	5–440	0–5.3	—
Lead	0.024–10.4	1–113	0.01–14.5	—
Mercury	—	0.029	—	—
Nickel	—	0.02–1.5	—	—
Zinc	0.02–4.9	1–15	0.01–5.23	—
Total coliform (ml^{-1})	—	—	240–99100	—
Total coliform (organisms km^{-1})	—	15.9×10^{10}	—	—
Faecal coliform (ml^{-1})	—	—	5500–11200	—
Faecal coliform (organisms km^{-1})	—	0.9×10^{10}	—	—
Faecal streptococcus (ml^{-1})	—	—	120–20000	—

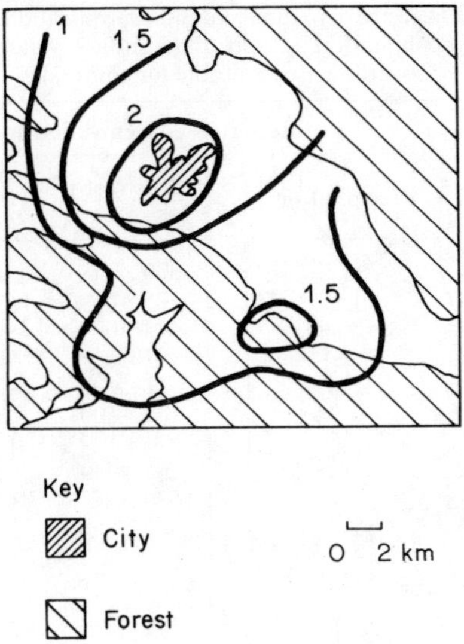

Figure 3.10. Distribution of sulphur (mg l^{-1}) in precipitation around Uppsala, Sweden, September 1962 (from Andersson, 1969; reproduced by permission of The Swedish Geophysical Society)

USA, which increased the concentration of suspended solids in the greater catchment by at least a thousand-fold (Figure 3.12), whilst elevating discharge by four times. The peaks are approximately coincident—and this is a widely reported phenomenon, the rate of run-off being governed by the rate of

---

Notes to Table 3.5.
* Unless otherwise indicated units are mg l^{-1}.
† Units are mg kg^{-1}.
‡ Units are kg curb km^{-1}.
Sources: Angino *et al.* (1972), Borch-Jensen (1978), Brownlee *et al.* (1970), Burt and Day (1977), Cordery (1976), Cowen and Lee (1976), Dauber *et al.* (1978), Ellis (1975), Field and Lager (1975), Goettle (1978), Hedley and Lockley (1978), Holbrook *et al.* (1976), Knight and Rimkus (1977), Koplyay and Buxton (1977), Lavalle (1973), Lloyd (1975), Mattraw and Sherwood (1977), Milaszewski and Osuch-Pajdzinska (1972), Mueller and Anderson (1978), Norton (1974), Oliver *et al.* (1974), Pecher (1974), Peukert (1974), Pope *et al.* (1978a, 1978b), Randall *et al.* (1977), Roberts *et al.* (1977), Sartor *et al.* (1974), Söderlund and Lehtinen (1973), Waller and Coulter (1974), Wanielista *et al.* (1977), Weibel *et al.* (1966).

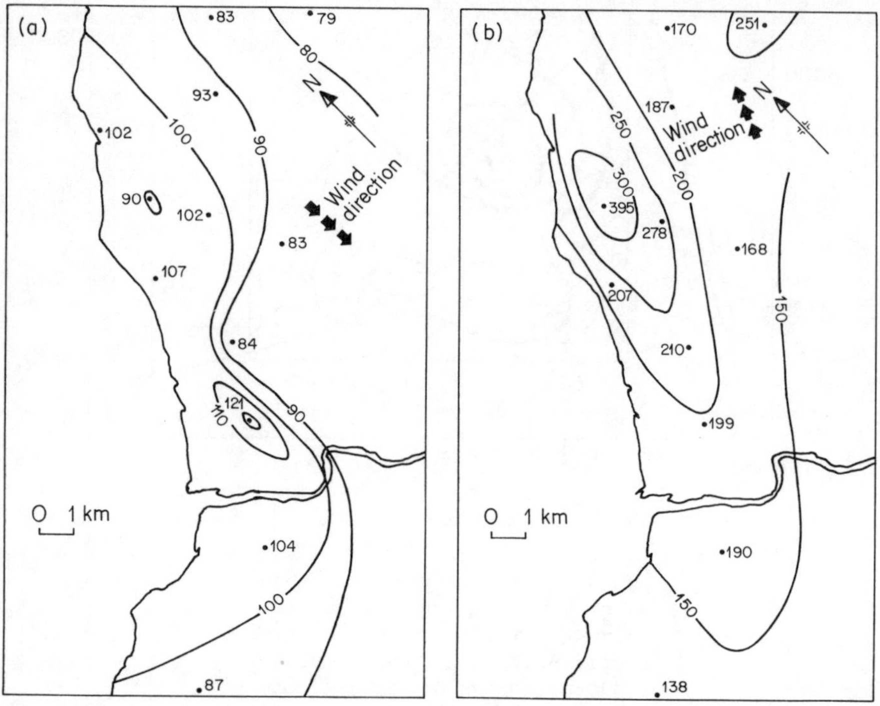

Figure 3.11. Electrical conductivity (μmho cm^{-1}) of precipitation samples around Avonmouth, Avon UK: (a) wind from north; (b) wind from south-west (from Burt and Day, 1977; reproduced by permission of Gordon and Breach Science Publisher Inc.)

overland flow, which in turn governs solid transport. Whipple *et al.* (1974), for example, report similar characteristics for Mile Run, New Jersey, USA, which has residential (39%), commercial (15%), and industrial (5%) areas; here the discharge hydrograph also appears largely to dictate the suspended solid load, as shown in Figure 3.13. The problem with this type of data is that not all the components are uniquely supplied by the urban run-off—the authors themselves state, 'The influence of industrial wastes was shown ... by direct observations of heated and coloured discharges ... and rapid changes in BOD at certain times of the day'. This amply demonstrates the difficulty and necessity of defining the nature and source of the sample more accurately.

In a recent study in Switzerland, Roberts *et al.* (1977) demonstrated that, in an area relatively free of traffic, the urban run-off (as sampled in a sewer separated from domestic and industrial wastes) was fairly clean. The run-off was at least up to secondary sewage standard, except in terms of suspended solids; despite regular and proficient maintenance of the streets and catchpits, the mean suspended solids concentration over a six-month period in 1975 was

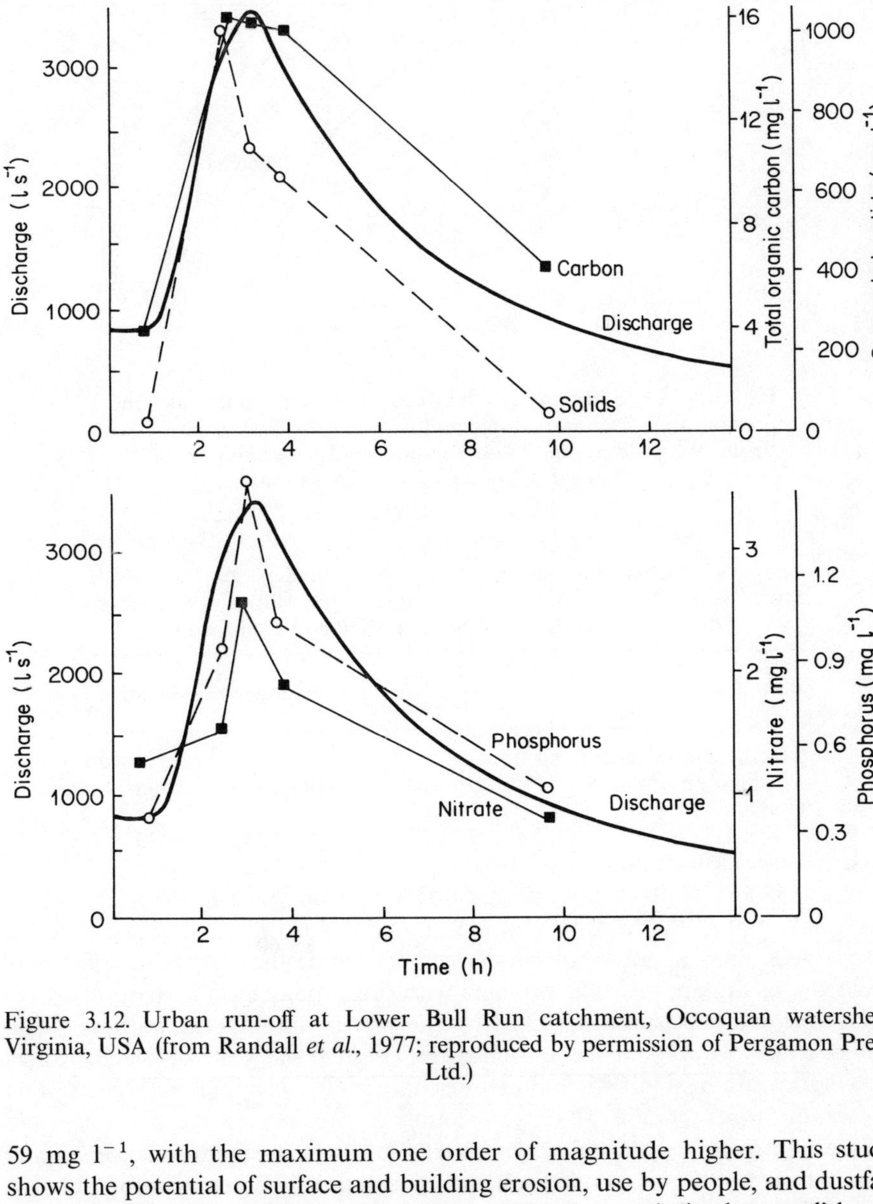

Figure 3.12. Urban run-off at Lower Bull Run catchment, Occoquan watershed, Virginia, USA (from Randall *et al.*, 1977; reproduced by permission of Pergamon Press Ltd.)

59 mg l^{-1}, with the maximum one order of magnitude higher. This study shows the potential of surface and building erosion, use by people, and dustfall to contribute to suspended solids loading. However, such 'background' levels can be grossly exaggerated in the heavily used, older urban areas. Rapid City, South Dakota, USA, has been the site of an investigation by Pirner and Harms (1978) on a primary catchment of 7 km^2, 70% developed with middle class, single family dwellings; 19% of this area is impervious. Table 3.6 shows

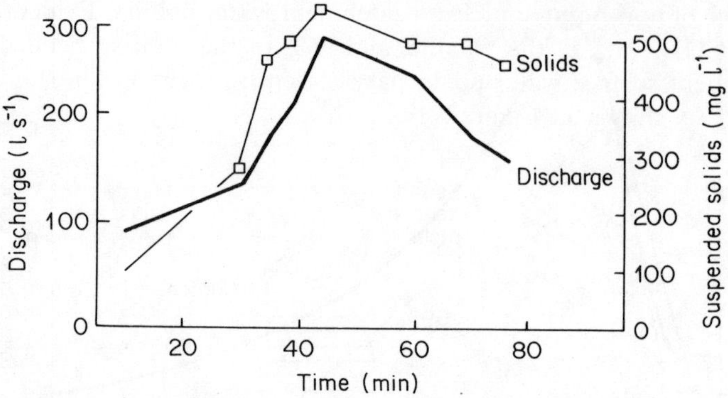

Figure 3.13. Relationship between sediment removal and precipitation—water quality data for Mile Run, New Jersey, USA (from Whipple *et al.*, 1974; reproduced by permission of the Water Pollution Control Federation)

Table 3.6. Urban run-off water quality parameters for the Meade Street basin, Rapid City, USA (from Pirner and Harms, 1978; reproduced by permission of Scranton Gillette Commun.)

Parameter	Number of samples	Range	Mean
Specific conductance (μmho cm^{-1} at 25°C)	58	140–2400	365
COD (mg l^{-1})	55	50–925	230
Suspended solids (mg l^{-1})	60	62–7518	2564
Volatile suspended solids (mg l^{-1})	60	12–603	225
Total phosphorus (mg l^{-1} P)	60	0.09–0.4	0.23
Soluble phosphorus (mg l^{-1} P)	59	0.02–0.17	0.11

the results from a number of completely sampled precipitation events, and clearly the mean for suspended solids is considerably greater than found in the Swiss study. The authors report a strong first flush for all parameters, citing, for example, one event which released 178000 kg of suspended solids (equal to seven months continuous discharge from the catchment sewage works) into a stream-flow of only 2000 l s^{-1}. Over 50% of the solids came in only 33 min, and the total in only 3 h, thus demonstrating the need for more concern over the importance of non-point sources to catchment water quality. Evidently, 31% of the total COD, 77% of the volatile, and 96% of the total suspended solids in the catchment surface water could have arisen from the urban run-off during June 1975, as shown in Figure 3.14.

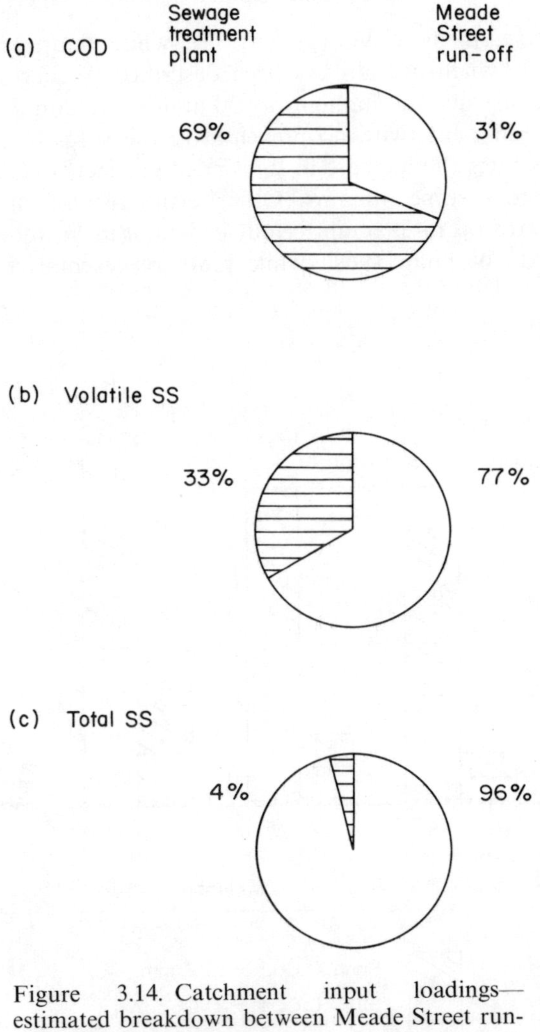

Figure 3.14. Catchment input loadings—estimated breakdown between Meade Street run-off and Rapid City wastewater treatment plant, June 1975: (a) COD 82600 kg; (b) volatile suspended solids 40000 kg; (c) total suspended solids 384500 kg (from Pirner and Harms, 1978; reproduced by permission of Scranton Gillette Commun.)

Urbanization may be divided into three phases, usually with some degree of overlap:

(1) initial construction;
(2) construction complete—use for original intention; and
(3) some degree of modification and reconstruction.

In terms of sediment yield, (1) and (3), while temporary, may be as significant as (2), which has already been described. Weber and Reed (1976) estimated that construction of a highway 30 m in width could contribute up to 650 kg m^{-1} of sediment (with no preventative measures to control erosion). Figure 3.15 illustrates the increase in suspended sediments resulting from road construction across some streams. Other estimates for areas under construction show erosion rates of up to four orders of magnitude above those of agricultural land (Wolman 1967), while more representative perhaps is the

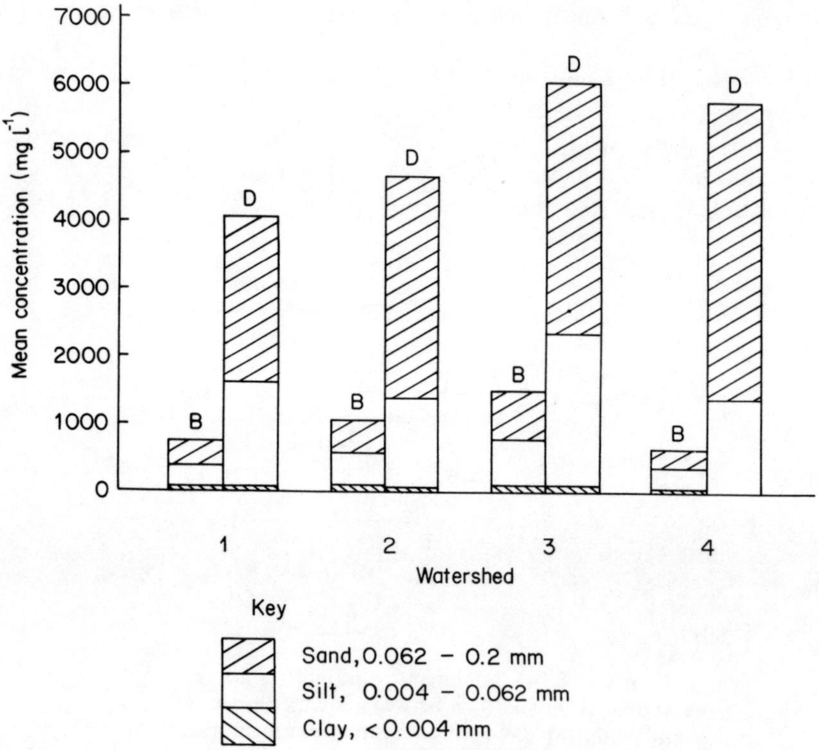

Figure 3.15. Sand, silt, and clay in water samples before (B) and during (D) highway construction crossing Conodoquinet Creek tributaries, Pennsylvania, USA (from Weber and Reed, 1976; reproduced by permission of the American Society of Civil Engineers)

increase of one order estimated by Walling and Gregory (1970) for a small catchment which the authors monitored prior to and during the onset of building activity.

Sediment erosion from road construction has been demonstrated to reduce both the diversity and density of fish and macrobenthic communities in receiving waters (Reed, 1976). Diversity indices indicated a long-term change in aquatic community structure, although fish reportedly repopulated the area of increased siltation within 12 months of construction finishing.

Nutrients

Phosphate and nitrate can readily reach concentrations of 2 mg l^{-1} in the first flush of urban run-off, with upper limits around 3.5 and 2.5 mg l^{-1}, respectively. Average concentrations during storms are generally lower, although Cordery (1976) reports work in Australia showing the nutrient concentrations to remain fairly constant, unless there is a high peak discharge, which may increase phosphate removal. The importance of sediments in transporting nutrients cannot be overemphasized, and a good example is that of the Lower Bull Run catchment (Figure 3.12), where there are good correlations between sediment transport and nutrient concentration. On the basis of this type of work, Cordery concluded that, for his urban catchment, the contributions from urban run-off were so great that investment in tertiary treatment of sewage effluent would not be justified. Table 3.7 shows the

Table 3.7. Comparison of sewage effluent, base-flow, and urban storm run-off additions to selected water quality parameters, Lower Bull Run, Virginia, USA (from Randall *et al.*, 1977; reproduced by permission of Pergamon Press Ltd.)

Parameter	Weight (kg) added during a 13 mm event by		
	Run-off*	Effluent	Base-flow
Nitrite N	1.4	0.0	0.0
Nitrate N	52.2	2.5	0.41
Ammonia N	43.1	0.34	1.0
TKN†	89.9	0.68	1.45
Total N	143.5	0.95	0.64
Total P	35.4	0.73	0.77
Ortho P	20.0	0.48	0.45
Suspended solids	46308.0	277.4	150.7
Total organic carbon	479.0	2.86	2.36
BOD_5	296.9	1.77	2.36

* 16.84 km^2 commercial – residential area.
† Total Kjeldahl Nitrogen.

conclusion drawn by Randall *et al.* (1977) that, for an event of 13 mm rainfall, the urban run-off added between one and three orders of magnitude more nutrients by weight to the catchment surface water than the river base-flow or sewage effluents.

On average, around 40% of the total phosphorus in urban run-off is associated with solids, but the range (13–90%) makes assumptions impossible (Cowen and Lee, 1976); a great proportion (about 50%) of this may be adsorbed onto particulates less than 43 μm in size (Sartor *et al.*, 1974). Removal by conventional settlement of such fractions is grossly inefficient, and increases the investment cost of urban run-off treatment. The picture is similar for nitrogen, with up to 65% being adsorbed onto sediments.

Conventional settlement techniques cannot remove great fractions of the mass flow of nutrients, and the importance of taking into account the adsorbed load is only recently being more widely accepted. Since adsorbed nutrients are available for release and dissolution, the traditional ideas of compounds being conservative or non-conservative dissolved substances may lead to excessive errors in water quality simulations. High-velocity flows, for example, can be shown both to move nutrient-laden sediment and to release nutrients directly from bed sediments (Harms, 1977).

Although the arguments are not for this chapter, it does appear that the nutrient load in urban run-off may be sufficient to effect the transition from oligotrophic to eutrophic conditions in receiving waters. If nitrogen or phosphorus is limiting phytoplankton growth in the microgram per litre range, as has been suggested (Stoerner *et al.*, 1978; Welch *et al.*, 1978), then it may well be upheld that urban run-off exceeds the threshold of algal nuisance.

Macro- and Micro-organisms

Although there is a qualitative change in macro-invertebrate diversity indices in streams running through urban areas, there is little documented evidence to show the impact solely of urban run-off. For streams in an urban area, the diversity profile for benthos drops to a minimum in the late summer/early autumn, compared to the healthy area which should remain stable and high (Pratt and Coler, 1976).

Von Kaenel and Tschumi (1978) have reported that motorway effluents draining into a marshy area of Switzerland, despite relatively sophisticated treatment techniques, altered the distribution and density of benthic fauna. Tubificid worms dominated in the road-affected area, whereas midge larvae (*Chironomidae*) dominated in unaffected areas.

Such are the effects in receiving waters—but what of urban run-off as a source of micro-organisms (see Table 3.5); much of the data are confused because of the accidental mixing with sewage, or the sampling of combined sewage overflows. Animal faeces should be the only source of micro-organisms in run-off, and this may be highly significant. For example, in New York, USA, an estimated 500000 dogs probably leave around 5000–20000 tonnes of faeces and 100000–200000 litres of urine on city streets (Ellis, 1975). All will be flushed to stormwater, either separate or combined.

Oliveri *et al.* (1977) found a consistently high recovery of both pathogenic and indicator micro-organisms in urban run-off from Baltimore City, USA. Faecal coliform-to-streptococcus ratios evidently indicated ingress into the stormwater sewers from sanitary sewers—yet again showing the problem of in-sewer sampling. It would also appear that micro-organisms can be expected to be scoured in a similar manner to the other adsorbed components (Matson *et al.*, 1978).

Organic Substances and Heavy Metals

As a measure of the organic load, the biochemical oxygen demand (BOD) is a technique of long standing, and is well reported for urban run-off (see Table 3.5). During dry periods BOD build-up can be expected from surface deposition, although recent work is tending to show that the nature of removal of accumulated material ("mechanism of removal") can be at least as important in dictating the load in any one event. Whipple (1975) estimated that a 50 mm precipitation event put 227000 kg of BOD into the Delaware estuary, USA, from which the river took up to six days to recover in terms of dissolved oxygen. The chemical oxygen demand, which incorporates some inorganic substances, can also be important—indeed, the COD of urban run-off due to the removal of persistent organics of low biodegradability is potentially great, up to two orders of magnitude higher than the corresponding BOD. Again, sediment association is predominant with organic substances (note the suspended sediment:total organic carbon ratio; Figure 3.12), with usually around 80% of total hydrocarbons adsorbed. Specific organic substances are dealt with in the next section (road surface water) due to their immediate association with highway traffic.

Of all the components of urban run-off yet described, heavy metals are best known to have high affinity for suspended solids. Sediments are effective in buffering the concentrations of soluble metals to a lesser or greater degree, depending on the metal and environmental conditions. As before, smaller sediment particles tend to assume a disproportionate importance, and another complicating factor is the tendency for the formation of organic—metal complexes with suspended material. Metal removal through a precipitation event exhibits a pronounced first flush, and concentrations of lead, for example, can reach around 12 mg l^{-1} (Bryan, 1974). Lead and zinc generally comprise about 80–90% of the total metal load by weight, and, because of this and the well documented output of lead from car exhausts, much of the literature is focused exclusively on the former metal.

Wisseman and Cook (1977) describe heavy-metal accumulation in the sediments of Wapato Lake, Washington, USA, and in particular the contribution from urban run-off. The lake is shallow (2 m) and urban; perched over glacial till, it receives the drainage water from its 4.2 km² basin (76% residential, 13% park, 8% highway, 3% lake). Surface water from 40.8 km of various types of roads enters the lake untreated, and Figure 3.16 demonstrates

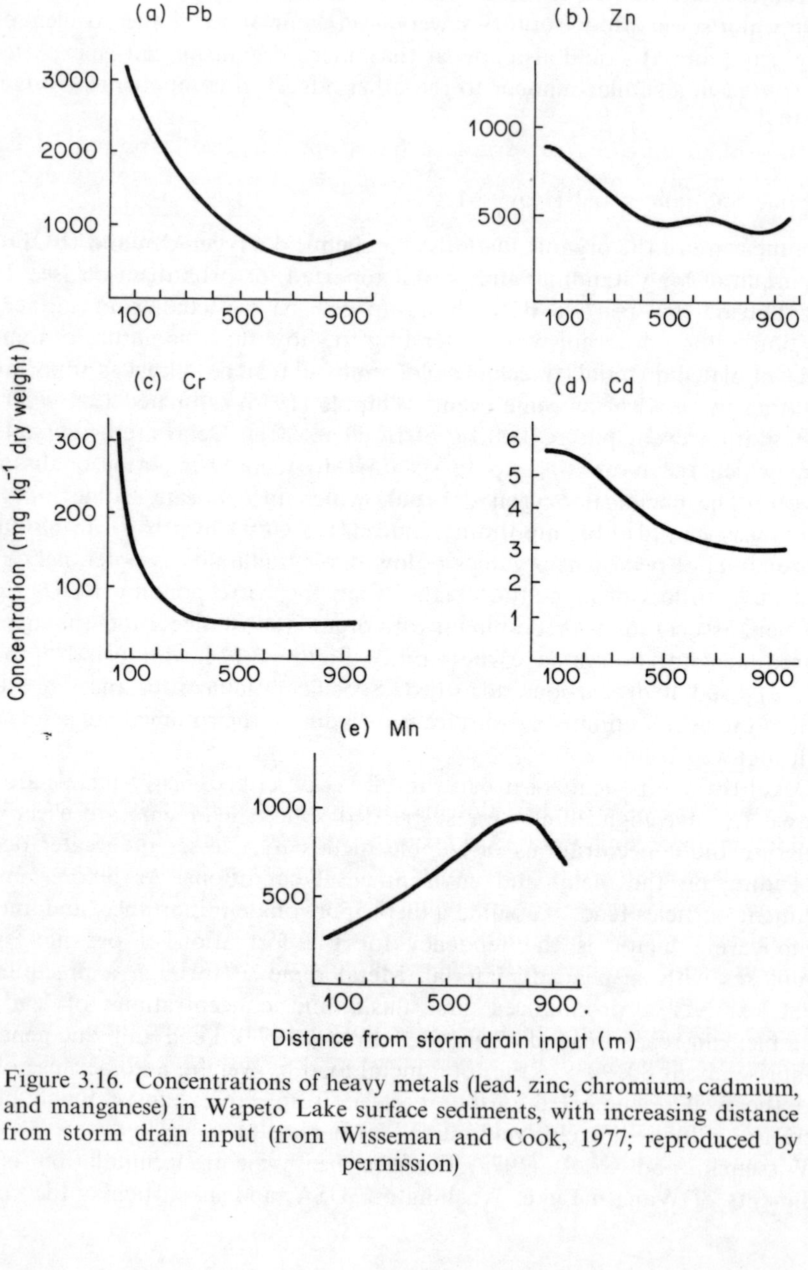

Figure 3.16. Concentrations of heavy metals (lead, zinc, chromium, cadmium, and manganese) in Wapeto Lake surface sediments, with increasing distance from storm drain input (from Wisseman and Cook, 1977; reproduced by permission)

how chromium, lead, zinc, and cadmium fall off in concentration in the bed sediments away from the input from highway and city streets. No significant correlation could be found for cobalt, nickel, and copper distributions, and the authors attribute the positive manganese correlation to the reducing conditions in summer increasing mobilization. It appears that the heavily trafficked highway and city streets contribute more heavy metals than the residential area, and are the major source of metals in the lake.

IMPACT ON WATER QUALITY—ROAD SURFACES

An individual unit of road drainage (i.e. from one grid) gives only a small increment in environmental loads of pollutants. However, the problems generated by urban conglomerations and their thousands of kilometres of roads draining to surface water at one or a few points may be significant. A similar type of problem may be created by lengths of motorway draining into rural watercourses that otherwise may receive no man-made effluents.

It is almost impossible to study the road in the urban area due to the physical constraints and inaccessibility of the drainage systems, and the use of a motorway as a model for a road surface generalization has been suggested and implemented (Pope *et al.*, 1978a). A motorway has the inherent advantages of a system with relatively open drainage, and a lack of interfering parameters such as pedestrians, animals, and parked and service vehicles. Figure 3.17 amply illustrates these advantages, where an elevated section of the M5, an uncongested system, passes the streets of Bristol, UK. Whatever the road under study, it has its own input/output system, and a generalization of this is illustrated in Figure 3.18. The relevant water quality parameters of this system are most pertinently dealt with as heavy metals, de-icing compounds, and organic chemicals, although others have been monitored (see Table 3.5).

Heavy Metals

Lead, which is by far the best documented heavy metal in road surface water, is added in an organic tetra-alkyl form to petrol as an anti-knock agent. Over the next 20 years, there is likely to be a phasing-out of this technique, but, for the time being, various western countries add between 0.15 and 0.55 g l^{-1}, although in the USA the situation is more complex since a certain amount of fuel is unleaded for use with engines fitted with catalytic exhaust emission control systems.

Tetramethyl and tetraethyl lead are volatile compounds, and while some undoubtedly escapes to the atmosphere by evaporation and (unburnt) in exhaust emissions, the levels reported in the atmosphere are low compared to inorganic combustion products (Harrison *et al.*, 1974). Although Potter *et al.* (1977) have identified organic lead compounds in road drainage grids, and

Figure 3.17. The M5 Motorway, near Bristol, UK, demonstrating the advantages of a motorway surface in modelling road surface contributions to urban run-off. (a) Comparison of local urban road (with associated interfering structures) with the super-elevated motorway in the background. (b) The super-elevated motorway, demonstrating definition and discrete nature of the road surface catchment. There is a singular lack of interfering structures, enabling assessment of vehicle generated pollutants

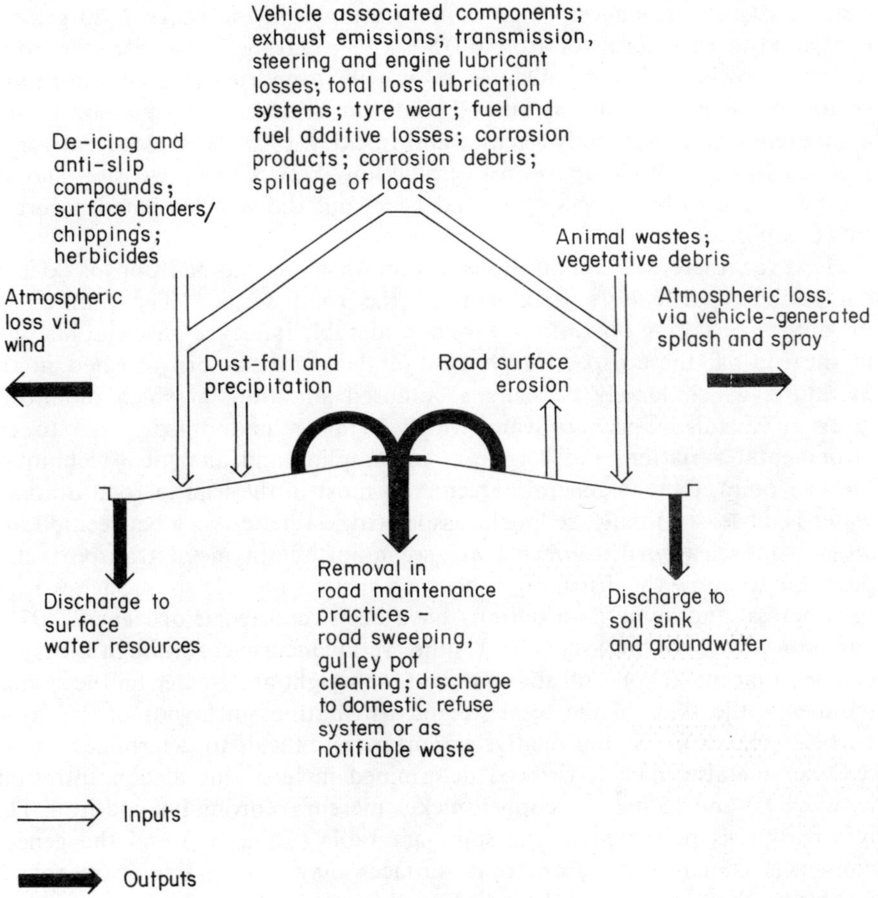

Figure 3.18. Input/output system of the road surface

Patterson (1965) has suggested that the major contribution to the levels of lead alkyls in rivers may be derived from stormwater discharges, the significance and occurrence of these compounds in watercourses has by no means been satisfactorily evaluated to date. Atmospheric wash-out probably makes little contribution to surface fresh waters.

In order to remove the lead from the combustion chambers of an engine, and to prevent it coating the internal surfaces with involatile lead oxide, scavengers (ethylene dibromide and ethylene dichloride) are added to the petrol, creating relatively volatile lead halides and oxyhalides on combustion. These are emitted in particulate form in car exhausts (although some is retained in the exhaust manifold and system, and in the engine oil), and their concentration builds up in road surface dusts. The lead in urban road and street dusts can reach very high concentrations—in the range 1–20 g kg^{-1}, compared to rural levels of around 0.8 g kg^{-1}. During a one-year study on a section of Swiss motorway, Dauber *et al.* (1978) reported lead concentrations of up to 1.8 mg l^{-1} in the run-off, but this represents a small mass load compared to a British study (Hedley and Lockley, 1975), where concentrations reached 12 mg l^{-1}. Comparison of traffic flow between the two sites showed the Swiss site to be far less used, and bears out the vehicle-related nature of this component.

However, there are complicating factors—such as the addition of de-icing compounds, particularly rock salt, to the road surface. The nature and magnitude of salting operations have considerable influence on metal loadings in the run-off; these aspects have not yet been satisfactorily treated in the literature, which largely presents a confused situation in which monitored levels of metals appear to vary widely. This is undoubtedly due to environmental variation, and variations in sampling and analytical techniques. On one point, there is general agreement—most of the lead in road drainage water is, at least initially, sediment associated; estimates vary between 50 and 99%. Quite clearly, if sediments are so important in metal transport, it is desirable to know the distribution between various particle sizes. Such work is in progress, and some distributions have been reported (Pope *et al.*, 1978b; Sartor *et al.*, 1974), but as yet the results are inconclusive. Although it may be claimed that metal concentrations weight for weight are greater on the smaller sediments, the skew in the total weight distributions in favour of the larger particles makes individual relative importances difficult to determine.

Other metals which have been determined include zinc at concentrations between 1.2 and 15 mg l^{-1}, copper, nickel, mercury, chromium, and iron. This information is presented in the summary table (Table 3.5) and the general consensus is that water from road surfaces may play a significant role in increasing the environmental load of all these elements.

De-icing Compounds

Application of de-icing salts may have the side effect of elevating metal levels. Complexation of Hg^{+} with Cl^{-}, and competition between Na^{+} and Hg^{+} for exchange sites on sediments, for example, may lead to the increased mobilization of mercury. Nickel and chromium are both reported as being present in rock salt, while cyanide is added to reduce agglomeration problems.

The use of rock salt itself generates its own problems. It is the most widely used de-icing material because of its efficiency and economic advantages. Its use may result in groundwater pollution, as well as affecting surface water quality. Immediately following a thaw, sodium concentrations in the road surface water may reach 2600 mg l^{-1} (Scott, 1976), and such levels may create saline layers in receiving lakes, preventing vertical mixing, and lead to a lack of oxygen in the lower layers of the lake. The environmental implications of road salting have been well reviewed by Field *et al.* (1974).

Organic Chemicals

The COD, which can reach up to 800 mg l^{-1} in road surface water, and the BOD, which is generally at least one order of magnitude lower, are useful general indications of organic pollution levels. However, probably the most interesting area is that of the potentially carcinogenic aromatic hydrocarbons. These are present both in vehicle exhaust emissions and bituminous road surfaces, but have only recently been seriously examined in road surface effluent.

Lead, around which so much work has centred, comprises less than 30% of exhaust emissions, and of the remainder, over 35% is carbon (Ter Hear *et al.*, 1972). A proportion of this carbon is in the form of polycyclic aromatic hydrocarbons (PAHs) which are produced by incomplete combustion of carbonaceous materials. PAHs are ubiquitous in the environment, being produced naturally by some plant forms and also by phenomena such as forest fires (Youngblood and Blumer, 1975). Such evidence has prompted scepticism over the importance of anthropogenic sources, but contrary reports based on sedimentary PAH appear to demonstrate that man has been a primary agent in PAH production and distribution (Hites, 1976). However, there is no dispute that PAHs are emitted in car exhausts (McKee and McMahon, 1967), and that the amount emitted increases with engine age and inefficiency. This constitutes a major environmental input.

The carcinogenic nature of certain PAHs has been clearly established, benzo(a)pyrene reputedly being the most potent of all four- and five-ring structures (Falk *et al.*, 1964). Others possess carcinogenic properties to a lesser or greater degree, of which benzo(b)fluoranthene, benzo(j)fluoranthene, benzo(a)anthracene, benzo(e)pyrene, and indeno(123-cd)pyrene (Figure 3.19) have all been reported in motorway run-off (Pope *et al.*, 1978b).

Although these compounds have a low solubility in clean water, in the presence of other dissolved organic material, solubility is increased (Andelman and Suess, 1970), such that only 85% may be solid associated (Pope *et al.*, 1978b). This presents problems for effluent quality standards, even with sophisticated settlement techniques. Where full treatment of the run-off is possible, removal of PAHs is low, due to their resistance to oxidation (Malaney *et al.*, 1967), such that photo-oxidation in the atmosphere and surface water is more important than biodegradation (Suess, 1976).

Automobile exhaust condensate washed from the streets directly to the river has been estimated to be the major source of PAH in the Charles River, Boston, USA (Hites and Biemann, 1972), and road washings have now been noted by a number of authors to contribute to levels of PAH in sewers.

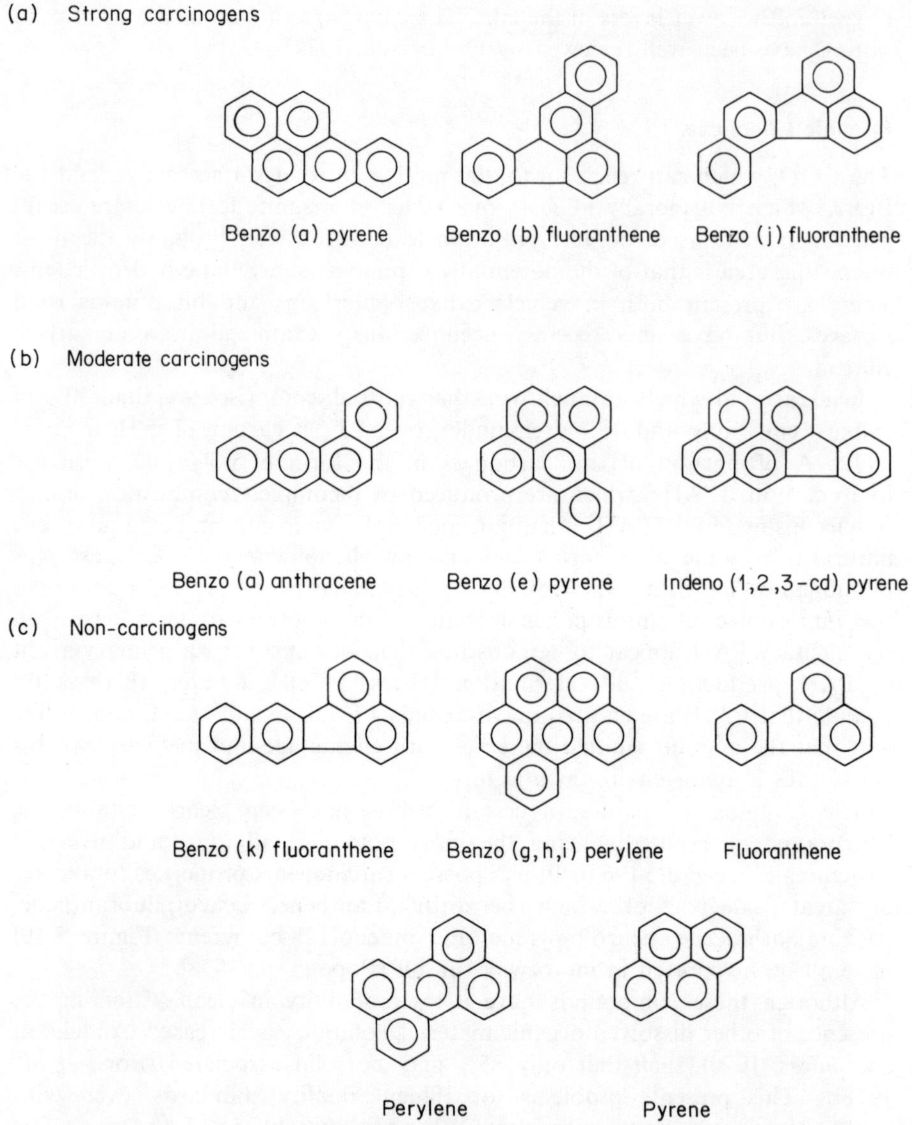

Figure 3.19. Polycyclic aromatic hydrocarbons reported in motorway run-off

Exhaust emissions are not the only source of hydrocarbons on the road surface. The loss of lubricants from moving vehicles is sufficient to coat road surfaces with an organic layer, reducing coefficients of friction between tyres and the road, clogging tyre tread and obscuring road markings (Figure 3.20). This material is removed by overland flow, and has a high PAH load.

Figure 3.20. The M1 Motorway, UK, demonstrating the build-up of organic debris on the road surface, and its removal by overland flow. (a) Before rain, note the obscuration of road markings in the lane frequented by heavy goods vehicles. (b) After rain, with a visibly cleaner road surface

Of engine lubrication oil, 10–20% is emitted in the exhaust, and 5% leaks to the road surface (Grossling, 1976). The remainder may be:

(1) used as fuel;
(2) used as road oil;
(3) re-refined; or
(4) released to sewage, surface water or ground water.

The expense of re-refining highly emulsified spent oils usually means that 30% of used crankcase oil alone is purposefully wasted, and of the 8.3×10^9 l of general lubrication oil produced, 3.8×10^9 l is wasted. The Environmental Protection Agency (EPA) estimate that up to 25% of this is discharged to sewers or dumped on open spaces (Myers, 1976), and so the road drainage grid becomes a route for quasi-intentional pollution. Grossling (1976) estimates that 2.9×10^9 l of used oil reaches the oceans every year by such routes. Although there is some evidence of the road surface itself being a source of PAHs, the motor vehicle is clearly of overriding importance in increasing these compounds in road surface water.

In general, the upper layer of the Earth's crust contains PAH in the range 0.1–1 mg kg^{-1} (Andelman and Suess, 1970), but motorway surface water sediments have been shown to contain around 20–80 mg kg^{-1} (Pope *et al.*, 1978b). The World Health Organization (WHO) has a limit of 200 ng l^{-1} for six PAHs in drinking water; motorway run-off was shown to contain up to 14900 ng l^{-1} of these PAHs, compared to local surface streams with less than 200 ng l^{-1}, and raw domestic sewage which rarely has more than 1000 ng l^{-1}. Blumer *et al.* (1977) report 'total PAH' levels of up to 300 mg kg^{-1} around a highway in Switzerland, but the background (mountain soil) levels are also high (4–8 mg kg^{-1}); however, if allowance is made for the different quantification procedure, then this would probably put this estimate in a similar range to those cited earlier. Nevertheless, the results are interesting in terms of their relative distributions. Table 3.8 shows the data portrayed as a percentage of the maximum, and this clearly demonstrates the influence of the highway on PAH levels in the local environment.

Thus, PAHs are readily released from road surfaces and appear to make up a significant proportion of the pollutional load of both urban run-off and highways crossing rural catchments. Some are carcinogens and add to chronic low-level exposure, which is readily accepted as being more dangerous than acute short-term exposure.

Mechanisms of Quality Control

While the replacement of the internal combustion engine would reduce PAH (and lead) output considerably, tyres and transmission/chassis lubrication systems would still lose PAHs and general hydrocarbons. The body/chassis

Table 3.8. Relative distributions of polycyclic aromatic hydrocarbons around a highway (initial data in mg kg^{-1} dry weight soil/sediment). (Reprinted with permission from Blumer *et al.*, 1977. Copyright by the American Chemical Society)

Sample designation	Quantity
Within town	
at highway, maximum	100.0
at highway, north	73.3
at highway, town centre	36.7
50 m from highway, town centre	33.3
at highway, south	28.3
400 m from highway, outlying, light use	7.0
250 m from highway, outlying area, dead-end road	2.0
Open country	
at highway, 750 m south of town	40.0
300 m from highway	5.0
700 m from highway	1.7
1000 m from highway	1.7
Alpine soils	
camp ground, 850 m elevation, no through traffic	2.7
mountain pass, 1200 m elevation, no road	2.0
alpine meadow, 1600 m elevation	1.3

could be totally sealed at great expense, but moving suspension/transmission/steering parts would have to be lubricated and, although the present autolube (i.e. automatic lubrication) system could be replaced with manual systems, it is unlikely that this could be encompassed within current economic and social objectives. Therefore, as long as we have, powered private transport, there will be problems from road surface pollution. In the very far future, we may be able to look forward to some form of 'slot car' mechanism using perhaps such remote propellant forces as linear induction. Such a technique would have in-built water quality and hydrological advantages using porous surfaces, but, for the present and foreseeable future, normal road effluents will have to be controlled and treated.

Creating a totally separate sewerage system is unlikely to reduce wet weather pollution by more than 50%, and is not the answer to water quality problems. For a variety of reasons, a 'partially separate' system would seem to be the most environmentally and economically acceptable solution. All sources of dirty water which would otherwise be environmentally deleterious should be treated, and the sewerage system should be increased in capacity to create more wet weather storage and treatment as necessary. Where physical and economic parameters prohibit such modifications, sources of dirty overland flow should be treated at discrete points. This is particularly applicable to motorway effluents. Drainage water must at least be settled, and oil emulsions removed by flotation and separation, in a more sophisticated manner than the current gulley pot system.

Pollutant Accumulation and Removal

Many data support the general point that there is a first flush of some magnitude for most pollutants after a dry period (e.g. Brownlee *et al.*, 1970). However, recent experience (Pope *et al.*, 1979) has shown that this may not be so fully defined, and is probably complicated by the material deposited in channels as the previous precipitation event ended, and by the rapidity of the initial overland flow. Although Whipple *et al.* (1978) failed to find any correlation between the antecedent dry period and pollutant loadings, this was probably due to faulty basic assumptions, for the longer the dry period, the more material will be deposited on the urban surface and so be available for removal. However, it may be that this 'mechanism of application' is overridden in importance by the 'mechanism of removal' (overland flow). This may be supported by the work of Söderlund and Lehtinen (1973) who found pollutant removal to be dependent on the intensity of the event.

In any event, given enough precipitation at sufficient intensity, following a normal distribution pattern, there does tend to be a decrease in concentration of certain constituents (most notably, soluble components) through time. However, this decrease does not necessarily reflect the full importance of the water body flowing at any one time, and despite a drop in concentration of

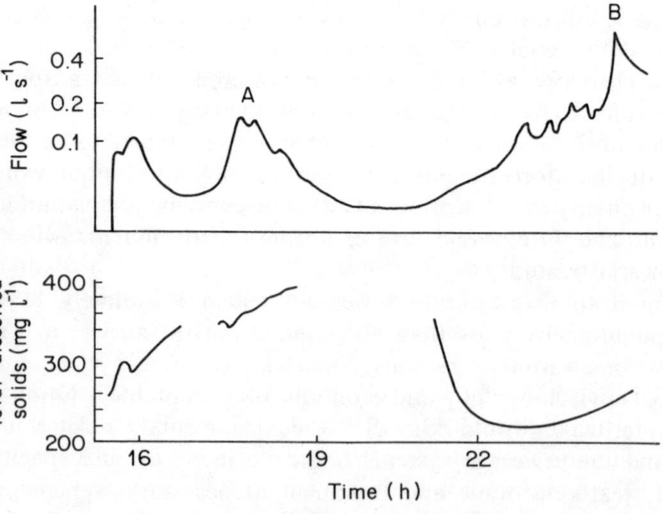

Figure 3.21. Hydrograph and total dissolved solids for a storm off a road surface—demonstrating the relationships between flow concentration and mass flow (from Pope *et al.*, 1978a)

dissolved solids concentrations from initial levels at the higher flow rates, more material may be removed from the surface. Figure 3.21 illustrates this; in two 20 min periods A and B the total weight of dissolved material removed during B is three times the level removed during A despite the concentration in A being approximately double that in B.

TRANSPORT OF HAZARDOUS LOADS

This creates a different type and scale of problem for the catchment ecosystem compared with the normal run-off dealt with so far. Potentially, a transporter of nominal 32 tonnes (UK maximum) weight will have a payload of around 20 tonnes and, in an accidental spill, this is the amount that may be drained from the road surface into either a receiving stream or the sewerage system. Of course, hazardous loads can also be moved by rail, pipeline and, less frequently, by an inland waterway. None of these transport systems are immune to spills, and in some instances effects may be more serious than with spillage on the road. However, the majority of commercial transport in the UK is by road, and the USA transports 2.5 times as much petroleum products and chemicals by road as it does by rail.

Currently, the movement of hazardous loads attracts much interest from the direct risks it poses to human life. Unfortunately, the pressurized tanker carrying propylene which exploded at San Carlos de la Rapita, Spain, in July 1978, killing 143 people, is an example of the type of disaster often necessary to provoke public demand for relatively safe but economically expensive legal controls.

The fact that roads drain to surface water, albeit often via the sewerage system, means that the potential for damage within any one catchment ecosystem is great. Should a toxic material drain to a sewage works and lower the capacity of that works to treat the wastewater it normally handles, then the effect of the accident may be magnified, since the normal waste will be disposed of untreated. This may be caused not only be removing the biolgoical capacity of the sewage works, but by physical disruption. For example, in the case of liquified natural gas, the normal load of 40000 l in a road tanker could fill about 160 km of 1.8 m sewer with an explosive air/gas mixture. In 1977, naphtha spilled into a drain in an accident in Akron, Ohio, USA, causing explosions 13 km away.

In the UK, legislative powers relating to road drainage are contained in the provisions of the Highways Acts 1959 and 1971, Public Health Act 1936, Local Acts, and Common Law (Hopkin, 1976). Section 103 of the Highway Act 1959, amended 1971, empowers a highway authority to construct and lay in the highway, or on land adjoining a highway, such drains as are considered necessary. Surface water may also be diverted into any existing drain. 'Drain' is defined as including a ditch, gutter, watercourse, soakaway, bridge, culvert, tunnel, and pipe. Local authorities are obliged under the Public Health Act 1936 to provide such sewers as may be necessary for effectually draining a district, and Section 22 of that Act provides for agreements between local and highway authorities in order to allow public sewers to be used for the drainage of highways.

Highway authorities are responsible, therefore, for the discharge of surface drainage into sewerage and land drainage systems, but, in practice, discharge of highway drainage has never been subjected to the consent of the Water Authorities. Under the provisions of Section 32, Control of Pollution Act 1974, however, any matter other than trade or sewage effluent to be discharged into relevant waters from a drain which a highway authority is entitled to keep open by virtue of Section 103 of the Highways Act 1959 should only be discharged with the consent of the Water Authority. The Water Authorities are therefore responsible for the effluent discharged. Effective control mechanisms must be rapidly developed to prevent such incidents as reported by Scheidt (1967), where accidental spills of pesticides, oil, and general chemicals caused fish kills and benthic disruption. One company evidently purposefully pumped diesel oil from its accident-damaged vehicle into the nearest channel.

Figure 3.22. Hazard warning system for transporters

There are around 12000 petrol tankers, 4000 non-pressurized general chemical tankers, and 300 pressurized tankers on UK roads. Many will go long distances on relatively safe motorways, but all must leave and join at some point via approach roads and normal trunk roads. Inevitably, spills will occur (though many could be prevented by better vehicle design). Spillages can only be dealt with effectively by a skilled and well informed approach to the accident as soon as possible after its occurrence. The major objective is to prevent the drainage system receiving a shock load of any material, but this has to be concurrent with the prevention of risk to the road user and the safety of those involved in the accident. Downstream water users may be warned in time to prevent harm, but the only way of preventing ecological damage once a spill has occurred is to clean up as efficiently as possible. To this end, solid and liquid removal techniques are available; the major difficulty in an emergency is identifying the chemical to be dealt with. Without this knowledge, the road surface is usually washed clean with water into the road drains.

In the UK, the Chemical Industries Association has introduced a voluntary scheme for marking vehicles carrying dangerous substances. This attempts to overcome errors in interpreting the nature of vehicle loads by the use of a numbering system, of which there are two types. The Hazchem code, which is

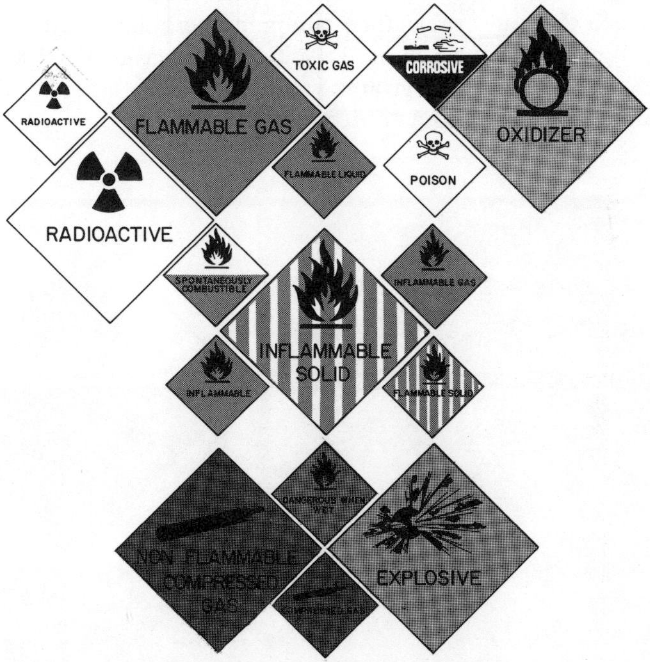

Figure 3.23. Warning diamonds—pictorial representation of the nature of a load

interpreted through reference cards held by the emergency services, gives advice on positive action for control, and the United Nations number, which is mainly for control during filling and emptying the tanker. These are usually combined on the UK Hazard Information Service label (see Figure 3.22) with an emergency telephone number for specialist advice, one of the internationally accepted warning diamonds indicating in visual terms the nature of the hazard (Figure 3.23) and a manufacturer's logo or name. Also in use are Transport Emergency cards (Tremcards), which contain information on particular substances and some explanation of the danger they present. These are carried by the driver, and their detail is expected to be particularly valuable in the carriage of waste materials, the exact nature of which is unlikely to correspond to a specific Hazchem or UN number.

Should a spill reach soil sinks and be a potential threat to groundwater, the geological structure will dictate any action which can be taken. Schmassman (1978) cites some Swiss experience in dealing with petroleum spills where action has to be efficient and fast. Within one hour of a spill of 8000 l of oil, the area affected began to be excavated, reaching 4 m depth after 6 h. Unfortunately, the infiltration had by this time passed this depth. It was not until 16 days and 9120 m^3 of soil later that the polluted ground had been removed. In some cases, percolation to aquifers has been reported and, if a spill does reach the water table, then pumping operations must be undertaken to prevent ingress into the aquifer. Despite such measures, a 10000 litre spill infiltrating a karst area raised groundwater oil levels to 6 mg l^{-1}. Such events may eliminate the aquifer as a source of supply.

CHAPTER 4

Impact of Man in Catchments (iii) Domestic and Industrial Wastes

M. S. Farrimond

INTRODUCTION

The effect of man's social and industrial activities can be seen in the extent to which river water quality changes as a river flows from its source to the sea. During this journey, water is abstracted for potable and industrial use and, after use, is returned to the river as effluent. Effluents are rarely the same quality as the raw water from which they are derived and are normally contaminated with some form of pollution. Use as cooling water, for instance, adds waste heat to the discharged effluent. In a large river system, water may be abstracted, used, and returned several times before it reaches the sea. This is especially so in the United Kingdom, where water demands are large relative to river flows and the population density is high.

Rivers in England and Wales have been classified into four categories as follows (Department of the Environment, 1971/2), and several surveys have been carried out in recent years to estimate the quality of the country's rivers:

class 1: rivers unpolluted or recovered from pollution;
class 2: rivers of doubtful quality and needing improvement;
class 3: rivers of poor quality requiring improvement as a matter of some urgency; and
class 4: grossly polluted rivers.

This classification of river water quality has recently been superseded (National Water Council, 1978a; see Table 7.5), but a national survey of rivers using this modified classification has not as yet been carried out. The NWC system is more objective than the DOE classification, as it sets class limits according to the statistical distribution of biochemical oxygen demand (BOD) and ammonia in the river. Table 4.1 shows the results of the 1975 DOE survey which revealed that 3.3% of all rivers in England and Wales are grossly polluted (Department of the Environment, 1978a). However, the length of river

Table 4.1. Quality of rivers in England and Wales in 1975

Water Authority area	Length of non-tidal river in various quality classes							
	1		2		3		4	
	(km)	(%)	(km)	(%)	(km)	(%)	(km)	(%)
North-West	3488	69.1	791	15.7	396	7.8	373	7.4
Northumbrian	2340	84.2	344	12.4	81	2.9	14	0.5
Severn–Trent	2351	57.0	1249	30.3	230	5.8	283	6.9
Yorkshire	4726	84.7	327	5.9	181	3.2	345	6.2
Anglian	2853	72.3	995	25.2	86	2.2	14	0.4
Thames	2346	81.0	475	16.4	72	2.5	5	0.2
Southern	1516	78.9	369	19.2	23	1.2	13	0.7
Wessex	1534	78.9	305	15.7	88	4.5	17	0.9
South-West	2937	90.3	260	8.0	28	0.9	28	0.9
Welsh	3947	85.1	345	7.5	257	5.5	87	1.9
Total	28037	77.6	5458	15.1	1449	4.0	1178	3.3

Source: *River Pollution Surveys of England and Walses, Updated 1975*. Reproduced by permission of the Controller of Her Majesty's Stationery Office.

in this class has fallen by 355 km since 1970. Of the grossly polluted rivers, 81% lie within the areas of the North-West, Yorkshire, and Severn–Trent Water Authorities. The major centres of population in these ares are located inland, and hence there is a smaller river flow available to dilute the effluents from these towns. Examples of this are at Leeds, Bradford, Sheffield, Manchester, Birmingham, and Leicester.

A total of about 11000 Ml (Ml = megalitre = 1×10^6 l = 1000 m^3) of sewage effluent is discharged daily into the rivers and canals of England and Wales. Approximately 17% of this volume is derived from trade effluents which are discharged to sewers for treatment at sewage works before final discharge to river. Many sewage works, mainly in rural areas, receive little or no trade effluent, whereas in other areas trade effluents may be as much as 60% of the total sewage effluent flow. The quality of the country's rivers is a function of background water quality as described in Chapter 1, and also of the quality of both sewage effluents and the direct discharges of trade effluent. Other factors affecting river water quality include discharges from sewerage systems, the almost inevitable number of accidental discharges that occur in industrialized areas, and drainage from tipping sites. The effect of agriculture and the urban environment are considered in Chapters 2 and 3, respectively.

Table 4.2 summarizes the numbers and flow of the different discharges to rivers and canals in England and Wales (Department of the Environment, 1978a). The total effluent flow of 20411 Ml day^{-1} is approximately four times

Table 4.2. Effluent discharges in England and Wales

Source of discharge	Number of discharges	Flow (Ml day^{-1})	Flow (%)
(1) Sewage works effluents	4843	11175	14
(2) Directly discharged industrial effluents	2077	9236	11
Total effluents	(6920)	(20411)	(25)
(3) Cooling water	882	59343	74
(4) Mine water	366	882	1
Grand total	8168	80636	

Source: *River Pollution Surveys of England and Wales, Updated* 1975. Reproduced by permission of the Controller of Her Majesty's Stationery Office.

the average flow of the river Thames at Teddington and the flow of cooling water is three times greater again. Most of the 59343 Ml day^{-1} of cooling water is derived from Central Electricity Generating Board (CEGB) generating stations.

Table 4.3 shows examples of the waste discharges which may cause chemi-

Table 4.3. Some sources of pollution

Origin of effluent	Type of pollution caused: Chemical	Physical	Physiological	Biological
Sewage	Organic matter	Suspended solids Turbidity	Smell	Bacteria Viruses
Coal mining	Dissolved inorganic matter	Suspended solids Bed deposits	Taste (iron)	
Food processing	Proteins Sugars Carbohydrates	Temperature		Bacteria
Metal industry	Metals Cyanide Phenols Ammonia	Colour	Taste (iron) Toxicity (cyanide)	
Organic chemical industry	Chlorinated organic compounds	Foaming	Taste (phenols)	
Electricity industry		Waste heat		

cal, physical, physiological, and biological pollution. Most discharges may cause more than one type of pollution; for instance, a steel works effluent may add inorganic compounds such as cyanide, thiocyanate, and ammonia to a river, but these in turn can be toxic and cause fish mortalities.

EFFLUENT VARIABILITY

The flow to a typical sewage treatment works with an average flow of 10 Ml day^{-1} is comprised of the components shown in Table 4.4. The dry weather flow (DWF) to such a works would be about 8 Ml day^{-1}, and under average flow conditions only about half the sewage entering the works comes from domestic sources. The factor between DWF and average flow varies between 1.10 and 1.25 depending on the particular sewerage system. Domestic water use has been shown (Thackray *et al.*, 1978) to be about 110 l head^{-1} day^{-1} and its components are shown in Table 4.5. This is expected to rise by about 40% by the end of the century. The domestic component of sewage flow comprises water used in the toilet, for waste disposal, personal washing, and laundering. These components together comprise about 90% of the water used in the home.

The trade effluent flow consists of that flow from trade premises which is discharged to sewer for treatment at the sewage works. A charge is levied by the Water Authority on the discharger for the reception and treatment of his effluent. The difference of 2 Ml day^{-1} between DWF and average flow is the flow from surface water which drains into the sewerage system after rainfall. This flow varies with the intensity and duration of the rainfall and is normally zero during dry periods. Infiltration is that water which enters the sewerage system through cracked joints and leaks in the fabric of the sewer.

The flow to a sewage works varies during any 24 h cycle and, at those works where a major proportion of the sewage is pumped to the works, flows can be zero during the night. On a dry day, the flow at a works varies between 0.5 and about 2.5 DWF, the lowest flows being received between 4 and 6 a.m. and

Table 4.4. Source of flow to a medium-sized sewage works

Source of discharge	Flow (Ml day^{-1})	
(1) Run-off from roads via combined sewerage systems	0.5–2.5	
(2) Infiltration	0.5–2.0	Dry weather flow
(3) Trade effluent	1.7	Dry weather flow
(4) Domestic sewage	3.8–7.3	Dry weather flow
Total	10	

Table 4.5. Water usage

Consumer	Amount ($l\ head^{-1}\ day^{-1}$)	Amount (%)
(1) Domestic use		
toilet/waste disposal	37	
drinking/cooking/washing-up/etc.	37	
baths/showers	20	
washing machine	9	
clothes rinsing	5	
garden watering	1	
car washing	1	
Total	110	43
(2) Other unmetered consumption	12	5
(3) Unaccounted-for water	55	21
(4) Metered consumption	81	31
Total	258	

the maximum flow about lunchtime. The extent of the flow variation at a works depends on the size and type of the drainage area. All other things being equal, a system covering a large area will tend to act as a buffer in that flow from the extremities will take longer to reach the works than flow from areas nearer the works. As a result, some equalization of the flows takes place and the ratio of maximum to minimum flow will be reduced. At smaller sewage works, the flow can change very rapidly in response to local rainfall. At a typical medium-sized works, the flow may vary through the day by up to a factor of 4 or more, whereas at a larger works this factor may be 2 or less.

Figure 4.1 shows the typical flow variations in crude sewage at a medium-sized works (average flow 8 Ml day^{-1}) and at a large works (98 Ml day^{-1}). Because of both the time lag through the works and the changing hydraulic loading on the various treatment units, better effluent qualities are produced during the night and early morning; the worst quality effluent is discharged in the late afternoon, and Figure 4.2 shows the variation in BOD load discharged from three medium-sized works. In each case, the minimum load is discharged at about 5 a.m. and the maximum load about 8 h later, the ratio between the maximum and minimum loads varying between 3 and 6. Activated sludge plants (see Chapter 7), which have longer retention times than percolating filter installations, may discharge the poorest quality effluents in the late evening.

Figure 4.3 shows the daily mean flow of both the River Trent at Nottingham and Nottingham sewage works; in addition, the ratio of the effluent flow to the upstream river flow is shown. The daily river flow varies between 1500 Ml and

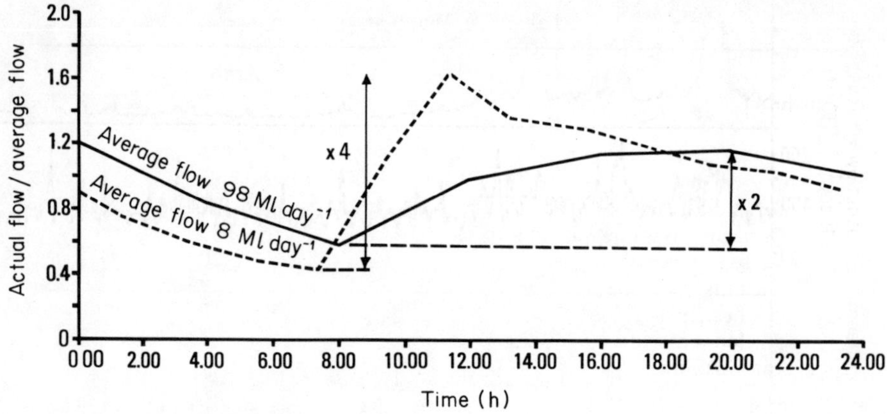

Figure 4.1. Diurnal flow variations at two sewage works

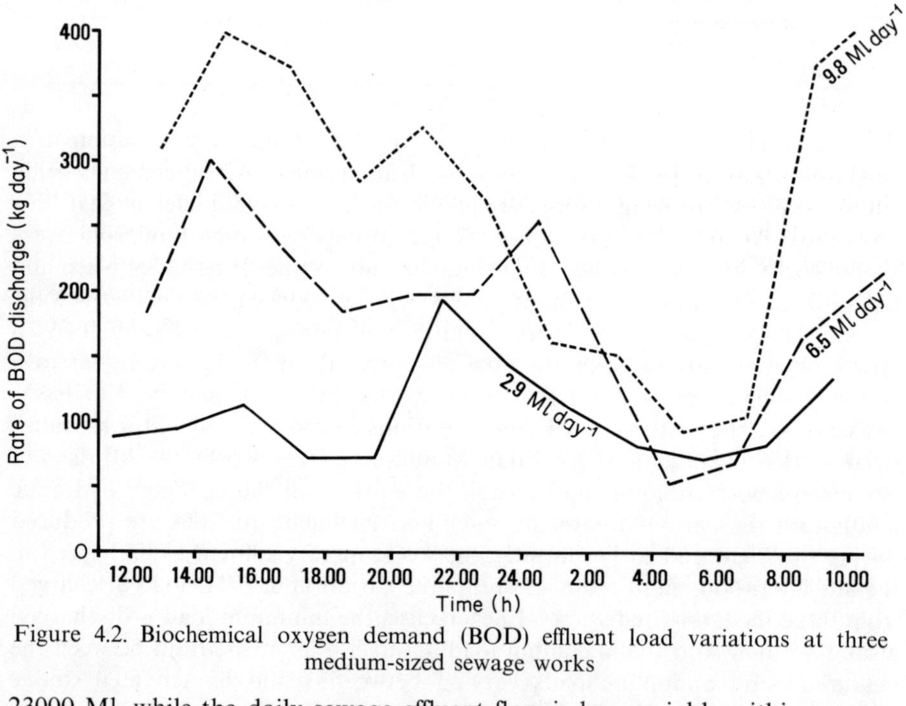

Figure 4.2. Biochemical oxygen demand (BOD) effluent load variations at three medium-sized sewage works

23000 Ml, while the daily sewage effluent flow is less variable within a range from 90 to 270 Ml. As would be expected, better dilution is provided for the effluent during the winter months when it represents as little as 2% of the river flow. However, during the drier months, this proportion rises to 10%. In themselves, these dilutions do not represent a significant effluent impact, but given similar river and effluent qualities, discharge at the lower dilution would

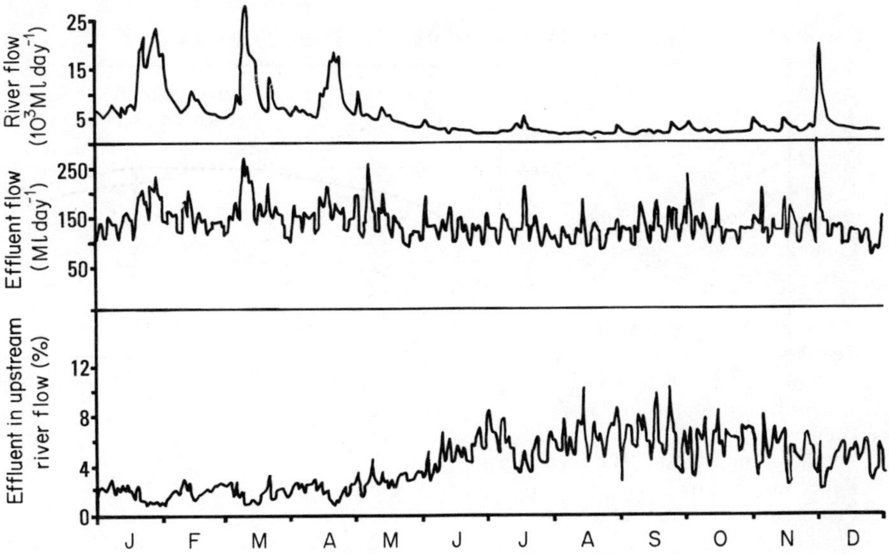

Figure 4.3. Flow contribution of Nottingham sewage works to the River Trent

increase the river BOD by an extra 1.5 mg l^{-1}. Because of the variability of both river and effluent flows and qualities, the setting of effluent consent standards has recently been placed on a more scientific basis (National Water Council, 1978a; see Chapter 8). The average flow of the sewage works is 137.7 Ml day^{-1}, but this disguises the significant difference between the weekday and weekend flows. This difference results from the discharge of large volumes of trade effluent into the sewerage system. About 30% of the flow to the works is trade effluent which is discharged by traders, who in the main work a five-day week. As a result, the average weekday flow is about 147 Ml day^{-1} whereas the flow at weekends is 115 Ml day^{-1}. The difference in flow of about 32 Ml day^{-1} is comparable to the estimated trade effluent flow of 40 Ml day^{-1} (based on 250 working days per annum). Some trade effluent discharges are seasonal and are only discharged to a works for part of the year.

Approximately 17% of the average flow to sewage works results from the discharge of trade effluent to sewers. A survey of major sewage works in the UK (Institute of Water Pollution Control, 1972) showed that only 48 (16%) out of 312 works received no trade effluent discharge at all. The type of trade effluent which was discharged to most sewage works was from the food processing industry; metal plating effluents were discharged to slightly fewer works. Table 4.6 shows the total flow of each type of effluent discharged and also the proportion of the works in the survey which received that type of effluent. Those industries producing wastes with high organic content include brewing, food processing, chemical, and textile industries, and, as a result,

Table 4.6. Discharges of trade effluent to sewers

Type of effluent	Total volume (Ml day^{-1})	Percentage of sewage works receiving this type of effluent
Food processing	166	56
Plating and metal finishing	195	53
General engineering	143	45
Laundering	25	40
General	72	39
Chemical and allied industries	184	31
Textiles		
wool	72	20
cotton and man-made fibres	101	20
Brewing	111	19
Leather tanning and fellmongering	20	19
Paper and board	53	13
Printing and ink manufacture	26	8
Plastics	25	8
Iron and steel industries	25	5

works may have been extended specifically to cater for the discharge of these effluents. Wastes from the metal finishing industries contain toxic metals which can inhibit biological treatment processes. Toxic metals, generally, tend to accumulate in the sludge produced at sewage works and, if their concentration becomes too high, it can limit the disposal of the sludge to agricultural land (Department of the Environment/National Water Council, 1977). The largest flow of trade effluent comes from the plating and metal finishing industries, and the chemical and allied industries, but, in the latter case, the discharges take place at fewer works and so the individual discharges are bigger.

The variability of about 500 sewage works effluents have been examined and, in general, all effluents, regardless of size and effluent quality, have a coefficient of variation of 0.5 for BOD and suspended solids (SS). For ammonia, the coefficient of variation is 0.7. Coefficient of variation is expressed as:

$$\text{coefficient of variation} = \frac{\text{standard deviation}}{\text{mean}}.$$

This relationship can be used to estimate the percentile values of the effluent quality distribution. For instance, for BOD and SS, the 95 percentile value is approximately twice the mean. Thus, a sewage works with a mean BOD of 12 mg l^{-1} will have a 95 percentile of about 24 mg l^{-1} and, if the total population of the distribution was sampled, only 1 sample in 20 would be expected to exceed this value. Similarly, a large works with a poor-quality effluent with mean BOD of 50 mg l^{-1} would have a 95 percentile value of 100 mg l^{-1}. This relationship (which is also dealt with in Chapter 7) is surprising considering the range of effluent flows considered (0.03–250 Ml day^{-1}) and the range of effluent quality (mean BOD 4–80 mg l^{-1}). Industrial effluents discharged directly to rivers tend to be more variable than sewage effluents. The flow from a dyeworks, for instance, varied by a factor of more than 30 during a typical 24 h cycle (Little, 1973) and the BOD fluctuated between 10 and 180 mg l^{-1}

Figure 4.4 shows the flow of the River Trent during 1976 and the amount of sewage effluent in the river. At times of average flow, the sewage effluent content at Nottingham (approximately 140 km from the source of the Trent) is

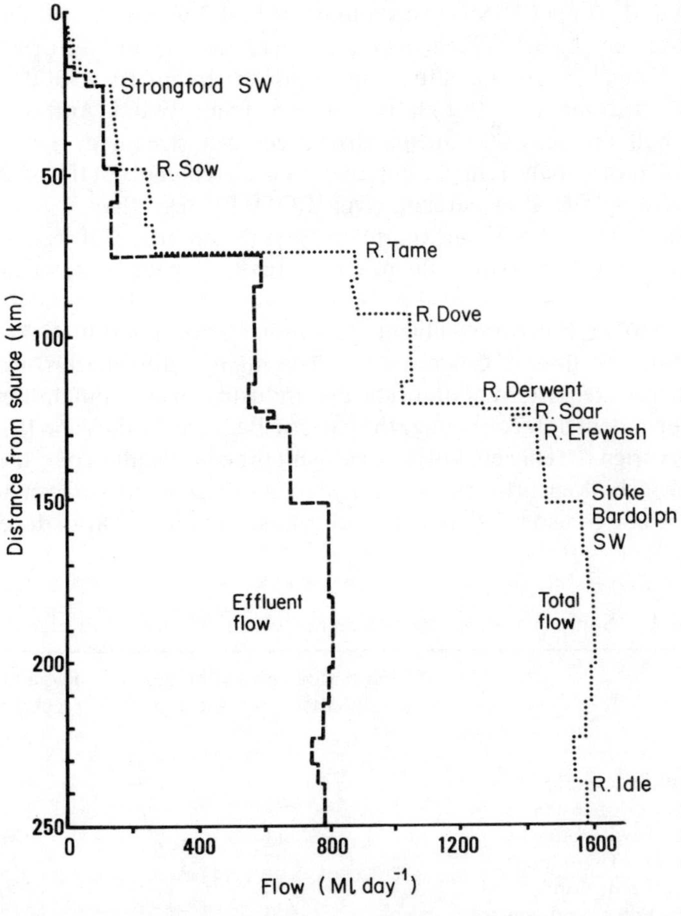

Figure 4.4. Flow of the River Trent (September 1976), showing effects of sewage works and rivers

about 28%, but during low summer flows this can increase to 50%. The River Tame, which drains Birmingham and the Black Country, and is a tributary of the River Trent, would in its natural state be a small river. However, because of the large population and industrial development in the area, parts of the river can comprise up to 80% sewage effluent at times of low flow. Similar situations exist in other areas where there is a large population centre discharging to the headwaters of a river system. Table 4.7 shows the average flows and percentage sewage effluent in some rivers in the UK.

The impact of effluent discharges on the receiving streams depends on the relative flows and qualities of the river and the effluent. Most sewage works are designed to produce what is termed a Royal Commission effluent (Royal Commission on Sewage Disposal, 1912); this is an effluent containing 20 mg l^{-1} BOD and 30 mg l^{-1} SS. The standard is based on an effluent dilution of 1 in 8 in summer (1 part effluent to 8 parts river water) and an upstream river BOD of 2 mg l^{-1}. In this situation, the downstream river BOD would be increased by 2 mg l^{-1}. Within the Severn–Trent Water Authority, approximately half the sewage effluents discharge to a stream giving a low flow dilution of more than 1 in 8, and therefore an effluent BOD of 20 mg l^{-1} would increase the downstream river BOD by less than 2 mg l^{-1}. The remaining effluents discharge to streams giving a dilution of less than 1 in 8 and so a Royal Commission effluent has a greater impact on downstream river quality.

To summarize, therefore, effluent quality varies as a result of the changing composition and flow of crude sewage. The main factor affecting flow during dry weather is the pattern of domestic and industrial usage, but, in wet weather, rainfall-dependent flow can change the pattern. Because the flow and load received at a works varies, the efficiency of the treatment processes is altered by the changing hydraulic and biological loads. The effluent is then discharged to a river, whose flow is rainfall- and season-dependent, and whose quality is also dependent on

Table 4.7. Some rivers and their percentage composition of sewage effluent

River and site	Estimated sewage effluent content at average flow (%)	Population in catchment ($\times 10^3$)
Severn at Shrewsbury	0.5	59
Severn at Gloucester	4	1140
Avon at Tewkesbury	11	820
Soar at Trent confluence	27	720
Tame at Trent confluence	34	2300
Dove at Trent confluence	3	125
Trent at Nottingham	25	4500

upstream discharges and abstractions. Direct trade effluent discharges may also affect river water quality. They may show a direct effect by contributing to the toxicity of the river water and so impairing the use of the river, or they may have an indirect effect by either overloading or reducing the efficiency of sewage treatment processes and so causing an inferior-quality effluent to be discharged.

SEWERAGE SYSTEMS

There are two types of sewerage system, the combined and the separate systems, and most areas contain some of each type. Most urban areas, especially in older town centres, are sewered by a combined system, whereas areas of newer development may be sewered by a separate or partially separate system. A combined system is one which receives run-off from roads and other paved areas as well as foul sewage. A separate system has two separate sewers, the first for foul sewage which is conveyed to a sewage works, and the second for surface water which is discharged directly to a stream. Traditionally, sewers had their overflows set at $6 \times \mathrm{DWF}$, so that only this flow was carried forward to the sewage works and any excess was discharged to stream. This has been revised (Ministry of Housing and Local Government, 1970) to a value of

$$Q = \mathrm{DWF} + 2E + 1360P \ (\mathrm{l\ day^{-1}})$$

where Q is the flow carried forward, DWF is the dry weather flow, E is the flow of trade effluent, and P is the population discharging to the sewer. This value is estimated to be slightly larger than $6 \times \mathrm{DWF}$. The discharge of excess sewer flows to streams is not, in principle, harmful to the receiving stream. In general, discharges should only occur when sewer flows are high as a result of rainfall and under these conditions river flows are also above average. The impact of the discharge of crude sewage is thus reduced by the greater dilution available. However, in many older urban areas, the storm overflows were set at a time when flows in the sewer were much lower. Because of increases in sewage flows, these overflows may not now be equivalent to $6 \times \mathrm{DWF}$, but to a smaller multiple of the present dry weather flow. Discharges, therefore, can occur at times of dry weather and the frequency of discharge is increased and may have a severe effect on the receiving stream. However, there are three ways in which the frequency and polluting potential of the overflows may be overcome;

(1) raising the level of the overflow;
(2) constructing larger-capacity sewers; and
(3) temporarily storing the overflow in storm retention tanks.

Of the flow passed down the sewer to the sewage works, approximately half ($3 \times \mathrm{DWF}$ or $P(1360 - 9G)$ where G is domestic consumption in $\mathrm{l\ head^{-1}\ day^{-1}}$) receives full treatment. The remainder is diverted to storm tanks which provide a minimum storage of 2 h. The tanks are often arranged sequentially such that the first one to fill does not overflow to stream, but a second tank fills and that tank discharges when full. By adopting this method of operation the first foul flush of sewage is stored and is not discharged. After the first foul polluting discharge during a storm, any further rainfall serves to dilute the sewage with clean surface water, and overflows from subsequent tanks are lower in BOD. In certain areas where the sewage works discharges to a potable supply river which gives little dilution to the effluent, the sewer overflow settings may be higher than $6 \times$ DWF in order to minimize the effect on the stream. In these circumstances, there would be correspondingly higher standards set on the overflows from the storm tanks at the sewage works, and on the works effluent. Grass plots have been used at some works to give additional treatment to storm sewage and so protect sensitive streams.

SELF-PURIFICATION

The BOD test was devised to simulate the conditions ocurring in a river, so that an estimate could be made of the oxygen-consuming capacity of the water. The test is standardized by sealing a sample and then storing it in the dark at 20°C. The BOD is the reduction in dissolved oxygen (DO) in the sample during the five-day incubation period. The dissolved oxygen is reduced because micro-organisms utilize carbon in the organic matter in the sample, to grow and multiply. This process uses up oxygen. Therefore, in theory, the more organic matter in the sample, the greater the reduction in dissolvd oxygen and the greater the BOD. However, appropriate bacteria may not be present in the sample, or toxic substances may be present, or some organic matter may not be degradable in this way. The process is further complicated by the presence of ammonia and nitrogen-containing molecules that may be converted to ammonia. The conversion of ammonia to nitrite and then to nitrate is also an oxygen-consuming process, and the presence of ammonia can therefore add to the total oxygen demand of the sample. By dosing samples with allyl thiourea (ATU), it is possible to suppress the conversion of ammonia to nitrate. This amended test is referred to as the BOD (ATU), and is a measure of the carbonaceous oxygen demand of the sample. In the absence of ammonia, the BOD and BOD(ATU) tests are equivalent.

Discharge of an organic waste such as sewage or cannery effluent will increase the BOD of the receiving stream. Figure 4.5 shows the impact of a discharge of sewage effluent on a river. Initially, the BOD is increased, and then it returns to its natural level; there is a corresponding initial reduction in dissolved oxygen (DO), but this also recovers to its natural level. The two processes, which operate concurrently, are termed *degradation* and *re-aeration* and the whole system of recovery is called *self-purification.* The process of degradation uses up dissolved oxygen, and if the rate of oxygen consumption

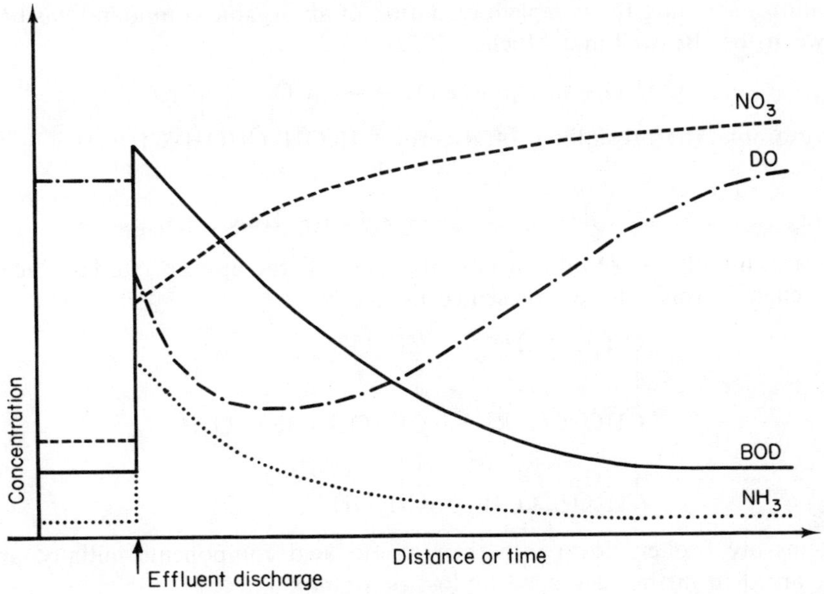

Figure 4.5. Impact of sewage effluent on river water quality

exceeds the rate of re-aeration, exceeds the rate of oxygen consumption, and then the dissolved oxygen level begins to rise. Eventually, the dissolved oxygen concentration approaches the saturation level and equilibrium is reached. These changes are responsible for the oxygen sag curve which has been represented mathematically as explained in Chapter 10.

In aerobic rivers, ammonia is converted into nitrate and this process is also oxygen-consuming. The nitrifying bacteria, *Nitrosomonas* and *Nitrobacter*, bring about oxidation in two stages—ammonia to nitrite, and nitrite to nitrate, respectively. The first stage is the rate-determining step; once formed, nitrite is rapidly converted to nitrate and the conversion of 1 mg l^{-1} of ammonia to nitrate requires 4.33 mg l^{-1} of dissolved oxygen. The rate of nitrification is also temperature-dependent and proceeds rapidly in a river like the Trent which contains large amounts of sewage effluent, and which has a higher-than-normal temperature (Garland, 1978).

The majority of organic compounds found in sewage and effluents, such as carbohydrates, fats, and proteins are utilized by bacteria as a source of food. In this process, these large molecules are broken down by enzymes to compounds of lower molecular weight. The most common pathways involve conversion to pyruvic acid ($CH_3 \cdot CO \cdot CO_2H$), lactic acid ($CH_3 \cdot CHOH \cdot CO_2H$), propionic acid ($CH_3 \cdot CH_2 \cdot CO_2H$) or acetic acid ($CH_3 \cdot CO_2H$). These acids are then further degraded to carbon dioxide and water or, in the case of pyruvic acid, are incorporated into cellular material for growth. The general chemical equation governing the complete oxidation of an organic compound has been shown to be (Buswell and Mueller, 1952):

$$C_aH_bO_c + (a + \tfrac{1}{4}b - \tfrac{1}{2}c)O_2 \longrightarrow aCO_2 + \tfrac{1}{2}bH_2O.$$

For example, the breakdown of glycerol, $CH_2OH \cdot CHOH \cdot CH_2OH$ is given by:

$$C_3H_8O_3 + \tfrac{7}{2}O_2 \longrightarrow 3CO_2 + 4H_2O.$$

Fats are initially broken down into glycerol and the appropriate fatty acids, and generally this can be represented as follows:

$$\begin{array}{l} CH_2O \cdot CO \cdot R \\ | \\ CHO \cdot CO \cdot R \\ | \\ CH_2O \cdot CO \cdot R \end{array} \longrightarrow \begin{array}{l} CH_2OH \\ | \\ CH \cdot OH \\ | \\ CH_2OH \end{array} + 3R \cdot CO_2H$$

Proteins are broken down into their amino acid components initially, and these are then further degraded by loss of ammonia:

$$R'CH \cdot NH_2 \cdot CO_2H \longrightarrow R'CHOH \cdot CO_2H + NH_3.$$

The rate of degradation of organic material is dependent on the size and structure of the molecules involved. For example, cellulose, a high-molecular-weight polysaccharide and a major constituent of plant material, is resistant to degradation. The human body does not contain enzymes which attack the $\beta-1,4$ link in cellulose, and so it passes through the body virtually unchanged. In natural water systems, there are few bacteria containing the necessary enzymes to attack and break this chemical bond, and therefore this compound is slow to degrade. Starch is also a polysaccharide, but of lower molecular weight than cellulose, and the glucose units are $\alpha-1,4$ linked, which allows the easier and more rapid degradation of starch. In addition to the mineral oils, there are other naturally occurring hydrocarbons such as carotenoids (naturally occurring pigments found in tomatoes, carrots, and other vegetables) and vitamins K_1 and K_2. The unsaturated hydrocarbons are easier to degrade than the saturated ones. Once oxygen has been introduced into the molecule, degradation follows a similar path to the one outlined above. These degradation processes are summarized in Figure 4.6. All of these processes utilize oxygen, either directly or indirectly, and thereby contribute to the reduction in dissolved oxygen in streams downstream of discharges of organic wastes.

Some of the solid matter discharged to rivers, and some of that precipitated in the chemical processes of degradation, settles to the bed of the river. Anaerobic reactions can take place in river beds and these are reduction reactions. The products of such reactions include methane, carbon dioxide, ammonia, and hydrogen sulphide.

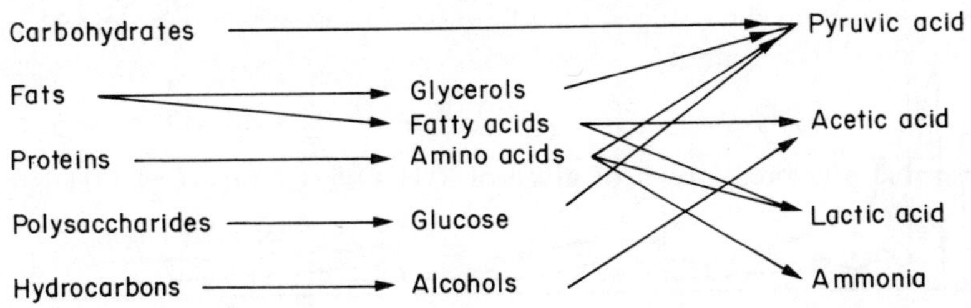

Figure 4.6. Summary of simplified degradation processes

IMPACT OF SEWAGE EFFLUENT

Table 4.8 shows the impact of a sewage effluent (BOD = 20 mg l^{-1}) on a clean river (BOD = 2 mg l^{-1}) with a flow of 10 Ml day^{-1}. The flow of the effluent varies from 0.5 to 50 Ml day^{-1} and the table also shows the minimum dissolved oxygen in the river and the time taken to reach the point of minimum dissolved oxygen. The change in downstream river water quality as the size of the effluent increases can be regarded as an historical record of the impact of the growth of an urban area on a local stream.

Figure 4.7 shows the average BOD and ammonia profile of the River Trent in 1977; increases or decreases resulting from additions of dirtier or cleaner water, respectively, are shown as vertical lines, and the effects of self-purification are shown as diagonal lines (upward sloping diagonals occur between points which are statistically indistinguishable). Decreases therefore can occur in two ways: first, by the addition of a cleaner tributary and, secondly, by self-purification. In the upper reaches of the river, which drain the Potteries conurbation, the Fowlea Brook and Stoke sewage works are the major discharges. During storms, surface water run-off from the paved urban area, and storm overflows from sewers, also serve to reduce river quality.

Table 4.8. Estimated impact of a discharge of sewage effluent*

Flow of effluent (Ml day^{-1})	River quality downstream of outfall		
	Maximum BOD (mg l^{-1})	Minimum DO (mg l^{-1})	Time to minimum DO (days)
0.5	2.9	8.1	2.9
1.0	3.6	7.5	3.1
2.5	5.6	6.2	3.5
5.0	8.0	4.6	3.6
10.0	11.0	2.6	3.7
20.0	14.0	0.5	3.7
50.0	17.0	0.0	3.7

* Flow of river 10 Ml day^{-1} at a BOD of 2 mg l^{-1} and DO of 9 mg l^{-1}.

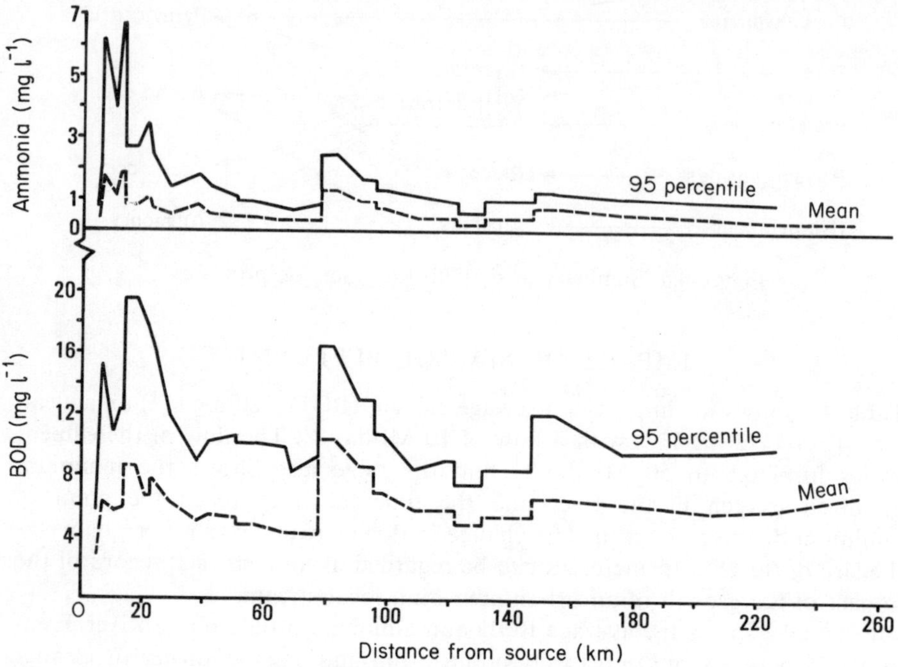

Figure 4.7. Biochemical oxygen demand (BOD) and ammonia in the River Trent, 1977

Dilution by the cleaner Rivers Sow and Blithe, and self-purification all ensure a good quality river by the time the Tame confluence is reached. At this point (approximately 80 km from source) the River Trent, under average flow, contains 30% sewage effluent from a population of 680000, but under low flow conditions the proportion of effluent rises to 60%. The River Tame, which drains most of the Birmingham conurbation, then joins the Trent, doubling its flow and its BOD, and increasing the ammonia to 1.7 mg l^{-1} from an average of 0.3 mg l^{-1}. However, as a result of dilution by the cleaner River Dove, and increased self-purification as a result of high river temperatures, the Trent has almost recovered by the Derwent confluence. The quality of the river then changes little until it reaches the Humber estuary. The self-purification capacity of the river is enhanced in its middle reaches by elevated river temperatures as a result of its use for cooling purposes by the CEGB. Because of these higher temperatures, the river is able to assimilate greater amounts of sewage effluent than might be possible in a colder river (Lester, 1975). The quality of the River Trent has improved dramatically over the last 20 years as a direct result of improved sewerage and effluent treatment. Table 4.9 shows the BOD, ammonia, and dissolved oxygen concentration for triennial periods since 1959.

Table 4.9. Quality of the River Trent at Nottingham, 1959–79

Year	Concentration (mg l^{-1})		
	BOD	NH_3	Minimum DO
1959–61	11.3	2.4	3.0
1962–64	13.0	3.5	2.6
1965–67	11.0	2.2	3.8
1968–70	8.6	1.1	4.2
1971–73	7.3	0.9	4.9
1974–76	6.5	0.6	4.2
1977–78	5.3	0.4	4.8
1978–79	5.0	0.5	4.8

SEWAGE EFFLUENT QUALITY

The BOD and SS contribution to sewage from domestic sources are both about 55–65 g head^{-1} day^{-1}, the COD being about 120 g head^{-1} day^{-1}. The quality of effluent discharged is dependent on the extent of treatment given, which, in turn, should be a function of the geographical location of the sewage works. For instance, coastal sewage works often discharge untreated sewage into the sea where there is ample dilution, whereas in inland areas effluents must be treated to a much higher standard in order to maintain river quality. Table 4.10 shows a comparison of effluent BOD loads discharged by the ten Regional Water Authorities (National Water Council, 1978b). The table also shows the effluent BOD discharged in g head^{-1} day^{-1}, and demonstrates the great difference in effluent treatment standards between a coastal Authority, such as the Northumbrian Water Authority, and a mainly inland Authority, like Severn–Trent. The former discharges 47.7 g head^{-1} day^{-1} of BOD, mainly into the North Sea, and the latter discharges about one-tenth of that into the rivers of the Midlands.

Phosphate in sewage comes mainly from the use of detergents, with smaller amounts from other industrial sources and domestic waste. Of the phosphorus arriving at a sewage works in crude sewage, about 1.92 g head^{-1} day^{-1} may be attributed to detergents and soaps and 1.34 g head^{-1} day^{-1} to other sources (Devey and Harkness, 1973). A recent survey had shown a domestic contribution of 1.8 g head^{-1} day^{-1} (Alexander and Stevens, 1976). Most detergents used are now termed ‘soft’ or biodegradable, and on average about 50% by weight of detergents is phosphate which is added to enhance the detergent properties of the product. During the sewage treatment processes, 50–60% of the phosphorus is removed, which results in an effluent containing about 6 mg l^{-1} phosphate (as P). In the River Great Ouse, nearly 95% of the

Table 4.10. Comparison of sewage effluent BOD loads (1977) (National Water Council, 1978b; reproduced by permission of the National Water Council)

Authority	BOD load (tonnes discharged)							Population served ($\times 10^6$)	Effluent BOD (g head^{-1} day^{-1})
	Non-tidal waters		Tidal waters						
	Crude	Effluent	Crude	Effluent	Total crude	Total effluent	Per cent removed		
North-West	502	50	214	178	716	228	68.2	6.80	33.5
Northumbrian	71	6	128	119	199	125	37.2	2.62	47.7
Yorkshire	401	30	94	77	495	107	78.4	4.55	23.5
Anglian	268	30	65	58	333	88	73.6	4.36	20.2
Thames	446	16	621	76	1067	92	91.4	11.28	8.2
Southern	70	4	82	32	152	36	76.4	3.48	10.3
Wessex	76	4	65	46	141	50	64.5	2.09	23.9
South-West	18	1	63	37	81	38	53.1	1.15	33.0
Welsh	65	6	135	100	200	106	47.0	2.72	40.0
Severn–Trent	616	31	9	7	625	38	93.9	7.85	4.8
Total		178		730		908		46.90	19.4 (average)

phosphorus load is attributable to sewage effluent (Owens, 1970) and a similar picture has been found for the Tyne estuary (James, 1976). Table 4.11 shows the sources of nitrogen and phosphorus in several different rivers in the United Kingdom. Phosphate concentrations in the River Trent have led to increased problems of calcium phosphate scaling in condenser tubes at power stations using the River Trent for cooling purposes. The increase in phosphate concentration appeared to be more related to reduced river flows than increased discharges of phosphate (Department of the Environment, 1978b).

A well treated sewage effluent contains upto 30 mg l^{-1} nitrate as N. A non-nitrified effluent will contain a similar amount of ammonia, the total inorganic nitrogen in an effluent normally totalling about 30 mg l^{-1} N. Some sewage effluents contain substantially less nitrogen, often as a result of receiving a large volume of trade effluent containing no nitrogen. Removal of nitrogen is now operationally possible at sewage works and this process may be used to reduce the nitrate content of public water supply rivers. However, sewage effluent may not be the major source of nitrate in the river and other sources must also be identified (Greene, 1978).

The enrichment of lakes and streams by nutrients can lead to a promotion of algal growth. This process is normally termed eutrophication and is dealt with in Chapter 5. The organic matter and nutrients in effluents can have this effect and lead to undesirable water quality changes. The production of algae utilizes carbon dioxide, nitrate, and phosphate (Stumm and Morgan, 1970):

$$106CO_2 + 16NO_3^- + HPO_4^{2-} + 122H_2O + 18H^+ \ (+\text{Other trace elements and sun, (i.e. energy)})$$

$$\text{Production} \downarrow \quad \uparrow \text{Respiration}$$

$$\underset{\text{Algae}}{(C_{106}H_{263}O_{110}N_{16}P)} + 138O_2$$

Table 4.11. Sewage effluent as a source of nitrogen and phosphorus in some UK rivers

River and site	Percentage derived from sewage effluents	
	Nitrogen	Phosphorus
Ouse at Tempsford	25	95
Tyne estuary	80	96
Thames estuary	80	—
Trent at Nottingham	66	95
Lough Neagh	—	78

Table 4.12. Boron concentrations in some Midland rivers

River and site	Total boron (mg l^{-1}) Mean	Range	Percentage of sewage effluent
Severn at Shrewsbury	0.06	0.02–0.28	0.4
Severn at Holt Fleet	0.16	0.09–0.23	3
Severn at Haw Bridge	0.22	0.03–0.41	4
Avon at Tewkesbury	0.54	0.05–0.85	10
Trent at Yoxall	0.44	0.25–0.71	15
Trent at Nottingham	0.42	0.13–0.69	25
Stour at Stourport	0.96	0.36–1.70	26
Tame at Chetwynd	0.69	0.31–0.90	34
Dove at Monks Bridge	0.19	0.05–0.41	3
Teme at Powick	0.16	0.02–0.47	0.2

Sewage effluents also contain boron, which is mainly derived from borates in washing powders. Sodium borate constitutes about 25% of the weight of domestic washing powders (Rosenfelder, 1978) to which it is added for its bleaching properties. The concentration of borates are little changed by sewage treatment, and so most is discharged in the effluent (Matthews, 1974). Although boron is an essential trace element for plants, continued application of irrigation water containing concentrations in excess of 0.5 mg l^{-1} may be harmful to some plants (Waggott, 1969). For instance, apples, plums, and peas are sensitive to boron (Wilcox, 1958), whereas sugar-beet, onions, and cabbages can withstand higher concentrations. The average boron concentration in sewage effluent is between 1 and 4 mg l^{-1} and Table 4.12 shows the boron concentrations in some rivers together with an estimate of the percentage sewage effluent. The maximum recorded values of 1.7 and 0.9 both occurred in rivers draining very urban and industrialized areas and containing high proportions of sewage effluent.

WATER IN INDUSTRY

Table 4.13 shows the water requirements of various industries expressed in terms of volume of water used per unit of product. For instance, between 6 and 10 litres of water are needed to produce 1 litre of beer and about 10000 litres of water are needed to produce 1 tonne of steel. Much of the water used is for cooling purposes, and so can be recycled within the system. It is the water which, after use, is discharged either to the sewer or directly to a stream which affects water quality. Table 4.14 summarizes the amount of process and cooling water discharged directly to rivers from different industries in England

Table 4.13. Water requirements of different industries

Industry or product	Approximate water requirement per unit product
Baked beans	30 $l\,can^{-1}$
Butter	20 $l\,kg^{-1}$
Cheese	2–27 $l\,kg^{-1}$
Sugar	15 $l\,kg^{-1}$ of sugar beet
Beer	6–10 $l\,l^{-1}$
Wine	3 $l\,l^{-1}$
Whisky	3–76 $l\,l^{-1}$
Paper and board	88 $l\,kg^{-1}$ (excluding cooling)
Petrol	7–10 $l\,l^{-1}$
Ammonia	250 $l\,kg^{-1}$
Soap	1–4 $l\,kg^{-1}$
Textiles	50–500 $l\,kg^{-1}$
Limestone	0.5–6 $l\,kg^{-1}$
Steel	3000–200000 $l\,tonne^{-1}$
Cement	0.9–2.5 $l\,kg^{-1}$
Cars	38000 $l\,vehicle^{-1}$
Coal	1000–3000 $l\,tonne^{-1}$
Electricity	200 $l\,kWh^{-1}$

and Wales. By far the largest discharge is from electricity generating stations which discharge 93% of all cooling water. Other major discharges are the chemical and petroleum industries and the iron and steel manufacturers.

Water is used for a variety of purposes by industry, for example:

(1) cleaning of raw materials (potato washing);
(2) rinsing water (milk bottles);
(3) direct cooling water (coke quenching);
(4) indirect cooling water (power stations);
(5) incorporation into product (beer);
(6) process water (dyeing); and
(7) transport of materials (vegetable preparation).

The most polluting waters are derived from process usage and the cleaning of raw materials, but the largest volumes come from cooling water usage. On a large industrial site, water may be used for several different purposes before it is finally discharged. For instance, cooling water is used until the dissolved solids content becomes too great. Part is then purged from the cooling circuit and the volume made up with fresh wash in order to keep dissolved solids at an acceptable level. The purge water can then be used as final rinse water, and then initial rinse water before being discharged.

Table 4.14. Volumes of effluents discharged to streams in England and Wales, 1975

Industry	Process water		Process and cooling		Cooling	
	Number	Volume ($Ml\ day^{-1}$)	Number	Volume ($Ml\ day^{-1}$)	Number	Volume ($Ml\ day^{-1}$)
Brewing	19	10.68	4	13.69	12	13.37
Brick making	15	6.61			5	0.94
Cement making	30	2.18	2	3.61	10	10.96
Chemical and allied	134	385.55	87	1280.74	142	522.96
Coal mining (trade effluent)	191	119.73	9	11.83	8	12.95
Electricity generation	71	1112.13	18	739.20	133	55124.47
Engineering	99	51.21	26	21.38	146	552.29
Food processing	115	279.51	32	105.12	101	463.73
Gas and coke	23	60.12	7	20.83	14	67.92
Glass making	5	6.72	1	0.55	14	16.94
Glue and gelatin manufacture	4	22.11	1	0.76	3	7.96
General manufacture	28	14.66	12	21.14	33	32.86
Iron and steel making	73	185.88	34	927.77	45	1190.05
Laundering and dry cleaning	5	0.31			4	3.54
Leather tanning and fellmongering	14	4.02	1	0.11	6	4.71
Metal smelting and refining	23	30.49	4	3.75	19	11.96
Paint making	3	5.8			5	3.08
Paper and board manufacture	80	446.57	25	4494.85	39	743.36
Petroleum refining	38	126.84	13	1721.81	12	716.62
Plastic manufacture	3	8.62	1	2.30	19	12.30
Plating and metal finishing	35	24.36	5	0.36	20	5.51
Pottery making	9	1.28			1	0.23
Printing ink and printing industry	1	0.23	1	1.49	5	0.58
Quarries and mining	207	482.67	3	10.52	7	2.24
Rubber processing	3	3.99	5	7.49	22	89.75
Soap and detergent manufacture	1	0.14	10	39.07	6	19.43
Textiles, cotton, and man-made fibres	62	86.15	11	77.71	29	695.96
Textiles and wool	38	7.96	2	0.40	13	8.67
General farming	169	15.91	3	10.02	2	0.19
Atomic energy	2	0.55	1	1.85	1	0.22
Water treatment plant effluent	211	181.05	4	4.63	5	7.57
Tip drainage	43	18.62				

Source: *River Pollution Surveys of England and Wales, Updated* 1975. Reproduced by permission of the Controller of Her Majesty's Stationery Office.

Discharges from Coal Mining

These discharges can come from several sources. First, mine water which is pumped from underground workings to prevent flooding—some or all of this water may be used for other purposes within the colliery if it is of satisfactory quality; in some cases mine water is used for public potable supply purposes after suitable treatment. Secondly, discharges of trade effluent from coal washing plants. Finally, surface water drainage from coal stocking sites, waste tips, and other areas.

In total, there were 366 mine water discharges in England and Wales in 1975; 828 Ml day^{-1} were discharged to inland waters, 37 Ml day^{-1} to tidal waters, and 16 Ml day^{-1} to canals. About 40% of the total number are discharged within the area of the Yorkshire Water Authority. Mine waters are often high in dissolved inorganic material and can contain up to 50000 mg l^{-1} of chloride. Other possible pollutants in mine water are iron, manganese, and low pH. During mining operations, pyrites (FeS_2) in coal seams is exposed to air and water and, as a result, iron and acidity are released into the water. When this contaminated water becomes oxygenated on discharge, ferrous iron is reduced to ferric iron and orange ferric hydroxide is precipitated (Glover, 1975):

$$2FeS_2 + 7O_2 + H_2O \rightleftharpoons 2Fe^{2+} + 4SO_4^{2-} + 4H^+$$

$$4Fe^{2+} + O_2 + 4H^+ \rightleftharpoons 4Fe^{3+} + 2H_2O$$

$$Fe^{3+} + 3H_2O \rightleftharpoons Fe(OH)_3 + 3H^+$$

Coal washing water and drainage from colliery yards is normally high in suspended solids and therefore is often passed through settlement lagoons before discharge, via oil interceptors. Two settlement lagoons may be needed in series, the first to remove heavy, easily settleable material, the second to remove finer suspended solids. Drainage from dirt tip areas varies in quality depending on the intensity of rainfall, heavy storms resulting in run-off containing high concentrations of suspended solids. Table 4.15 shows the dissolved inorganic content of several streams, half of which contain significant amounts of mine water, the remainder containing no mine water discharges. Figure 4.8 shows the changes in the annual average chloride concentration in the River Derwent and its tributary the River Amber, between 1955 and 1978. The increased chloride concentration is due to increases in both flow and chloride in mine water discharges to a tributary of the River Amber. Discharges of mine water can also affect the balance of inorganic ions in a river (Hughes and Edwards, 1977).

Mine waters and other discharges from collieries can reduce the potential value of a river. Because of their high inorganic content, such rivers can be unsuitable for public supply because of high treatment costs associated with demineralization. High mineral content is not desirable in water used for irrigation of certain fruits and vegetables (Doneen and Henderson, 1957).

Discharges from Electricity Generating Stations

In 1975/6 the CEGB operated 177 generating stations with a capacity of about 58000 MW. About 45% of the stations used river water for direct cooling purposes and a similar number used cooling towers (Central Electricity Generating Board, 1976). The remainder mainly used sea water for cooling. About 25% of the generating capacity in England and Wales is situated within the catchment of the River Trent and most of these stations use the river as a source of cooling water. The stations in the Trent catchment abstract about 4000 Ml day^{-1} on average, of which 95% is returned, the remainder being lost by evaporation. The returned water is warmer than that abstracted. Figure 4.9 shows a temperature profile of the River Trent and Table 4.16 compares the temperature of the Trent at several sites to the temperature of several other rivers which are not used for cooling water abstraction. Figure 4.9 also shows an estimated mean temperature profile of the Trent if there were no cooling water discharges; there would still be some slight rise and fall in temperature due to the addition of warmer sewage effluent and colder upland tributaries. It has been estimated that the power stations on the Thames estuary increase river temperatures at low flows by about 2°C, and the development of three

Table 4.15. Quality of some rivers receiving mine water discharges (means and ranges)

River and site	Chloride ($mg\ l^{-1}$)	Alkalinity ($mg\ l^{-1}\ CaCO_3$)	Conductivity ($\mu S\ cm^{-1}$)
Mease at Croxall	173	207	1240
	158–410	146–265	650–1800
Alfreton Brook	756	318	3300
	292–1850	145–465	1580–4950
Erewash at Trowell	219	167	1313
	106–360	100–230	820–1800
Leen at confluence	469	246	2126
	153–840	120–295	880–3200
Poulter at Elkesley	625	228	2380
	354–960	185–260	1800–3200
Dove at Mayfield	20	181	432
	13–28	125–220	360–490
Derwent at Matlock	28	135	409
	17–60	80–155	330–530
Soar at Sileby	82	182	943
	39–125	160–245	640–1260
Avon at Tewkesbury	71	204	913
	22–255	145–236	318–1169

The first five streams receive significant volumes of mine water and spoil tip run-off.

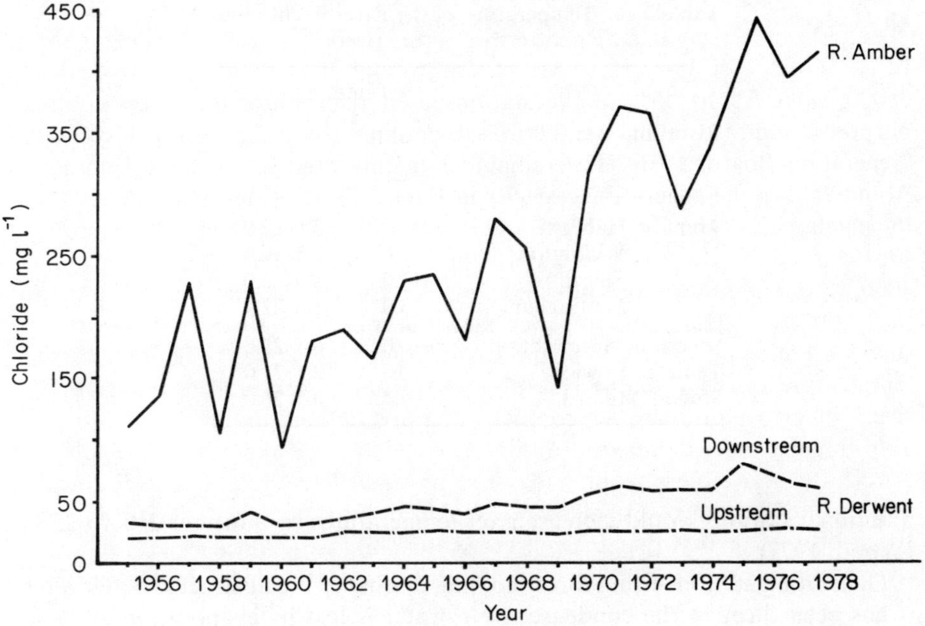

Figure 4.8. Annual average concentration of chloride in the Rivers Derwent and Amber, 1955–78

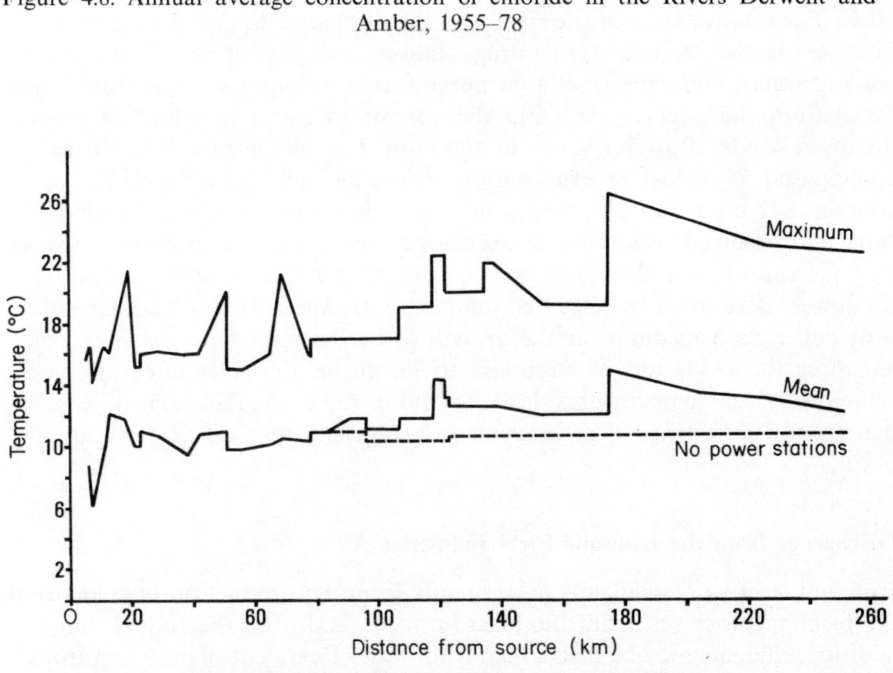

Figure 4.9. Temperature profile of the River Trent

Table 4.16. Temperature of the River Trent compared to three other rivers

River and site	Temperature (°C)	
	Mean	Range
Trent at Hanford	11.0	3.5–22.0
Willington	11.0	3.5–17.0
Shardlow	14.0	5.0–23.0
Nottingham	13.0	1.0–22.0
Severn at Shrewsbury	10.0	0.5–22.0
Teme at Powick	10.0	0.1–21.0
Sow at Milford	9.4	0.5–20.0

additional stations would increase river temperatures by a further 3°C to 25°C (Wood, 1977).

Generating stations using direct cooling discharge the abstracted water after it has been through the condensers. No water is lost by evaporation at these sites, and the temperature increase in the abstracted water can be as much as 10°C. The relative sizes of the river and the discharge dictate the temperature increase in the stream. Generating stations using cooling towers recycle cooling water, and only discharge purge water, an equivalent amount being taken from the river to maintain the volume in circulation and to control dissolved solids. If a river water of 800 mg l^{-1} of dissolved solids is used for cooling and 5% is lost by evaporation, then after one cooling cycle the water contains 842 mg l^{-1} of dissolved solids. In addition to increasing the temperature and dissolved solids of the abstracted water, the use of cooling towers may increase dissolved oxygen and reduce BOD as the water cascades down the tower. Because of the elevated temperatures, water used in cooling towers provides a good medium for the growth of the bacteria *Nitrosomonas* which enhances the oxidation of ammonia to nitrate in the river after discharge. Increased river temperatures lead to more rapid degradation of organic matter, and if the rate of re-aeration is insufficient, then de-oxygenation can occur.

Discharges from the Iron and Steel Industries

Iron and steel works effluents arise mainly from the production of coke, from the pickling processes, from the blast furnaces, and from the rolling mills. In addition, effluents may be discharged from water treatment plant, laboratories, and workshops, and surface water may be contaminated in oil storage areas and in other parts of the site. Water usage in steel works varies considerably depending on the availability of water and the conditions attached to its disposal after use. For works in which all water is only used once, water consumption can be as much as 150000 to 200000 litres per tonne of steel. Alternatively, on works practising a high degree of re-use, water consumption may be reduced to 3000 l $tonne^{-1}$ of steel.

Within a steel works, the effluents produced by different processes are often mixed together to neutralize each other before discharge. Cooling waters can be recycled and any purge water may be used for quenching. The impact of a steel works effluent on the River Ebbw Fawr (Williams *et al.*, 1975) and on the Tees estuary (Prater, 1975) have recently been discussed.

Discharges from Food Processing Industries

Effluents from food processing industries can be very variable in both quality and quantity, and will also vary according to the growing season. The BOD of the untreated effluent can vary between 100 and 100000 mg l^{-1} and the pH can be acidic (milk wastes) or alkaline (waste from the caustic washing of potatoes).

Waste waters from fruit and vegetable processing usually consist of wash water, fruit peelings and other waste matter, and wash-down water from the cleaning of machinery. Fruit processing wastes can be treated by the activated sludge process to produce good-quality effluents (Esvelt and Hart, 1970). The BOD of potato processing effluents can be about 2000 mg l^{-1}, but this can be reduced to levels acceptable for discharge to stream by the use of high rate filtration. Sugar-beet wastes are produced between November and February and contain large amounts of dissolved organic matter, mainly sucrose, and suspended material, mainly soil from washing the beets. If the waste is discharged to a sewage works for treatment, it often has to be stored in lagoons and discharged over a longer time period than the processing season. This is done to avoid either overloading the works for part of the year, or having treatment capacity underutilized for the rest of the year.

Dairy wastes consist mainly of whole milk or its derivatives, and they rapidly become acidic because of the conversion of lactoses to lactic acid. Another major component is the alkaline wash water from cans, bottles, and churns. These wastes have a BOD of about 1000 mg l^{-1} and can be treated by conventional biological processes, particularly alternating double filtration (ADF) and high rate filtration. Slaughterhouse wastes contain blood, animal excreta, dirt, and hair, and can have a BOD of up to 3000 mg l^{-1} with a total nitrogen content of about 500 mg l^{-1} (Tookos, 1977). These wastes are best treated at a sewage works in admixture with sewage, but can be treated on site with protein recovery.

Of the water used in a brewery, about half is discharged to the environment, the remainder either being recycled in the works, incorporated in the product or evaporated. Most of the untreated effluent (which has a BOD of about 900 mg l^{-1}) is wash water from process and ancillary equipment and the cleaning of bottles and casks (Isaac, 1966). Because of their volume and strength, brewery wastes are normally treated at sewage works which have often been specially extended to treat the extra organic load by alternating double filtration. Effluents from the food, drink, and paper industries when discharged to rivers can cause the growth of sewage fungus as described in Chapter 5. Poorly treated sewage effluent can cause similar growths, which result from the presence of easily degradable organic matter in the effluents.

Discharges from the Chemical Industry

The waste discharges from the chemical industry are many and vary greatly in quality. The effluents may contain the raw process materials, the reactants, products and by-products of the process, and some cooling water which may or may not be contaminated with process water. Some of the chemicals in the effluent are biodegradable, some are not, and some are toxic. It has been suggested (Thom and Agg, 1975) that there are about 250 important synthetic organic chemicals which may affect water quality. Of these, less than 16% are easily degraded by conventional sewage treatment and many of the remainder could result in a public health hazard if present at a public water supply intake in sufficient quantities. Effluents vary in quality as production schedules change, and this may lead to problems at sewage treatment works. Biological treatment processes are vulnerable to shock discharges of toxic substances or changes in influent quality, and they need time to acclimatize to the discharge of different substances. As a result, treatment plant efficiency may be reduced and inferior-quality effluents may be discharged. Organic sulphur compounds can cause treatment problems by interfering with nitrification.

Effluents from the manufacture of epoxy resins can be successfully treated by the activated sludge process, the BOD being reduced from about 1800 mg l^{-1} to 2.6 mg l^{-1} (Singleton and Hosden, 1973). The paint industry uses large amounts of titanium and silicon dioxides and solvents, which are also contained in the effluent (Huang and Ghadiran, 1974). The effluents from the carpet industry contain pesticides which are used to mothproof carpets, and other pest control compounds used on sheep fleeces. The wastes can also contain dyes and surfactants, and the presence of organochlorines can be toxic to biological treatment processes. Treatment at a sewage works results in reduced pesticide concentrations, but problems can still arise (Holden and Marsden, 1966). Table 4.17 summarizes the pollutants found in domestic sewage effluent and different types of industrial effluent.

MICROPOLLUTANTS

More than 1000 organic compounds have been found in rivers and effluents (World Health Organization, 1975), and doubtless there are many more

Table 4.17. Some sources and causes of water quality deterioration

Effluent	Factors affecting water quality deterioration
Domestic sewage	BOD, suspended solids, ammonia, nitrate, phosphate
Vegetable processing	BOD, suspended solids, colour
Chemical industry	BOD, ammonia, phenols, non-biodegradable organics, heat
Iron and steel manufacture	Cyanide, phenols, thiocyanate, pH, ammonia, sulphides
Coal mining	Suspended solids, iron, pH, dissolved solids
Metal finishing	Cyanide, copper, cadmium, nickel, pH
Brewing	Suspended solids, BOD, pH
Dairy products	BOD, pH
Oil refineries	Heat, ammonia, phenols, oil, sulphide
Quarrying	Suspended solids, oil
Power generation	Heat

remaining to be identified. Toxicological and carcinogenicity information is not yet available for all these compounds and, therefore, their significance cannot be assessed. Between 10000 and 20000 chemicals are available in the UK (Central Water Planning Unit, 1976) and about another 1000 are added to the total each year. Because of the great variety of chemicals and their differences in biodegradability, toxicity, and carcinogenicity, and the need to protect the environment and reduce health hazards, these chemicals are best controlled at source. Many substances are readily degradable, but industries do not always know exactly what amount of chemicals they discharge to the environment (nevertheless, availability of this information is important for the procedure of catchment quality control as described in Chapter 7). Biodegradability may be defined as the complete degradation of a compound such that the final products are biomass and the oxides of the elements constituting the original compounds (Wuhrmann and Meschner, 1974). Because a compound which is discharged to a river cannot be detected in that river, does not mean that it is rendered harmless. Some compounds may be broken down to simpler molecules, but these may then be resistant to further degradation and may have harmful effects not related to the original compound. In the future, to protect public health and the general environment, and to facilitate the most economic and efficient disposal of effluents, it may be necessary to classify our rivers as (1) those to which discharges are limited and industrial activity is closely controlled and (2) those which certain industries are directed to use. The first type would be used for public water supplies and the second would not.

SOLID WASTE DISPOSAL

The disposal of solid waste and strong industrial effluents at tip sites can also lead to water pollution problems. Table 4.18 shows the sources and amounts of solid waste disposed of annually in England and Wales. Household and domestic sources only account for about 10% of the total. The biggest source is coal mining, which produces 50 million tonnes of waste each year, most of it being stored on colliery spoil tips. Colliery, quarry, and general building wastes consist mainly of inert materials, whereas domestic waste is organic and, if not tipped in an organized way, can produce leachates of high BOD and ammonia content. Industrial waste covers many different types of material, from blast furnace slag to concentrated acid solutions and oils. Disposal of these materials should be carried out in geologically safe sites such that there is no chance of contamination of either underground or surface waters.

Another problem in solid waste disposal is the contamination of water by metals from lead mine spoil tips (Jones and Howells, 1975). Weathering of the metal sulphides produces soluble lead and zinc sulphates which are then washed into the river. The presence of lead deposits in acid and soft water areas can result in increased lead concentrations and increased toxicity. The impact of mining and mining waste on river water quality has been recently reviewed (Purves, 1975).

About 30% of sewage sludge is tipped, 44% is applied to agricultural land, 23% is dumped at sea, and the remainder is incinerated (Department of the Environment, 1976). All sludges contain some metals derived from domestic sources, but many contain higher metal concentrations as a result of trade effluent discharges to sewer. Analysis of the sources of metals to New York sewage showed that 49% of the cadmium came from domestic sources and

Table 4.18. Solid waste production in England and Wales

Source	Annual amount ($\times 10^6$ tonne)
Household and domestic	18
General industrial	23
Building waste	3
Power stations	12
Mining	60
Quarrying	50
Sewage sludge (dried)	1.2

33% from electroplating discharges, whereas 62% of nickel discharged was derived from electroplaters and only 25% came from domestic sources (Klein *et al.*, 1974).

The disposal of pulverized fuel ash (PFA) from power stations may cause pollution of groundwater. The metal content of the ash is a function of the type of coal burned, and a recent survey in the USA showed PFA to have zinc, lead, and arsenic concentrations of 3300, 1120, and 1200 mg kg^{-1}, respectively, some of which could be released into aqueous solution (Theis *et al.*, 1978).

ACCIDENTAL DISCHARGES

Many discharges of an 'accidental' nature occur each year. They can vary from a burst or leaking oil storage tank to the discharge of pesticides from storage areas following a fire. Table 4.19 shows the impact of a discharge of oil to a watercourse. Unless adequate precautions are taken, discharges of oil which may not increase river BOD or decrease dissolved oxygen, can result in the closure of water supply intakes and can damage fish and birdlife in and around the river. During 1977/78, 1895 incidents were reported within the Severn–Trent Water Authority, 36% of them being spillages of oil, 10% chemical spillages, 15% from sewers and sewage works, and 8% from agricultural activities (Seven–Trent Water Authority, 1978).

Accidental discharges are much more likely to cause fish mortalities than the continuing discharge of treated effluents. The quality of a river reflects the size and quality of the discharges to it, and much effort may be expended in treating effluents in order to improve river quality. However, the beneficial effects of such work may be wiped out by a single illegal discharge of a toxic substance.

ACKNOWLEDGEMENT

The author wishes to thank the Chairman and Chief Executive of the Severn–Trent Water Authority for permission to publish this chapter and for access to

Table 4.19. Oil on water

Approximate thickness of oil film (cm)	Appearance	Approximate amount (l) of oil covering 1 km^2
0.0000038	Barely visible	37
0.0000076	Visible as silvery sheen	74
0.000015	First traces of colour	146
0.000030	Bright bands of colour	293
0.00010	Colours begin to turn dull	974
0.00020	Colours much darker	1948

the information contained in it. Thanks are also due to Dr N. Harkness for helpful advice and to the editor for constructive suggestions. Nevertheless, the views expressed are those of the author and do not necessarily represent the views of the Severn–Trent Water Authority.

CHAPTER 5

Ecological Effects of Changes in Water Quality

A. M. Gower

ECOSYSTEM INTERACTIONS

One of the striking features of lotic waters (e.g. rivers and streams) and lentic waters (e.g. lakes) is the way in which they interact with, and are affected by, the surrounding land. Drainage and discharge from the surrounding land affect the quality and quantity of water flowing downstream, and it is thus essential that water quality is assessed in the context of the catchment ecosystem.

A catchment ecosystem (Figure 5.1) consists of a non-living abiotic component, and a biotic component made up of assemblages of living organisms including man. The two components interact through the flow of energy and cycling of nutrients, and one result of the interaction is seen in the quality of the water. Although a watercourse or lake may be regarded as an ecosystem in its own right, for the purpose of understanding the factors governing the quality of fresh water, the watercourse or lake is treated here as a subecosystem of the catchment. This is particularly important when one considers the extent to which lotic and lentic waters have their characteristics imposed on them by the surrounding land. Catchments vary in their characteristics such as geographical location, geology, geomorphology, biogeochemical processes, and the extent of human activity including resource exploitation, transportation, and urbanization. Such variations are reflected in the catchment's water, the final adjustment to water quality occurring within the aquatic subecosystem and resulting from interactions between aquatic organisms and their surrounding medium.

It is important to understand more about aquatic ecosystems to appreciate the ecological consequences of man's impact on water quality. Although lotic systems are complex, much can be learnt from those reaches which exhibit an alternation of shallow and deep water areas (riffles and pools, respectively). Except in heavily wooded catchments, the riffles are the main areas of plant

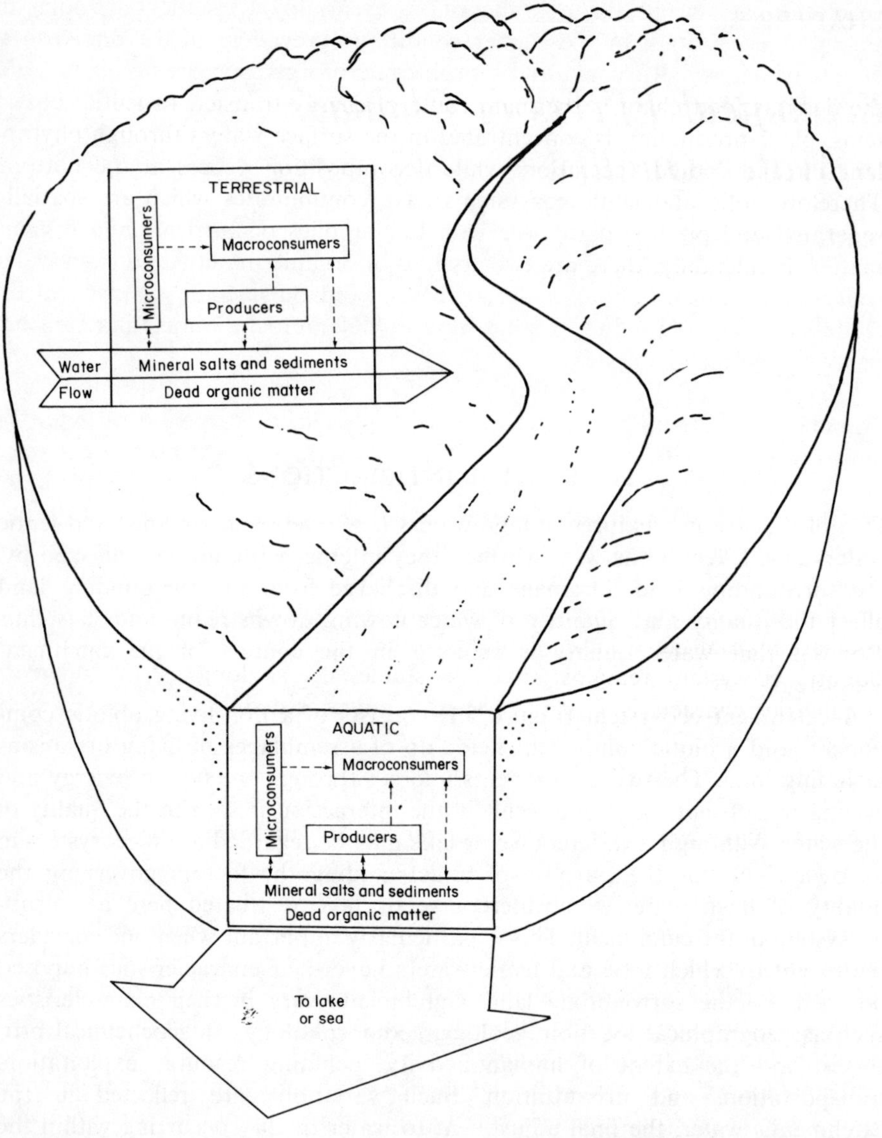

Figure 5.1. A catchment ecosystem showing compartmentalized terrestrial and aquatic subecosystems

production and at least some of the energy is transferred through the grazing food chain. In contrast, pools are important sites for decomposition of the accumulated detritus. In temperate areas in autumn, it is certainly striking to observe how most of the dead leaves entering a stream from terrestrial vegetation are swept through the riffles, eventually to settle out, often in enormous quantities, in the calmer conditions prevailing in the pool zones; here scavengers, detritivores, and saprobic micro-organisms are dominant, and the detritus food chain is the main route of energy transfer. In lentic ecosystems, plant production is concentrated in the surface waters through phytoplankton and rooted vegetation, while decomposition occurs at the bottom. Therefore, lotic and lentic ecosystems have communities which are spatially separated and organized to cope with the supplies of nutrients and organic matter. Incidentally, there are two sources of organic materials in fresh water ecosystems. Allochthonous materials enter from outside the system and include leaf litter and man's organic wastes, while autochthonous materials are produced within the system.

Of course, there are many factors which determine the distribution and abundance of organisms in fresh water. Abiotic factors operate through the substratum and the surrounding medium, whereas biotic factors result from the activities of the organisms themselves; the observed aquatic communities are the product of interactions between these factors.

Seasonal fluctuations, for example in the composition of communities and density of individuals, may occur in response to the variation in the nature and quantities of materials entering rivers and lakes, but this is part of the natural oscillation resulting from climatic and other changes. At the same time, the aquatic ecosystem remains relatively stable, i.e. as long as the range of oscillations characteristic of the system is not exceeded.

EUTROPHICATION

Many of man's activities in a catchment tend to increase the fertility of the water. There are point sources of enrichment such as sewage effluents, and diffuse sources which include run-off from agricultural land and enrichment from the atmosphere in disturbed and industrial areas. Figure 5.2 shows that an increase in nutrients would facilitate the flow of solar energy through the primary producers. Biological production would increase, resulting in larger biotic compartments. This in turn would affect the size of other compartments, with implications for water quality.

The whole process, involving the change of an unproductive body of water into a productive one through enrichment with plant nutrients is known as eutrophication. This term used to be applied to the natural successional changes characteristic of the maturation of lakes. However, human activity has considerably accelerated this natural process in many catchments through the discharge of excessive amounts of nutrients, resulting in cultural eutrophication of rivers as well as lakes.

Natural lake eutrophication also involves sedimentation, reducing the depth of the lake and the volume of water, thus helping to lead to the increased

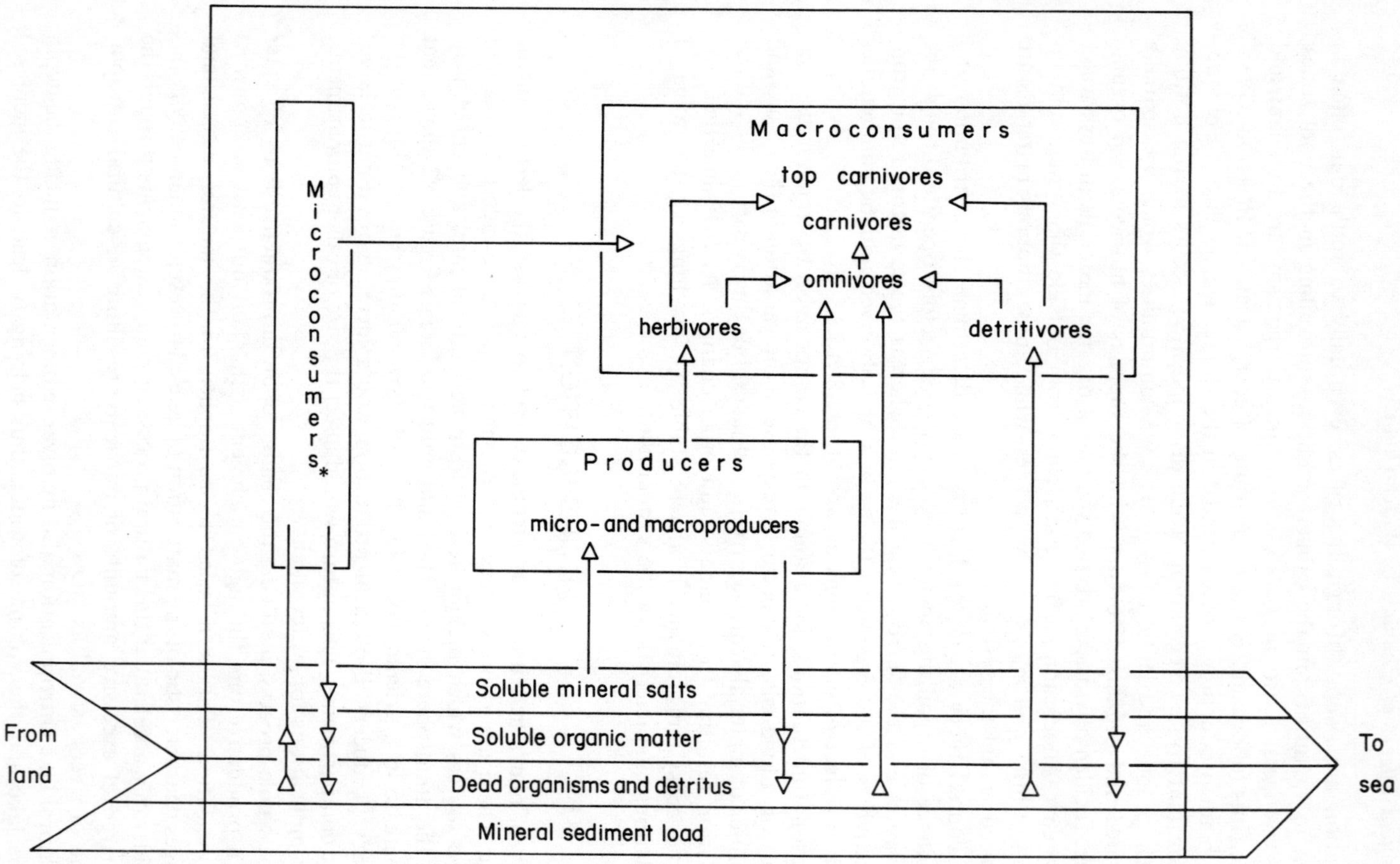

Figure 5.2. Compartmentalization of a freshwater ecosystem showing main interactions (*microconsumers include bacteria, fungi, and protozoa)

nutrient concentration. However, in many of the culturally eutrophic lakes of the world, there is an increase in nutrient concentration, without significant infilling of the lake basin. Since a lake is a bulge in a drainage system, where there is reduced or no through-flow, materials accumulate. In contrast, lotic systems are characterized by a more rapid and continuous through-flow, so that eutrophication is only brought about by a continuous supply of nutrients which is, for example, frequently provided by sewage effluents.

An aquatic ecosystem which has become culturally eutrophic is unstable partly because of the intensity of the phenomenon; natural eutrophication occurs on a geological time scale, but cultural eutrophication takes only decades or years or less. As a result, changes occur so rapidly that the community as a whole cannot adjust, and the oscillations exceed those experienced by the system on a normal time scale. There may be disproportionate increases in micro- and macro-consumers and producers at different times, and ultimately the amounts of plant and animal tissue may exceed the decomposing capacity of the bacterial population. Substantial accumulation of organic matter then occurs in the bottom deposits.

Eutrophication can be better understood by determining the quantity and types of nutrients responsible for increased growth of aquatic vegetation. Man's activities often add a variety of nutrients to aquatic systems such as carbon, nitrogen, phosphorus, silicon, and trace metals, most of which are used to varying degrees by different plant species. Nitrogen and phosphorus are generally of prime importance, and phytoplankton blooms can occur in lakes having concentrations as low as 0.3 mg l^{-1} of nitrate nitrogen and 0.01 mg.l^{-1} of phosphorus (Sawyer, 1947). It would seem that phosphorus, in particular, plays a key role, especially in the early stages of eutrophication (e.g. Vollenweider, 1968), perhaps partly through stimulating bacterial production of growth factors used by algae (Thomas, 1969).

Plants inevitably modify the nutrient content of the surrounding water, although in lotic systems this may sometimes amount to only a small proportion of the total quantity of nutrients moving downstream. Indeed, plants are one of the principal agents (together with denitrifying bacteria) in the removal of oxidized nitrogen and are of primary importance in the removal of phosphorus, especially in lotic waters (Owens, 1978).

Unlike organic pollution, where plants may be virtually eliminated, with cultural eutrophication they occupy a central position in the structure and functioning of the ecosystem. They can, for example, become the source of enormous quantities of food, both as living plants and subsequently as autochthonous detrital matter. Plants can also act as a substratum and a site for the activities of a whole range of organisms. For example, many invertebrates need the support of plants whilst feeding, breathing, or ovipositing. Coarse fish such as perch, roach, and rudd spawn in-amongst vegetation. Again, plants can modify the abiotic component. As well as adjusting nutrient concentrations, they alter the amount of dissolved oxygen through photosynthesis and respiration, secrete toxins, and can affect light penetration. Even the nature of the substratum can be altered, particularly by macrophytes which reduce water movement in the littoral zone of lakes and stream riffles. This allows sedimentation of particulate mineral and organic matter which will have further effects on depth of water, re-aeration rates, and oxygen demand.

The greatest effects of man-induced eutrophication are seen in lentic systems, and the resulting problems are well documented for lakes such as the Zürichsee in Switzerland and Lake Erie in North America, both of which receive large amounts of sewage effluent. In Britain, comparatively few lakes receive significant discharges of sewage effluent, and the main eutrophication problems occur in storage reservoirs when nutrient-enriched river water is impounded, thus providing favourable conditions for growth of phytoplankton. In Britain, natural lakes are more likely to suffer eutrophic effects from agricultural run-off containing fertilizers.

An important characteristic of temperate lakes and reservoirs is stratification. When this occurs in summer, a warm upper layer (the epilimnion) is separated from a cold lower region (hypolimnion) extending to the bottom. At first, light, temperature, and nutrients are favourable for algal growth in the epilimnion, but, as nutrients are used up, surface organisms die and fall into the hypolimnion. Decomposition and mineralization proceed, and nutrients accumulate which do not become available to the surface waters until destratification in the autumn. Decomposition of the organic matter reaching the hypolimnion results in oxygen depletion which may be considerable under eutrophic conditions, especially if there are also quantities of allochthonous materials reaching the lake bottom from the surrounding catchment.

Anaerobic conditions at the bottom of the hypolimnion can affect the redox conditions in the mud–water interface, which in turn governs the retention or release of nutrients. Investigations by Mortimer (1941, 1942), in the English Lake District, showed that phosphate and other ions are adsorbed on to ferric iron complexes in the oxidized mud surface. However, if the mud surface becomes anaerobic, ferric iron is replaced by the soluble ferrous iron and the complexes break down, releasing the nutrients which circulate through the hypolimnion and become available to the surface water on destratification. A lake undergoing cultural eutrophication loses homeostatic control partly because of continuous positive feedback operating at the mud–water interface, i.e. more nutrients cause increased synthesis of organic matter, and subsequently more detritus which leads to more oxygen being used at the mud–water interface and to more nutrients being released. However, the significance of this chain of events to the eutrophic process in a lentic system will depend on various factors: nutrients may be lost with the outflowing water and by precipitation onto the bottom sediments. As another example, anaerobic conditions and enrichment may be further exaggerated if there are large quantities of allochthonous organic matter reaching the lake botton from the surrounding catchment.

During eutrophication, the biotic compartments change in composition as well as in size. For example, desmids (a group of green algae), such as *Cosmarium* and *Staurastrum*, disappear from the phytoplankton. Diatoms (brown algae with silica shells), especially *Asterionella*, become conspicuous, and are then replaced by Cyanophyceae (blue-green algae). Dense populations of the latter may appear at the surface of lakes and reservoirs in calm, warm weather, forming 'blooms'. There are many factors responsible for 'blooms', including their tolerance of conditions near or at the water surface and the fact that they are not grazed so much as other algae. One interesting property of some blue-green algae (e.g. *Anabaena*, *Aphanizomenon*, *Gleotrichia*) is their ability to fix atmospheric nitrogen, and this contribution of nitrogen to the annual lake income may be significant, e.g. in Clear Lake, California, it is between 30 and 50% (Horne, 1977). So this may be an important route of nitrogen enrichment, and is another example of loss of homeostatic control, especially in a lake or reservoir where there is adequate phosphorus but inadequate nitrogen for increased algal growth. Nitrogen fixation by blue-green algal blooms can introduce substantial amounts of nitrogenous compounds into the water which ultimately become available for other species including non-nitrogen-fixing blue-green algae such as *Microcystis* and *Oscillatoria*. Therefore, even with enrichment deficient in nitrogen, the route is established for positive feedback resulting in increased eutrophication.

Zooplankton also becomes more abundant; there are changes in the relative dominance of some of the species and there is evidence, for example, that the cladoceran *Chydorus sphaericus* increases in association with 'blooms' of blue-green algae, perhaps using the latter as a substrate. As organic detritus accumulates in the hypolimnion, relatively few organisms remain which can tolerate the low oxygen tension. These include burrowing tubificid worms and the tube-dwelling red chironomid (midge) larvae *Chironomus plumosus* and *C. anthracinus*, all of which contain haemoglobin. With further eutrophication, the benthic community may be restricted to tubificids. Eutrophication can also cause a dramatic change in the quality of the lake fishery. Game fish such as salmonids are replaced by coarse fish. Factors responsible include oxygen depletion in the hypolimnion, and the smothering of gravelly and sandy substrates (necessary for spawning and development of eggs) with organic detritus.

In lotic waters, the entry of well oxidized sewage effluents and run-off from agricultural land results in direct enrichment; indirect enrichment can occur when plant nutrients are released downstream of a zone affected by organic pollution. In both cases, the nutrient compartment (see Figure 5.2) is increased but is continually moving downstream, and it is therefore the continuous enrichment that maintains the high concentration of nutrients. There is a responding increase in growth of macrophytes (e.g. *Ranunculus fluitans*) and attached algae (e.g. *Cladophora*) in flowing water, and macrophytes together with phytoplankton in sluggish reaches and backwaters.

Cladophora (commonly known as blanket weed) is a benthic filamentous green alga which is attached to stones and other hard surfaces. In eutrophic waters, it grows rapidly and becomes conspicuous for much of the year (except winter in temperate rivers). Pitcairn and Hawkes (1973) were able to correlate the growth of *Cladophora* (measured as standing crop) with the concentration of phosphorus. They also found that growth is probably restricted in waters with a phosphorus concentration of less than 1 mg l^{-1}, and that growth can also be affected by the interaction between the amount of nitrogen and phosphorus in the water. The establishment of dense growths of this alga can cause a number of changes in the surrounding environment. For example, its presence in the riffle zone increases habitat diversity. The alga itself harbours a variety of organisms such as chironomid larvae (especially Orthocladiinae) and epiphytic diatoms. The algal filaments trap silt and detritus, creating favourable conditions for silt-dwelling detritivores such as the water louse, *Asellus*. In the autumn or at times of spate, masses of *Cladophora* filaments may become detached and washed downstream, resulting in pockets of de-oxygenation wherever they settle and decay. The changes brought about by *Cladophora* are to some extent seasonal, with the impact in British rivers being least during winter when only the basal tufts of filaments usually remain. Much of the collected silt may then be scoured and washed downstream. *Ranunculus fluitans* (water crowfoot), a submerged macrophyte with divided leaves, may also thrive under enriched conditions and produce many of the effects described for *Cladophora*. Large growths of *R. fluitans* may reduce or prevent the growth of *Cladophora* through diminishing the high light intensity upon which the alga depends.

In slower flowing waters, dense populations of phytoplankton are symptomatic of eutrophication. For example, Moss (1977) showed that large phytoplankton populations were a reflection of very eutrophic conditions in the Norfolk Broads and the lowland rivers of eastern England. The phytoplankton increases turbidity, and consequently there is a decline in macrophytes and the associated invertebrates. In the Broadland rivers, the Broads themselves serve as culture banks for plankton which are then inoculated into the main river channels.

The excessive growth of plants can result in a pronounced fluctuation in dissolved oxygen over a 24 h period (Figure 5.3). Photosynthetic activity produces high levels of dissolved oxygen during the late afternoon followed by oxygen depletion at night, due to continued respiratory activity. De-oxygenation may be severe, with dramatic effects on those organisms whose oxygen requirements are not satisfied during the period of low oxygen concentration. It is interesting to note that Simonsen and Harremoës (1978)

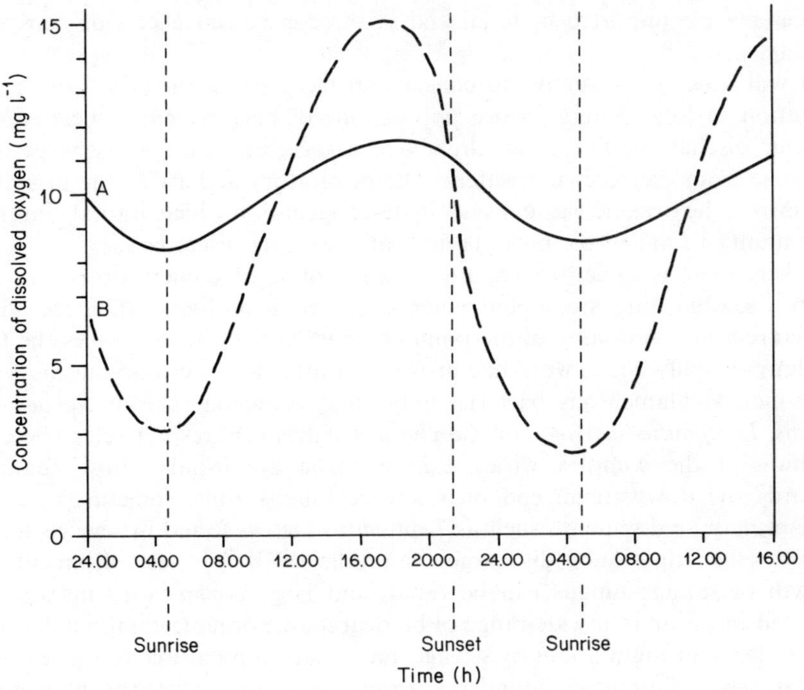

Figure 5.3. Dissolved oxygen profiles for comparable reaches of a nutritionally balanced stream (curve A) and a eutrophic stream (curve B)

have found diurnal oxygen fluctuations caused by vegetation to be the dominating ecological problem in Danish rivers. Exaggerated diurnal fluctuation in pH is another consequence of increased photosynthetic activity, and this, together with oxygen depletion, can cause the death or emigration of a number of organisms including fish.

ORGANIC WASTES

Since rivers and lakes are the receptacles of materials from the surrounding land, they have developed the ability to cope with significant inputs of allochthonous organic matter. However, if excessive amounts of energy-rich organic matter enter a watercourse, e.g. from food and drink processing, paper manufacture, and other industries, organic pollution may result. The detritus food chain within the aquatic community becomes exaggerated while the primary producers and the grazing food chain, for the most part, are bypassed. Much has been learnt about the trophic pathways from organic pollution episodes, and, in contrast to eutrophic waters, it is saprotrophic micro-organisms and macro-invertebrates that occupy the key position in processing the imported material and in the consequent effects on ecosystem quality.

It will now be instructive to consider in more detail the effects of organic pollution in lotic waters. Dense populations of bacteria are associated with organic discharges. They may already be associated with the discharge itself, or come from the receiving waters. The respiratory activity of the organisms feeding on the organic waste results in de-oxygenation which, if total, limits the community to anaerobic bacteria and air-breathing insect larvae.

Where there is some oxygen, a few species of aerobic micro-organisms may form a sewage–fungus complex which can exert a profound influence on the structure and functioning of the polluted zone. Sewage–fungus takes the form of slimy or fluffy cottonwool-like growths of attached micro-organisms which may include filamentous bacteria, fungi, and protozoa such as *Sphaerotilus natans*, *Leptomitus lacteus*, and *Carchesium polypinum*, respectively. The composition of the complex varies, e.g. protozoa are usually most abundant towards the downstream end of a sewage–fungus zone; sometimes one organism may be dominant, such as *Leptomitus lacteus* found in sewage–fungus below refuse tip liquor discharges (Nuttall, 1973). The establishment and growth of sewage–fungus can be rapid, and large conspicuous masses may occur in response to a wide range of biodegradable organic compounds. Some of the effects brought about by sewage–fungus are comparable to those caused by excessive plant growth under eutrophic conditions, i.e. the filamentous nature of the complex entraps silt and detritus and reduces water flow. From time to time, detached pieces of the complex are carried downstream, and exert a local oxygen demand where they settle out. Sewage–fungus may become attached to plants, invertebrates, and fish as well as to the substratum. It can also colonize riffles, and this is where it has the most dramatic effect—through smothering vegetation and facilitating accumulation of silt and detritus. Silt-dwelling detritivores, including tubificid worms and chironomid larvae, become relatively abundant, and thus conditions are created which are quite unique to the fast-flowing reaches of a river.

Moving progressively downstream, as the organic matter is used up by saprotrophic organisms, there is a decline in oxygen demand and a corresponding increase in dissolved oxygen. The changes in oxygen concentration downstream of an organic discharge produce the familiar oxygen sag curve and this is shown in Figure 5.4 together with the diurnal variation in dissolved oxygen at selected stations. These spatial and temporal changes in dissolved oxygen play an important part in determining community structure. For example, in spite of abundant food, there may be insufficient dissolved oxygen for certain organisms; the replacement of *Gammarus* by *Asellus* is a good example, as both are scavenging crustacea, but *Asellus* is more tolerant of oxygen depletion.

Figure 5.5 illustrates the effects of organic pollution on riffle communities

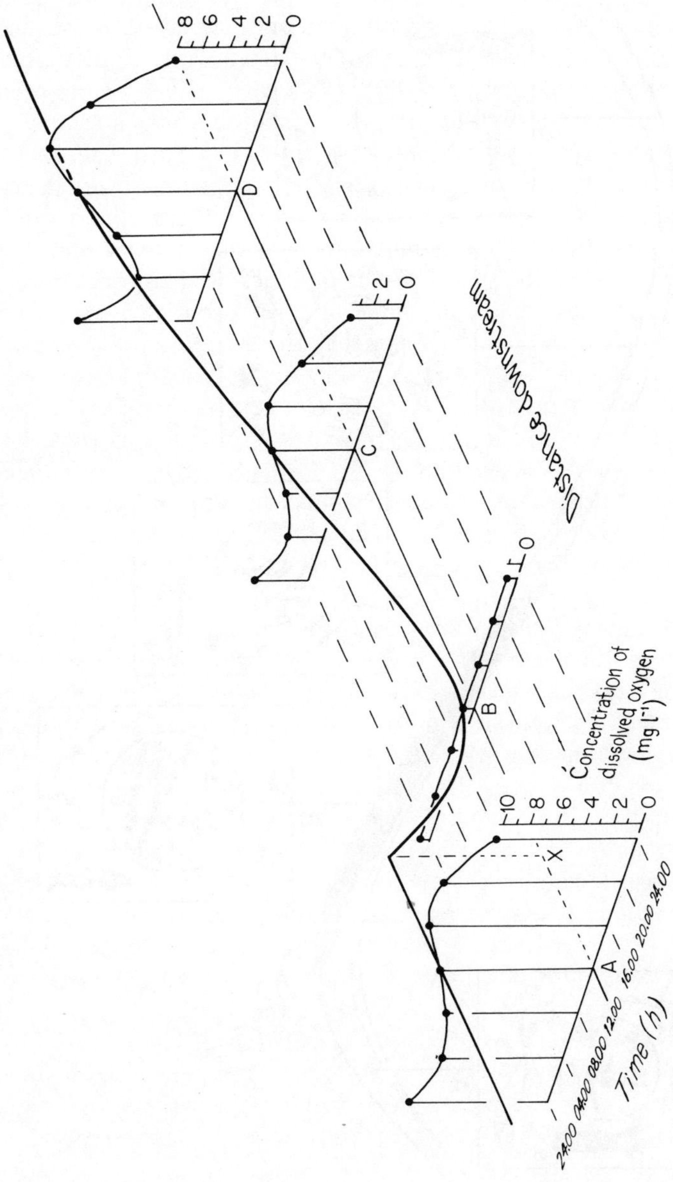

Figure 5.4. Diurnal variation in oxygen concentration at four stations along a river receiving an organic discharge at point X, between Stations A and B. The oxygen sag curve at 12.00 h is shown

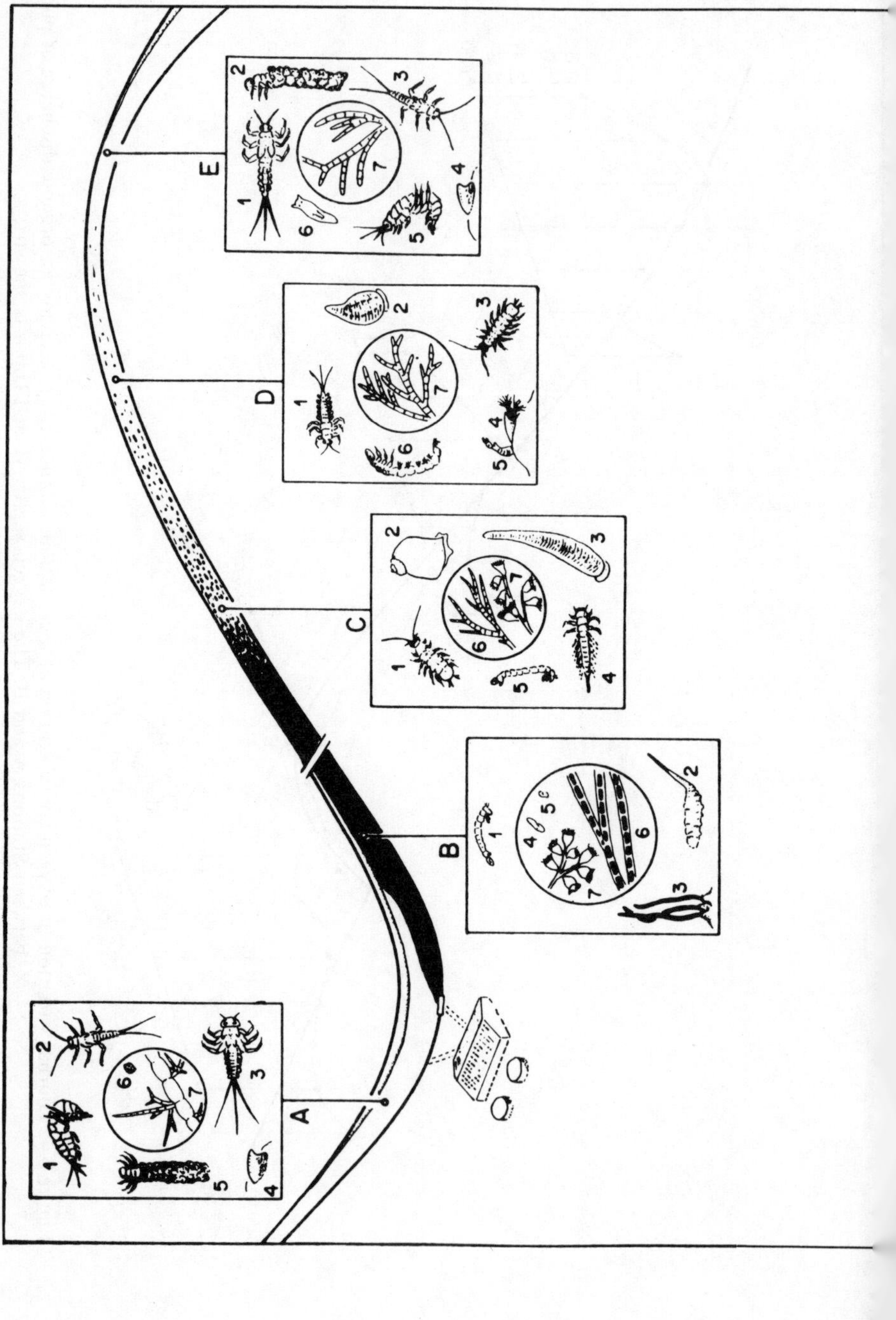
A
B
C
D
E

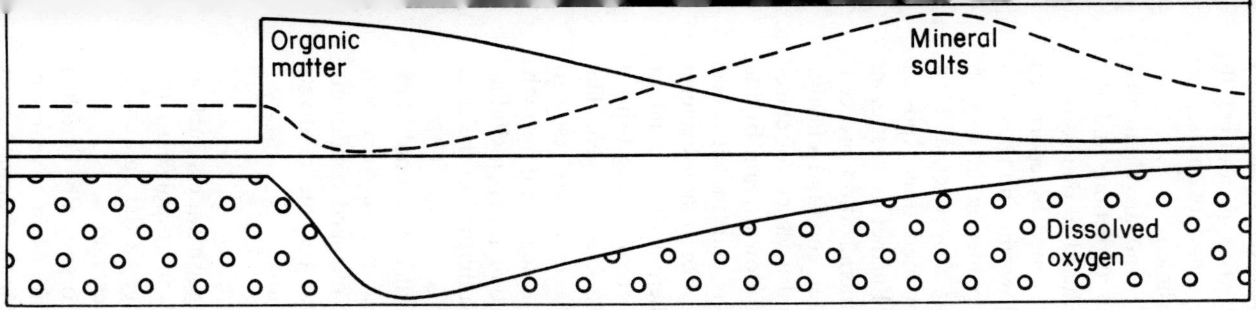

Figure 5.5. Representative organisms of riffle communities at five stations along a river receiving an organically polluting discharge. (Modified after Hawkes, 1963, *Effl. Water Treat. J.* Reproduced by permission of Effluent and Water Treatment Journal.)

Key to organisms

A	B	C	D	E
1 *Gammarus pulex* (freshwater shrimp)	1 *Chironomus riparius* (blood worm)	1 *Asellus aquaticus* (water slater)	1 *Baetis rhodani* (mayfly nymph)	1 *Ephemerella ignita* (mayfly nymph)
2 *Nemoura cinerea* (stonefly nymph)	2 *Eristalis* (rat-tailed maggot)	2 *Lymnaea peregra* (wandering snail)	2 *Glossiphonia complanata* (leech)	2 *Potamophylax* (cased caddis larva)
3 *Ecdyonurus venosus* (mayfly nymph)	3 *Tubifex tubifex* (sludge worm)	3 *Erpobdella testacea* (leech)	3 *Asellus aquaticus* (water slater)	3 *Nemoura cinerea* (stonefly nymph)
4 *Ancylus fluviatilis* (river limpet)	4 *Paramecium caudatum* (free-swimming ciliate)	4 *Sialis lutaria* (alder fly larva)	4 *Simulium ornatum* (blackfly pupa)	4 *Ancylus fluviatilis* (river limpet)
5 *Limnephilus* (cased caddis larva)	5 *Colpidium* (free-swimming ciliate)	5 *Chironomus riparius* (blood worm)	5 *Simulium ornatum* (blackfly larva)	5 *Gammarus pulex* (freshwater shrimp)
6 *Cocconeis* (diatom)	6 *Sphaerotilus natans* (sheath bacterium)	6 *Stigeoclonium tenue* (filamentous green alga)	6 *Hydropsyche angustipennis* (net-spinning caddis larva)	6 *Dugesia* (flatworm)
7 *Draparnaldia* (filamentous green alga)	7 *Carchesium polypinum* (attached ciliate)	7 *Carchesium polypinum* (attached ciliate)	7 *Cladophora* (blanket weed)	7 *Cladophora* (blanket weed)

where the organisms most sensitive to changes in water quality are found. However, this figure only gives a generalized picture of the effects of an organic discharge in a river; the precise details vary from one river to another in relation to the nature of the discharge and receiving waters, and, of course, the prevailing conditions in the catchment ecosystem as a whole.

Below an organic discharge (Station B in Figure 5.5), plants are eliminated by the turbid conditions. There may be severe oxygen depletion, and the normal diverse stream riffle community (Station A) is reduced to one consisting of relatively few species, each of which may be represented by enormous numbers of individuals. For example, burrowing tubificid worms and the tube-dwelling *Chironomus riparius* are able to survive the reduction in oxygen concentration, which incidentally can remain uniformly low over a 24 h period, as shown in Figure 5.4. Both organisms disturb the mud, thus exposing a larger surface area for oxidation and bacterial activity. Further downstream, as the organic matter is used up and oxidation and mineralization proceed, helped by the continued re-aeration from the flowing water, conditions favour increasing community diversity. For example, *Asellus* and a variety of chironomid larvae may become relatively abundant, followed downstream by the omnivorous net-spinning caddis larvae of *Hydropsyche* and the attached larvae of the blackfly *Simulium* which use their mouth brushes to filter bacteria and small particulate detritus from the flowing water. There is thus abundant invertebrate food for predaceous leeches and also for bottom-feeding fish—if they are able to survive the low concentration of dissolved oxygen. The continued mineralization and the improvement in water clarity may result in an oscillation towards eutrophic conditions with increased algal growth, especially of *Cladophora*. Eventually, a more balanced community, including higher plants and a resident fish population, will be re-established in the river as seen at Station E in Figure 5.5.

Many of the clean water organisms associated with riffle and pool communities are eliminated from an organically polluted zone, but may reappear in the recovery zone—in their order of tolerance to the changed conditions. The organisms which replace the intolerant species and become relatively abundant in the polluted zone come from three main sources:

(1) those organisms already associated with the discharge;
(2) species which are more typically associated with sluggish waters such as pools and backwaters, and which increase in numbers in response to the heavier organic load; furthermore, they extend their range as the amount of detritus and associated bacteria increase in the riffle zones; and
(3) insect larvae whose aerial flying adults oviposit in organically enriched waters; and, of course, those species with several generations in a year will have repeated opportunities to colonize waters suitable for their development (Gower and Buckland, 1978).

The processing of an excessive input of organic matter in a river, and recovery from its effects, is a good example of the homeostatic control of a lotic system. Negative feedback is realized through the biotic shuffling within the aquatic community to cope with the organic load at its different stages of breakdown. Thus, a river is able to purify itself, provided there is a sufficient stretch of water in relation to the size of discharge.

INERT SUSPENDED SOLIDS

Organic discharges are frequently laden with biodegradable solids in suspension. In contrast, there are other discharges, e.g. from coal and china clay mining, in which the load of suspended particles may be almost entirely inert; such discharges serve to increase the mineral sediment compartment of a fresh water ecosystem (see Figure 5.2).

Where mining activity occupies upland sites in a catchment, inert suspended solids will be discharged into fast-flowing waters with an eroding substratum. As the load of particles is deposited, the original (heterogeneous) substratum is transformed into a more uniformly depositing one. Particles of inert suspended solids can range widely in size and density so that the heavier ones settle out first and the lighter ones travel furthest and settle last. Of course, from time to time, and especially during spates, the settled particles may be scoured out and resuspended, thus providing an unstable shifting substratum. The primary ecological effect of this type of discharge is essentially physical in nature, in which the particulate load in suspension is abrasive, the effective light penetration is reduced, and sedimentation occurs. Secondary effects acting on the biota include changes in the food supply and reduced competition. The net result is a sparse community with reduced diversity.

Plants (together with many herbivores) are generally eliminated with only isolated tufts of moss, or algae such as *Stigeoclonium* and *Ulothrix* surviving on the surfaces of large stones and boulders. Allochthonous organic matter is therefore the main source of energy available to the aquatic community, and the prevailing conditions favour organisms which are associated with the detritus food chain and which can also cope with the 'rain' and accumulation of mineral particulate material. The filter feeding larvae of *Hydropsyche* and *Simulium* are not successful. However, burrowing detritivores such as tubificid and naidid worms, and also chironomid larvae, become more prominent, although their densities are related to the food supply available, and they do not increase on anything like the scale seen with organic pollution where the pollutant is also a source of food.

In upland streams in mining areas, the brown trout, *Salmo trutta*, may be severely affected both directly and indirectly by inert suspended solids. Although able to migrate through affected reaches, their longer-term survival in these waters is jeopardized by the reduction in food supply, the effects of the physical impact of particles on sensitive surfaces such as gills, and also the silting of suitable gravelly spawning grounds. According to Alabaster (1972), concentrations of inert suspended solids exceeding 100 mg l^{-1} are generally detrimental to river fisheries.

POISONS

Heavy Metals

Mining activities in mineralized areas, and industrial activities such as smelting and electroplating, sometimes provide us with another form of water quality deterioration—namely, excessive amounts of heavy metals. The impact of a toxic metal discharge on the biotic community is dramatic, with reduced diversity due to elimination of most species, overall reduction in the number of individuals within the community, and survival of only those species tolerant of the contamination.

How metals exert their toxic effect is not always clear, but the implications for the biotic community are better understood if it is appreciated that heavy metals can exist in soluble, colloidal, and particulate phases, and consequently can reach every part of an aquatic ecosystem:

(1) They may be associated with the surface film where they are complexed with allochthonous organic compounds leached out from the surrounding land.

(2) They can occur in the water column itself: (a) in solution (as free metal ions and as inorganic and organic complexes); (b) in the form of colloids; (c) as suspended particles; and (d) adsorbed on, or complexed with, suspended particulate matter.

Generally, it is the metal ion in solution which is toxic to aquatic organisms, but the association of metals with organic and inorganic particles is likely to be a significant source of contamination in filter-feeding molluscs and insects.

(3) The eventual fate of many heavy metals is with particulate matter and as precipitated material, both of which accumulate in the bottom sediments; this is why stream and lake sediment analysis is sometimes used as a geochemical technique in search for metal-rich areas. Aston and Thornton (1977) have further used the technique to provide an indication of the concentrations of trace metals likely to be achieved or exceeded in the associated waters. Such a relationship is useful, not only from the water quality standpoint, but also for predicting the likely impact on aquatic organisms.

Although sediments act as a sink for heavy metals, the latter may still be potentially mobile. For example, release of these metals into the overlying waters may be effected by changes in the physicochemical conditions, such as pH and oxidation potential (Eh). Movement of metals out of sediments may also be achieved by biotic factors such as uptake through rooted plants (Harding and Whitton, 1978). From here, the metals may enter food chains, or be released into the water by the plants themselves. It should also be remembered that micro-organisms are associated with river and lake sediments. As a result of their absorptive and adsorptive capacity, bacteria compete very effectively with the sediments in accumulating metals, and have a significant effect on their mobilization into food chains (Ramamoorthy *et al.*, 1977; Patrick and Loutit, 1976, 1978). For example, tubificid worms have been shown to concentrate metals from ingestion of contaminated bacteria and the metal-enriched worms may then be eaten by other animals, including fish.

Clearly, then, it would be erroneous to assume that it is only the metal in solution which is available to aquatic organisms. Metal contamination of aquatic organisms may be derived from one or more sources, the relative importance of which varies with the organism as well as with the distribution of the metal in the different compartments of the ecosystem.

So far, we have briefly considered the occurrence of metals and their entry into the biotic component within aquatic ecosystems. We shall now examine some ecological effects. When there is direct uptake of heavy-metal ions in solution, the toxic action eliminates many plants, with only resistant species surviving, e.g., in running waters, *Lemanea* (a red alga) is resistant to zinc and lead and *Stigeoclonium* (a filamentous green alga) is resistant to zinc and copper. The effect on invertebrates can also be dramatic, and the reader is referred to the investigations of Carpenter (1924) and Jones (1940a, 1940b) on some Welsh rivers polluted by lead and zinc from mine wastes; also the study by Butcher (1946) of the River Churnet, polluted by a discharge from a copper works. In each case, fish were absent, no doubt largely due to the direct uptake by the gills of heavy metals from solution.

When the effect on animals is more severe than on plants, there may be excessive plant growth, partly caused by the absence of herbivores. In the River Churnet, there was total elimination of the fauna from waters with a high concentration of copper (0.5–1.0 mg l^{-1}), which allowed the establishment of the green alga *Chlorococcum* and the diatom *Achnanthes affinis.* Further downstream, as the copper became diluted, species diversity of the algae increased, but invertebrates were slow to recover with only chironomid larvae being recorded at a concentration of 0.12 mg l^{-1}.

The invertebrates most tolerant to metal pollution in fresh water appear to be tubificid worms and insect larvae, especially certain species of the Chironomidae, Ephemeroptera, and Plecoptera, and some species of campodeiform caddis larvae. In the fast-flowing reaches of lotic waters, there is no replacement, and the surviving species consist of those riffle organisms tolerant of the changed conditions—for example, nymphs of the stonefly, *Leuctra*, which even became more abundant in contaminated Welsh rivers (Jones, 1940b), possibly because of the absence of competition.

To summarize, the organisms within an aquatic community are affected differently by heavy-metal pollution, and this is reflected in the recovery zone of a metal-contaminated river. Species reappear in their order of tolerance to the poison which gradually declines in concentration downstream, due very largely to physical and chemical processes such as dilution and precipitation.

One recent investigation which demonstrates the complexities of heavy-metal contamination of an aquatic ecosystem is that of Södergren (1976). He was able to associate a significant reduction in the juvenile salmon population of the River Rickleå in northern Sweden with cobalt discharges from a diamond factory. The discharges contained precipitated cobalt hydroxide which dissociated in the acid receiving waters of the river, with the release of cobalt ions. However, the increased concentrations recorded in the river water (maximum of 0.047 mg l^{-1}) were still below levels known to be toxic to fish and their eggs. There was a seasonal variation in the cobalt concentration in the river water, with the highest measurements (0.010–0.043 mg l^{-1}) at times of low flow in the autumn and winter, lower concentrations (0.002–0.005 mg l^{-1}) during May and June due to the spring flood, and zero values when the factory was closed for its summer holiday in July. Although there was some increase in the cobalt content of the surface sediments below the discharge, the concentration factor in the moss *Fontinalis* was significantly greater, with 19 times the concentration of cobalt being found in clumps collected downstream of the factory.

Associated with the moss were a number of invertebrates of which the mayfly nymphs of *Ephemerella mucronata* (which grow throughout autumn and winter), *E. ignita* (which grow in the summer), and *Baetis rhodani* (which has summer and winter generations) are important food organisms for salmon, especially when the insects appear in drift. In the summer, it was found that nymphs of *E. ignita* and *B. rhodani* were abundant, but in the winter *E. mucronata* and *B. rhodani* disappeared from the river below the factory. The elimination of mayfly nymphs during winter occurred at a time when cobalt concentrations were at their highest. Södergren demonstrated that a cobalt concentration of 0.032 mg l^{-1} reduces growth and increases the mortality of nymphs, and suggests that the sensitivity of *Ephemerella* could also be due to their consumption of accumulated cobalt in the *Fontinalis*, especially as they were shown to tolerate high concentrations of cobalt in the water when prevented from eating.

The change in water quality in the River Rickleå acting through the trophic levels was probably the cause of decline of the salmon population. As a result, salmon parr which are more dependent on drifting mayfly nymphs than any other species of fish, were denied this source of food for much of the year—and especially in spring when large nymphs of the overwintering species are normally available—with consequent reduction of the juvenile salmon population. This investigation provides us with an example of the need to think in terms of the whole community and of the temporal changes characterizing the dynamic nature of an aquatic ecosystem. Too often in the past, effects of toxic materials have been assessed solely in terms of short-term toxicity experiments on single species.

Long-term detailed observations are required to identify sublethal effects such as reduced growth and rates of reproduction and changes in behaviour patterns. Of course, behavioural changes in an organism may be difficult to detect by man, but are recognized by other organisms. Even slight changes in an animal's behaviour may make it more vulnerable to predation. For example, Sullivan *et al.* (1978) found that sublethal levels of cadmium in fathead minnows cause subtle alterations in their behaviour pattern including schooling, with the result that they were more heavily preyed upon by largemouth bass. In this way, sublethal concentrations in the prey could prove to be lethal to predatory organisms through the accumulation of poison from susceptible prey. Therefore, both ecologically and toxicologically, such effects are as important as acute lethal effects.

Again, if an organism is subjected to stress by a toxic substance, it may fail to compete successfully with other organisms. For example, high concentrations of heavy metals in lake sediments retard larval growth of the midge, *Chironomus tentans* (Wentsel *et al.*, 1977a, 1977b). In a metal-contaminated lake studied by these authors, larvae of *C. tentans*, which feed primarily on organic detritus, were found to be numerous in sediments with low levels of heavy metals, but their ability to compete effectively with a more resilient detritivore—the oligochaete, *Limnodrilus*—was reduced with increasing metal concentration. This allowed the greatest densities of the worm to occur in sediments with the highest metal contamination. Incidentally, the sediments of aquatic systems, acting as sinks for poisons such as heavy metals, often yield examples of metal-tolerant organisms. *Limnodrilus* has already been mentioned, and in the estuarine situation there is the burrowing polychaete worm, *Nereis diversicolor*.

One invertebrate which has been shown to be particularly tolerant to heavy metals in rivers receiving mine drainage in south-west England is *Asellus meridianus* (Brown, 1976). In the River Hayle, with 0.1 mg l^{-1} of copper and trace amounts of lead, this crustacean is tolerant to both these metals, whereas in the River Gannel, with 0.04 mg l^{-1} of copper and a high concentration (0.19 mg l^{-1}) of lead, it is tolerant to only lead. Brown (1977) further demonstrated that both tolerant and intolerant specimens of *Asellus* can accumulate metals from solution, and tolerant animals can also accumulate copper and lead from metal-enriched food. Copper-tolerant animals accumulated copper, and lead-tolerant animals accumulated lead—to tissue levels which proved lethal to the intolerant animals.

The tolerance mechanism for copper was demonstrated to be associated with its removal from the blood, and storage in the hepatopancreas. Storage sites for lead are believed to include the hepatopancreas, the cuticle and underlying tissues. Sulphur–metal complexes associated with copper granules which were found in the hepatopancreatic cells of *Asellus* from the River Hayle are probably partly responsible for the dual tolerance shown by these animals. Finally, adaptation to heavy metals in *Asellus* is probably reinforced by a genetic influence, as resilience to lead continued in the F2 generation cultured from *Asellus* collected from the River Gannel.

Pesticides

Heavy metals are just one of a range of poisons that may affect water quality. Pesticides present us with an example of toxic substances manufactured and deliberately applied to the environment to poison pest organisms. It is interesting to reflect on the ways in which pesticides reach aquatic ecosystems. Some (e.g. lampricides, some herbicides, molluscicides, and insecticides) may be introduced directly into water to control any unwanted aquatic organisms. Some have point sources, and enter through effluents discharged from textile and other industries using moth and other pest-proofing pesticides. In addition, significant quantities have been recorded entering rivers and lakes (directly, and also by contaminated run-off) from aerial spraying used in the control of agricultural and forest pests. Again, since there is often considerable volatilization of many pesticides both during and after application, they may get carried for enormous distances, eventually to contaminate catchments far removed from the original point of release. If the pesticides are persistent chemicals such as organochlorines (e.g. DDT, dieldrin, and BHC) which do not break down readily, they are likely at some stage to find their way into fresh water.

Like heavy metals, pesticides can contaminate virtually every part of an aquatic ecosystem. The main route of uptake and accumulation (e.g. of organochlorines) for aquatic organisms has been the subject of some debate in recent years. Uptake from the surrounding water and the organism's food are both important; the main route will surely vary with a number of factors, including the formulation in which the pesticide occurs, its distribution in relation to the organism's mode of life, and the characteristics of the receiving waters.

When a pesticide enters an aquatic ecosystem, one or more populations of organisms are poisoned with repercussions for other species. For example, entry of a herbicide (such as Paraquat or Diquat) causes the death of aquatic plants *in situ*, and this, in turn, will have a number of ecological consequences. First, the nature of the habitat changes through the destruction of plants (sometimes accompanied by substantial delay in regrowth) and the resultant effect on the substratum. Dependent populations would also be affected, with survival only of the more adaptable species. Then there may be changes in water quality, not only through the addition of the chemical but also through oxygen depletion and increased concentration of plant nutrients resulting from the microbial decomposition and mineralization of the dead plants.

The accidental or deliberate contamination with insecticides is known to have differential effects on the various plant and animal populations due, for example, to lethal and sublethal effects, accumulation, resistance, and, of course, competition. Insects may be severely affected even by low concentrations of an insecticide which may cause increased agitation and loss of coordination. This is of particular significance in running water, where the disturbed behaviour of actively predaceous larvae, e.g. of stoneflies and caddis flies, exposes them to the mercy of the current. Many of these larger insects take one or more years to recover and re-establish themselves because of the longer duration of their life cycle. In contrast, some of the smaller insects which have two or more generations per year, e.g. many Chironomidae and Simuliidae, are able to recolonize more quickly after the initial elimination, and may become more abundant while competition and predation are reduced. Thus, the presence of a pesticide in a fresh water system can result in an abnormal ecological oscillation, an example of which is afforded by Davies (1950) who recorded the use of DDT to control the larvae of the blackfly, *Simulium*, in a stream. Although good control was achieved at first, with drastic reduction in numbers, it was followed by an even more spectacular increase, with densities reaching 17 times the prespraying figure.

Heavy metals and pesticides have been used here to provide insight into the impact of poisonous substances in aquatic systems, but the list of toxic substances is forever growing and includes cyanides, phenols, chromates, and PCBs. The toxicity of such substances will vary with other aspects of water quality such as pH, dissolved oxygen, calcium content, temperature, and presence of other poisons. It is often stated that poisons exert a comparatively simple effect on aquatic ecosystems by reducing the variety and abundance of life. Ecologically, however, there are repercussions not only for the intolerant species but also for the community and the functioning of the ecosystem as a whole. Even an apparently simple deterrent effect may bring about the redistribution of organisms in an aquatic system as a result of poisoning a part of it.

COMPLEX POLLUTANTS

The emphasis so far has been on single-character pollutants with specific changes in water quality. However, point and diffuse sources of contamination are frequently of a complex nature. Again, aquatic systems are likely to receive a variety of effluents, and this means that ecological effects frequently reflect changes brought about by mixtures of wastes.

Organic discharges may result not only in the addition of solids with an associated high BOD but also the release of poisons such as ammonia and sulphides. In any case, toxic and organic wastes are frequently found together—indeed, they may occur in the same discharge, as in pulp and paper mill effluents and in effluents from sewage works treating combined sewage and industrial flows. If such effluents are of a poor quality, resulting in oxygen depletion in a river, the oxygen sag curve may only take on a shallow profile, but will extend for some distance downstream. This is due to delayed recovery because of an inhibitory effect of the toxic material on microbial activity. These conditions are likely to prove lethal to fish; generally, most poisonous substances are more lethal to fish at low oxygen concentrations, the main reason being that fish pump more water over their gills to compensate for low dissolved oxygen, and therefore bring a relatively large amount of toxic substance in contact with the gills.

Refuse tip liquors are pollutants, the complexity of which varies with the nature of the wastes from which they are derived. Leachate from tips used for domestic refuse generally includes organic matter, ammonia, and iron, and has an acid pH. If this discharges into a river, there may be total de-oxygenation resulting in septic conditions with black ferrous sulphide in the mud. Following the waters downstream, sewage–fungus usually becomes established, but only a sparse fauna develops. The main reason for this is the formation of precipitated ferric hydroxide—by chemical oxidation, and also by bacterial oxidation forming gelatinous-like growths. The ferric hydroxide appears as a rusty red 'ochre' over the substratum; it may give the sewage–fungus an orange tinge and also encrusts organisms living under these conditions. Suspended particles of ferric hydroxide appearing in the flow increase the turbidity of the water. The combination of these conditions proves hostile to most organisms, including some members of the typical replacement community associated with organic pollution, and it is some way downstream before the normal river community reappears. Clearly this is a complex pollutant, but fractionation of its individual components and their associated effects is sometimes seen downstream of a liquor discharge.

Ferric hydroxide is also an important constituent of acid mine drainage which has seriously affected the quality of surface waters in many parts of the world. The effects of acid mine drainage include de-oxygenation resulting from oxidation of ferrous compounds, establishment of an 'ochre' over the substratum, inhibition of bacterial activity by the low pH, and the toxic and physical action of metals in solution and suspension. When metals such as zinc, cadmium, and copper occur in acid mine drainage, they may act synergistically at low concentrations with deleterious effects on algal growth and community structure.

In many ways, the most complex forms of water quality deterioration result from urban stormwater and run-off which carry a wide range of contaminants including grit, greases, oils, detergents, combustion by-products, and plant and animal debris (see Chapter 3). At times of heavy rainfall, the formidable force of the increased flow charged with suspended particles exerts a significant abrasive effect, and this removes many organisms which may have become established in the receiving streams during an intervening quiescent period. The receiving streams in urban areas are sometimes canalized with reinforced concrete banks, which result in homogeneous conditions and a reduced capacity for self-purification. As a result, deposits of gelatinous sludge with a high oxygen demand may accumulate. It is evident that as a catchment develops, then, even in a relatively clean suburban area, the pattern and quality of run-off will bring about a reduction of diversity in the aquatic community; this results from the inhospitable environment created by the extremes in quantity and quality of the water in the receiving stream.

BIOLOGICAL SURVEILLANCE AND MONITORING

Since aquatic organisms integrate information about their surrounding environment, an analysis of the biotic component will yield information on the quality of the aquatic ecosystem as a whole, and careful scrutiny of the data is necessary before correlating biological or ecological change with variation in water quality. However, provided the necessary caution is taken over the collection of data and their interpretation, the information obtained from the biotic component augments data from water and sediment analyses. For example, an examination of aquatic organisms will show any additive or synergistic effect of the micropollutants referred to in Chapter 7, especially as the impact of these is likely to vary from one aquatic system to another, depending on factors such as hardness, pH, and dissolved oxygen. The collection of biological information on a regular basis is thus an essential part of water quality assessment and management, especially in view of the multiplicity of demands on aquatic systems (including aesthetic and recreational, as well as domestic, agricultural, and industrial).

There is a range of organisms and communities which can be employed in the biological assessment of water quality. While bacteriological analyses are used for monitoring potable water supplies (see Chapter 7), plankton studies are important in assessing the quality of lentic systems, and benthic macroinvertebrates are favoured for lotic systems. Although natural fish populations are not used widely in routine monitoring, they are nevertheless important and can yield much information about the water quality characteristics of their environment (Brown, 1978; Price, 1978). Fish act as bio-accumulators for a range of contaminants; it is also becoming increasingly feasible to use fish in automatic monitors in the field, where physiological and behavioural changes in response to water quality are recorded.

As already mentioned, benthic macro-invertebrates are widely used in water quality monitoring, especially in lotic ecosystems. In contrast to fish, these animals have limited mobility, so they are relatively easy to sample and may be assumed to reflect conditions at the place of collection. They have life cycles extending to months or years, so any sporadic changes in water quality will remain evident until the benthic community has undergone complete recovery. Furthermore, benthic invertebrates are represented by a large number of species (belonging to a range of phyla) which are differentially sensitive to water quality. A change in water quality will therefore affect the variety of species and their relative abundance within the community (for example, see Figure 5.6).

A useful account of methods for the assessment and presentation of summarized biological data on water quality in rivers is given in Hellawell (1978). A number of indices have been devised as a means of condensing biological information into a numerical value, and two widely used examples are the diversity index (such as that of Shannon and Weaver (1963)) and the score system (Chandler, 1970). The diversity index is calculated using the formula:

$$\text{Diversity index} = -\sum_{i=1}^{s}\left[\left(\frac{n_i}{N}\right)\log_e\left(\frac{n_i}{N}\right)\right]$$

where s is the number of species in the sample, n is the number of individuals in each species, N is the total number of individuals in the sample, and $e = 2.78$ or simply 2. This index is a measure of the number of species and the evenness of distribution of individuals among the species. However, caution has been urged over the use of this index on its own, as the value may be affected by factors other than environmental quality. Unfortunately, the index does not incorporate information on ecological requirements or tolerances of the various species, and differences in water quality may be marked by changes in dominance of the species within a community without markedly altering the diversity.

In the score system (Table 5.1), which was originally devised for samples from riffle zones of a river, points are awarded to each species according to its level of abundance and to its position in a list of organisms arranged in their order of tolerance to organic pollution. Thus, intolerant species are awarded a higher score and tolerant species a lower score with increasing abundance. The points are summed and a high index value (but rarely exceeding 2000) would be obtained for a good-quality river. The system has the potential of wider applications regarding the type of habitat sampled and the quality of water to be assessed. Useful modifications of the orginal scheme have been suggested by Balloch *et al.* (1976), Cook (1976), Bryce *et al.* (1978), and Hargreaves *et al.* (1979).

While ecological data make a valuable contribution to water quality assessment and management needs in the short term, their greatest contribution might be seen in the future, as information is accumulated to show the direction of more subtle long-term changes occurring in aquatic ecosystems. It

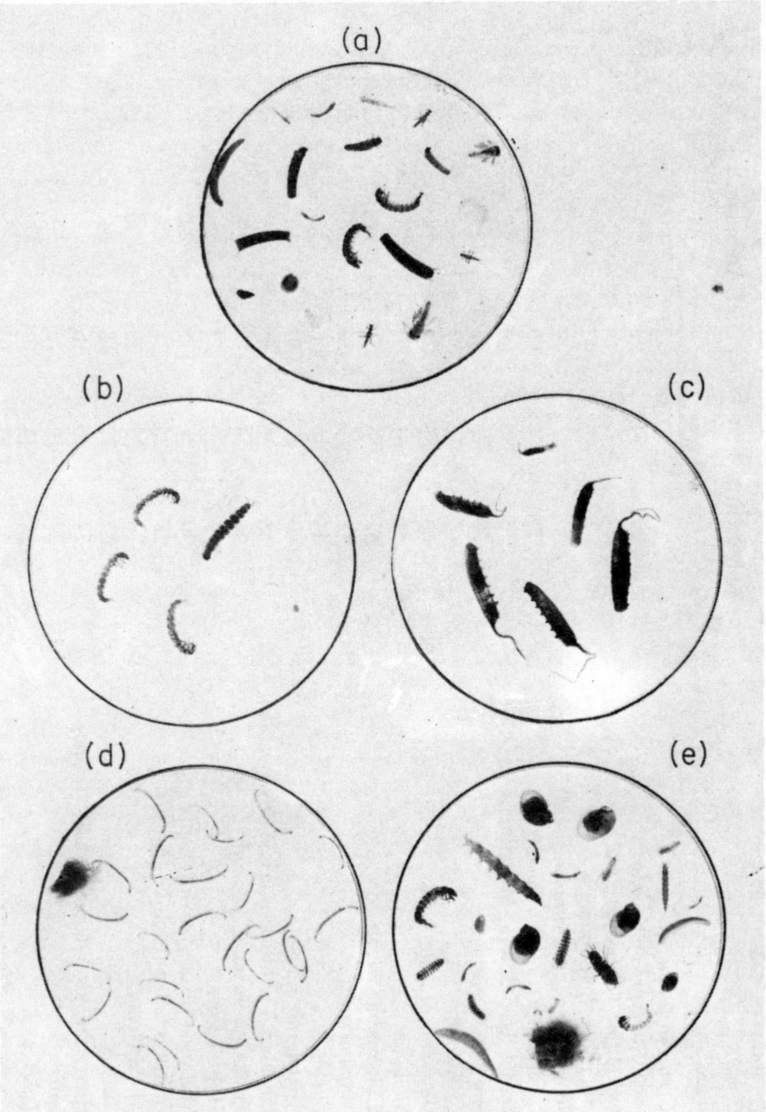

Figure 5.6. Riffle macro-invertebrate communities collected from rivers of varying water quality in south-west England. Sample (a) is from a clean river. The remaining samples are from rivers or streams receiving: (b) drainage from a copper mine, (c) refuse tip liquor, (d) slurry liquor from agricultural land, and (e) paper mill effluent

Table 5.1. The biotic score system—the numbers give the points scored. (From Chandler, 1970, *Water Pollut. Control.* Reproduced by permission of The Institute of Water Pollution Control.)

Groups present in sample		Increasing abundance				
		Present	Few	Common	Abundant	Very abundant
—	*Crenobia alpina*					
Each species of	Taeniopterygidae, Perlidae, Perlodidae, Chloroperlidae	90	94	98	99	100
Each species of	Leuctridae, Capniidae, Nemouridae (excl. *Amphinemura*)	84	89	94	97	98
Each species of	Ephemeroptera (excl. *Baetis*)	79	84	90	94	97
Each species of	Cased Trichoptera, Megalptera	75	80	86	91	94
—	*Ancylus fluviatilis*	70	75	82	87	91
—	*Rhyacophila*	65	70	77	83	88
—	*Dicranota, Limnophora*	60	65	72	78	84
—	*Simulium*	56	61	67	73	75
Genera of	Coleoptera, Nematoda	51	55	61	66	72
—	*Amphinemura*	47	50	54	58	63
—	*Baetis*	44	46	48	50	52
—	*Gammarus*	40	40	40	40	40
Each species of	Uncased Trichoptera (excl. *Rhyacophila*)	38	36	35	33	31
Each species of	Tricladida (excl. *C. alpina*)	35	33	31	29	25
Genera of	Hydracarina	32	30	28	25	21
Each species of	Mollusca (excl. *Ancylus*)	30	28	25	22	18
—	Chironomidae (excl. *C. riparius*)	28	25	21	18	15
Each species of	*Glossiphonia*	26	23	20	16	13
Each species of	*Asellus*	25	22	18	14	10
Each species of	Hirudinea (excl. *Glossiphonia, Haemopis*)	24	20	16	12	8
—	*Haemopis*	23	19	15	10	7
—	*Tubifex*	22	18	13	12	9
—	*Chironomus riparius*	21	17	12	7	4
—	*Nais*	20	16	10	6	2
Each species of	Air-breathing Diptera (e.g. *Eristalis*)	19	15	9	5	1
	No animal life			0		

is therefore essential that biological data are carefully obtained and recorded, and identification should be taken to species level to be at all meaningful (Resh and Unzicker, 1975). Finally, it should be remembered that biological data are complementary to chemical and physical data, and their interrelationships might usefully be examined by techniques such as multivariate analysis. The impact of man's activities on these interrelationships also needs to be monitored, as does his impact on the functional characteristics of aquatic ecosystems such as energy flow and nutrient cycling.

CHAPTER 6

Water Quality and Health

Richard C. Ballance and Betty H. Olson

HISTORICAL PERSPECTIVE

Recognition of the health benefits associated with a good and abundant domestic water supply can be inferentially traced to the Egyptians, the Romans, and the Greeks of 1000 B.C. and earlier. Marvellously constructed aqueducts were built to conduct clean, clear water by gravity from distant catchments to serve the needs of major cities and population groups. In all likelihood, the only water quality criteria were aesthetic ones—colour, taste, odour, and turbidity—and a prime objective was convenience of supply. Since the ancient domestic supplies came from lightly inhabited upland areas or from springs, the water was probably of excellent bacteriological and chemical quality. Adequate quantities of water permitted ample—indeed lavish—use, with the result that personal and public cleanliness was easily possible, an almost accidental effect of the desire for convenience.

Ancient systems of water supply deteriorated and became inadequate as populations increased and system maintenance was neglected. Unsanitary living conditions were the norm for hundreds of years after the fall of the Roman Empire, and it was not until the urbanization association with the industrial revolution that piped water supplies began to appear in European cities. These brought their own problems in the form of epidemics of cholera, newly arrived from the East, and typhoid, an older resident on the European scene. The newly invented pumping machines were able to pump water from nearby rivers—the same rivers that received urban drainage conveying human excreta and other wastes—thus recycling the causative organisms of waterborne disease. The sciences of bacteriology and epidemiology were still in their infancy. Water quality criteria had not changed since Roman times and some water companies had installed sand filtration equipment to improve the clarity of their product. Epidemiological investigations in London showed that users of unfiltered water suffered a greater incidence of cholera than did those who were supplied with filtered water. Proof that cholera is water-borne was obtained conclusively—albeit empirically—by Dr John Snow when he removed the handle of the Broad Street pump and halted further spread of the disease (Skeat, 1969). This was in 1854, more than 20 years before the discovery of the bacteria which causes cholera, and at a time when the miasmic theory of disease was a popular concept.

The understanding of communicable disease has progressed greatly since 1854, together with an appreciation of the need for good-quality water for domestic purposes. Cholera and typhoid, the two classical water-borne diseases, are virtually unknown in the industrialized countries, but their periodic appearance in the less developed countries is disturbingly frequent. New hazards are now being detected in the domestic water supplies of those countries that are highly industrialized, and these will be reviewed in a later section of this paper.

SURFACE WATER

The quality of water in a catchment will be indicative of the natural and artificial conditions and processes occurring upstream of the point that is sampled for a quality determination. As explained in Chapter 1, the composition of the precipitation falling on the drainage area sets the initial chemical quality of the water, but this is soon changed by interactions with vegetation, soils, and impervious surfaces. In a completely natural situation, the changes can be detected by increased concentrations of dissolved and suspended materials resulting from:

(1) the solution of aerosols previously deposited on vegetation and impervious surfaces;
(2) the solution of calcium, magnesium, and sodium (occurring as chlorides, carbonates, bicarbonates, and sulphates) from weathered rocks or limestone deposits;
(3) the solution of organic and other inorganic substances usually at very low or trace concentrations;
(4) evaporation; and
(5) the suspension of particulate matter, both organic and inorganic, picked up by the water as it flows over the surface.

The geological and vegetal characteristics of the catchment will be the primary determinant for the type of chemical and physical change that water undergoes prior to its entry into the stream. The magnitude of the change will be more dependent on the natural factors of climate and land form. For example, the combination of frequent heavy rainfalls on steeply sloping land will lead to higher rates of erosion, and thus higher concentrations of suspended solids, than when rainfall is moderate or the land is relatively flat.

Normally, the natural addition of dissolved and suspended materials does not adversely affect future beneficial uses of the water in the receiving stream. In fact, many of these 'impurities' enhance the desirable qualities of the water for specific end uses. Some natural influences may, however, impair stream water quality for domestic use. For example, when surface run-off passes through densely wooded areas or marshy lands, the water may acquire a putrid odour, or may be highly coloured and acidic because of the presence of humic acid (UNESCO/WHO, 1978).

The interventions of man in the catchment usually has profound detrimental effects on the quality of water in the drainage system (see Chapters 2, 3, and 4). Activities based on the resource industries—mining, forestry, and agriculture—invariably add contaminants in great concentration and of a more polluting character than those that result from natural run-off.

The presence of manufacturing industries brings a more critical dimension to stream water quality. These industries are notorious producers of wastes, many of which are contained in wastewater discharges. When such wastes reach a stream—advertently or inadvertently—its water quality undergoes a noticeable change, because a stream has a limited capacity of self-purification. Some wastes—particularly toxic chemicals and heavy metals—are not readily amenable to self-purification forces and may even inhibit natural purification processes. Domestic sewage and treated effluents are, in principle, of a nature that the contaminants they contain can be absorbed by the stream without causing serious long-term damage. However, few municipal sewages are strictly domestic. Most contain some industrial wastes and virtually all contain evidence of the wide array of chemical products used in the modern home. Also, the capacity of the stream to accept a waste discharge is highly dependent on the strength and quantity of the wastes in relation to the quantity of stream-flow and the flow regime, as well as the physical, chemical, and biological conditions of the stream water.

The occurrence of sulphates in stream waters is a pertinent and appropriate example of anthropogenic deterioration of stream quality. Sulphur may enter environments from a variety of sources: domestic sewage, industrial wastes, mine wastes, urban and road drainage, and fertilizers. Approximately 40% of the sulphur in an average river system can be attributed to the activities of industrialized society (Friend, 1973). The sulphur content in Lake Erie increased by 2 mg l^{-1} each decade over the period 1850–1967 (Beeton, 1965; Nriagu and Coker, 1976).

The quality of the water in a stream will be a limiting factor in the number of beneficial purposes for which the water may be used. Generally, when a source for drinking water is sought, the best possible natural quality will be a major objective, along with adequate quantity and reasonable accessibility. Few surface waters close to cities and towns are of sufficiently high quality that they can be used as drinking water without some form of treatment to remove contaminants that would endanger human health or cause the water to be aesthetically unacceptable.

Contaminants can be carried over very long distances in the stream water. Undissolved solids in rivers are made up of bottom deposits and a much larger quantity of suspended material in transport. Except in very fast flowing streams, the suspended material has a particle size smaller than about 100 microns (μm) and is composed of fine sand, silt, and clay particles. These particles are inorganic, but will often be coated with organic materials such as plankton and detritus. Suspended materials will settle in quiescent zones of the stream where velocities are low, to become resuspended seasonally when increased run-off results in higher stream velocities. At these times, erosion forces in the catchment and along the stream banks add to the load of suspended solids.

The natural concentration of suspended solids is widely variable from a few milligrams per litre to as much as several grams per litre in extreme cases. In general, small, intermittent streams display the widest variations in suspended solids concentration; larger streams with low variation in quantity of flow show variations of about one order of magnitude. Thus, suspended solids concentration is almost solely dependent on the physical considerations of stream-flow velocity and the nature of the stream bed.

Dissolved solids and their concentration present a more complex picture. The initial quality of the water at upstream locations and the nature of the material with which the water comes into contact will control the nature and amount of material that is dissolved. Contact time between the water and the solute adds a further variable dimension, so that, with longer contact times, dissolved solids concentrations can be expected to increase. Temperature will also affect the dissolved solids concentration. Higher temperatures first of all permit higher solution rates of soluble materials. This can be further aggravated in downstream areas of tropical rivers where precipitation is low and evaporation is high, leading to increases in the concentration of dissolved materials. While this is most evident in the concentrations of dissolved salts, other soluble materials present can also show a tendency to increase.

Unlike suspended material, dissolved solids will not be precipitated in quiescent stretches of the stream as a result of velocity reduction alone. Other changes in conditions—temperature, carbon dioxide, pH or dissolved oxygen—can decrease the solubility of some types of dissolved solids and result in their transformation to suspended material which may then come out of suspension. More dangerous, from the point of view of health, is the possibility that some harmful substances—principally the heavy metals—may be absorbed and concentrated by aquatic micro-organisms and plants and thus enter the food chain (see Chapter 5). Heavy metals may sometimes accumulate or become concentrated at hazardous levels in the tissues of fish or aquatic plants used by man for food, thus causing such diseases as Minamata (organic mercury poisoning) and Itai–Itai (cadmium poisoning). It is highly improbable that dissolved inorganic or organic solids could be completely removed solely by natural processes. Most solids, after being dissolved in the water, will persist at more or less the same concentration until diluted by further inflows to the main stream from tributaries.

Surface waters invariably contain a rich variety of micro-organisms. Specific types present will depend on the environmental conditions in the water, and an analysis of the micro-organisms and other biota will give valuable insights into the quality of the water.

Discharges of sewage from municipalities contain pathogenic bacteria along with very large numbers of bacteria which are normal inhabitants of the human intestine. Optimum conditions for growth of intestinal bacteria—body temperature, low oxygen, and adequate food—are rarely present in surface waters; consequently, these bacteria tend to disappear due to natural death processes and predation by other organisms. The presence of easily identifiable species of intestinal bacteria in water is a certain indication of the entrance into the stream of wastes containing human excreta, and thus of the potential presence of pathogens. When large numbers of these indicator organisms are detected in a water sample, it is interpreted to mean that contamination of the water has been recent and or great. The organism most commonly sought as an indicator of faecal contamination is *Escherichia coli*. There is some evidence that the test for *E. coli* may suffer from interference by other organisms which occur in water and that, under certain environmental conditions, there may be regrowth (Dutka, 1973). The test for *E. coli* is nevertheless, at present, the most sensitive and specific indicator of faecal contamination.

GROUNDWATER

Groundwater is often utilized as a source for domestic water supply. In England and Wales, approximately 30% of water demand is met through the use of groundwater (Young *et al.*, 1977). In arid regions, such as Israel and Southern California, groundwater can account for up to 70% and 50%, respectively, of the total water supply (Saliternik, 1976). Over the 35 years 1935–1970, the use of groundwater in the United States increased on a yearly basis from 28 km^3 to 90 km^3 (Hodges, 1977). This dramatic rise in groundwater use has created several concerns for the protection of the quality and quantity of this source. Which issue is most pressing will depend on the geological and climatic characteristics of the catchment basin. In the arid south-western part of the United States, overdraft, with the resulting problem of sea-water intrusion in coastal areas, is of crucial concern. In the eastern United States, the major problem confronting users of groundwater sources is chemical and/or biological contamination through infiltration processes (Anon, 1978).

Overdraft—the withdrawal of water from an aquifer at a rate greater than the rate of natural replenishment—may be alleviated by artificial recharge. In this practice, water which would otherwise be discharged to a river system is retained in a percolation basin from which it penetrates the soil and eventually reaches an underlying aquifer. In order to maximize the rate of recharge, a percolation basin should be located in an area where the soil is highly permeable. The technique has been practised in Africa and North America (Walker, 1978) with stormwater drainage serving as the recharge water. Other naturally occurring 'excess' water can be used equally well; for example, some portion of peak river flows during periods of heavy rain or snow melt.

In densely populated areas, the quantities of groundwater withdrawn may exceed the amount of replenishment possible with naturally occurring water, and water managers have turned to the use of treated wastewater for the replenishment of groundwater supplies. This process usually involves spreading of wastewater in percolation basins or river beds. Alternatively, the wastewater may be injected directly into the aquifer. In the former case, wastewater is usually secondary treated effluent from domestic sewage sources which has received additional treatment such as filtration. Direct injection as practised in the United States requires that the wastewater be subjected to advanced treatment following the usual secondary treatment processes. Water Factory 21, located in Orange County, California, is an example of this process. Advanced treatment in this plant includes lime treatment, ammonia stripping, breakpoint chlorination, filtration, activated-carbon adsorption, reverse osmosis, and final chlorination.

Artificial recharge methods have accompanying health concerns. The use of stormwater presents risks of both biological and chemical natures. Storm runoff can often contain myriad pollutants ranging from viruses to toxic inorganic and organic chemicals from industrial processes. The potential for movement of many of these substances through the soil column into a groundwater aquifer is invariably present. The type of soil and the depth of the underlying aquifer are important factors in determining the magnitude of the potential problem. In compacted strata, the movement of pollutants is limited, but, if fissures are present, contaminants can travel great distances (Butler *et al.*, 1954). In the case of direct injection, the interface with the soil column and its resulting purification capabilities have been removed. Therefore, advanced treatment processes are utilized to produce an adequate product water. The two major health concerns with the use of this reclaimed water are viruses and trace organic compounds.

Groundwater, free from pollutant contamination, is usually of higher ionic strength than surface water, and is characterized by low concentrations of suspended solids and pathogenic organisms. Groundwater is usually high in minerals such as iron, manganese, sulphate, chloride, calcium, and magnesium. However, there are certain naturally occurring problems. The water may be coloured if organic acids from decaying plant materials enter the aquifer. Nuisance bacteria can cause taste and odour problems. Most notable among these are *Crenothrix*, *Leptothrix*, and *Gallionella* (iron bacteria), and sulphate-reducing bacteria, *Desulphovibrio desulphuricans* (Walker, 1978). These conditions must be alleviated through disinfection.

The major contaminants of groundwater which are currently of interest are nitrates, organic compounds, biological pathogens, and inorganic salts. The susceptibility of a well system to contamination is largely dependent on its depth, since shallow wells are more accessible to infiltration from surface sources. Shallow wells are particularly susceptible to contamination from landfills and refuse tips, especially if they are located on a down gradient from the filling operation. Usually, deep wells draw water from below a deep stratum, thus being protected from surface contamination. However, when water is drawn from great depths, it is difficult to determine the origin of the water, and also the origin of any pollutants which may be present. When an aquifer lies below more than one catchment basin, practices outside the immediate area of the catchment in which the well is located can affect the quality and quantity of the entire aquifer.

Contamination of groundwater within a catchment basin will depend upon agricultural (fertilization and irrigation) practices, land disposal of industrial wastes, storage of hazardous wastes (chemical or sewage), and infiltration from sewers, as well as on the geological formations unique to the aquifer. In Dade County, Florida, 342 samples from 42 wells, ranging in depth from 3 to 18.3 m, indicated that the quality of the water reflected the land-use patterns of the overlying catchment basin (Pitt *et al.*, 1976).

Irrigation causes salt build-up in soils and, in time, these salts are leached out of the soil, eventually reaching the groundwater. Application to the land of water containing 750 mg l^{-1} of dissolved salts has resulted in the addition of 11 tonnes ha^{-1} of salt (Eckholm, 1975). Another cause of increased salinity or chloride content in groundwater in the eastern United States and Europe is road salting practices during snowfall periods. In certain areas, the sodium levels are 20 mg l^{-1}, which is the suggested maximum for individuals on low sodium diets. Further, the chloride content has increased ten-fold and hardness has tripled between 1964 and 1977 in these areas.

Increases in the level of nitrates in groundwater presents a hazard to health—especially of bottle-fed infants and young children. In the human intestine, nitrate is converted into nitrite, which combines readily with haemoglobin and reduces the absorption of oxygen into the blood. The resulting clinical condition is called methaeomoglobinaemia; an infant suffering from this condition is sometimes referred to as a 'blue baby', although a blue condition may result from other causes.

In California's central valley, one of the United States' largest agricultural areas, the major sources of nitrates in groundwater are feed lot wastes and fertilizers (Miller and Smith, 1976). On Long Island, New York, nitrogen fertilization of potato crops applies 222–278 kg ha^{-1} N of which 83–187 kg ha^{-1} N is taken up by the vegetation. Therefore, 55–194 kg ha^{-1} N can be leached into groundwater (Baier and Rykbost, 1976). Increases in nitrate content of groundwater have also been found in the United Kingdom, and in certain instances have exceeded the World Health Organization (WHO) lower recommended limit. Levels up to 60 mg l^{-1} NO_3-N have been recorded in unsaturated aquifers in agricultural regions. This figure contrasts greatly with 4 mg l^{-1} NO_3-N in unfertilized grassland sites (Young *et al.*, 1977). In Israel, the question of increased nitrate levels is of primary importance, due to the heavy dependence on groundwater for water supply and the presence of intensive agricultural efforts. It is estimated in Israel that fertilizer contributes 40640 tonnes yr^{-1} nitrogen to the soil, while all other sources total 10970. The amount of nitrogen leached through the soil amounts to 24230 tonnes yr^{-1}. A central question in Israel is the fate of organic nitrogen in the soil column and its influence on the amount of nitrate eventually reaching the aquifer (Saliternik, 1976).

Pathogenic micro-organisms present a potential threat to the contamination of groundwater. Micro-organisms of concern include *Salmonella*, *Shigella*, *Vibrio cholerae*, enteroviruses, reoviruses, adenoviruses and hepatitis viruses, *Entamoeba histolytica*, and *Giardia lamblia*. All of these organisms have the potential to move through the soil column vertically and horizontally. The ability of the soil column to remove such organisms depends on its physical and chemical characteristics; bacteria are generally removed from low-sand-content soils at a depth of 40–45 cm. The crucial factors involved in bacterial or viral contamination of groundwater are the extent of treatment that wastewater receives before coming into contact with land, and the nature of soil system involved. Problems can occur with septic tanks or storage tanks which develop leaks. Most reported incidents of enteric illness from groundwater have resulted from infiltration of sewage waste products. A notable exception occurred in Riverside, California, in 1965. Although the etiological agent was determined to be *Salmonella typhimurium* (Greenberg and Ongerth, 1966), the source of the contamination in the groundwater system was never identified. Interestingly, in a south-eastern section of the United States, there have been incidents of coliform bacteria appearing in groundwater from storm run-off and not from domestic sewers or septic tank effluent (Pitt *et al.*, 1976). With more emphasis on the use of wastewater for groundwater recharge, there is a growing concern over the contamination of groundwater from viral agents. The movement of viral particles is highly influenced by the type of virus and nature of soil particles.

Metals have been a problem in contaminating groundwater as a result of disposal of industrial wastes in ponds or landfills (Schiffman, 1978). In 1942, an outbreak of brain disease (lead poisoning) occurred in a town near Tokyo,

Japan. The lead source was a landfill containing 400 discarded dry-cell batteries (Walker, 1978). In another example, a waste disposal pond system which had received plating wastes containing Cr, Pb, Cd, Ni, CN, and As for a number of years had produced little effect on the water quality in the immediate area. However, metals could be detected 300 m from the disposal site, and the threat of further metal infiltration into the groundwater appeared imminent from the detection of metal-coated sand grains at depths of 6.8–13.4 m.

Organic compounds are the most baffling and potentially dangerous health threat. Myriad organic compounds are contained in domestic and industrial wastewater, thus establishing the potential for groundwater contamination. Waste lagoons can contain such organic chemicals as *N*,*N*-dimethylformamide, ethyl alcohol, acetone, dichloromethane, 2-ethoxyethanol, diacetone alcohol, 4-methyl-1.2-pentanone, and 2-ethoxyethylacetone. The movement of many of these compounds in the soil column is not well understood, nor is it known whether they are rapidly decomposed or remain unaltered. Further, it is not known whether they are absorbed by the soil or adsorbed onto the surface of soil particles.

Examples of a few recorded groundwater contamination events serve to illustrate some of the dimensions of the problem. The discharge of the weed killer 2,4-D into a sewer system, and its release into a stream, resulted in the contamination, 6–9 km downstream, of the public supply wells of the city of Montebello, California. Eleven wells were affected and a medicinal taste was noted 4–5 years after the accident, even though it was estimated that the chemical had been diluted by a factor of 10^7 (Swenson, 1962). Creosol used in treating wood was disposed of in a lined ponding system. After 10 years of operation, a leak was discovered. Further testing revealed that the chemical had migrated into the groundwater which was flowing in the direction of a fresh water pond. The pond water contained 2.1 mg l^{-1} of phenolic compounds and the groundwater contained 14.4 mg l^{-1} (Schiffman, 1978).

Recently, there has been concern over the closure of wells in New York and Connecticut due to the migration of organic chemicals from disposal sites into the groundwater. In certain instances, this contamination has been blamed as the cause of various adverse health effects, including miscarriages and increased disease rates. Organic pollutants are of special concern because of their suspected ability to persist unaltered for long periods in the environment, to move through the soil column, in many instances to provide no obvious evidence that contamination has occurred, and, finally, because of the long time period required for many of these compounds to reach groundwater. When this time lag is 10–20 years, it is possible that the source of the pollutant would have been abandoned or forgotten many years earlier. Thus, health officials and water managers must exert particular care in dealing with this problem.

Groundwater is a fragile and important resource which must be carefully managed to maintain its purity. Today, most groundwater sources are used with only minimal treatment, such as chlorination. Many of the aforementioned pollutants could alter this situation drastically and cause substantial increases in the cost of supplying safe potable water.

DRINKING WATER QUALITY

Water intended for human consumption must be free from organisms that are the causative agents of disease, and must not contain chemical substances at concentrations that may be hazardous to human health. In addition, drinking water should be aesthetically acceptable—free from unpleasant or objectionable taste, odour, colour, and turbidity. Temperature is not too critical but, for drinking, a cool temperature is generally preferred.

These general remarks comprise a general standard for water quality which allows a wide latitude in interpretation and application. For some of the parameters, it is relatively simple to set a more precise standard. For example, drinking water should be free of organisms—bacterial and viral—whose presence would indicate faecal contamination. Similarly, certain chemicals are known to be toxic and their concentrations must therefore be below some value which has been determined from experience, taking into account that drinking water is but one of several pathways by which substances enter the body. A second variable is that the toxicity of a chemical will usually be related to the size and susceptibility of the person who ingests it. Consequently, in order to provide a high degree of safety, the allowable upper limit of concentration of a toxic chemical should be determined on the basis of the health effect on the smallest and most susceptible members of the population group. While reasonable accuracy in setting the allowable upper limit may be obtained from animal experiments, there is often very little reliable epidemiological evidence to support this. In practice, it is normal to err on the safe side, and most standards include a large factor of safety against the toxic effects of chemicals.

In recent years, attention has been increasingly directed to the long-term effects on health from the chronic ingestion of low concentrations of toxic substances, particularly those such as lead, mercury, and cadmium which have cumulative toxic effects at relatively low dose levels after long-term exposure. The issue has become further complicated by the existence of hundreds of thousands of manufactured chemical compounds, many of which are toxic, which are discharged to rivers along with the effluents from industries and municipalities. Of particular concern are the polycyclic aromatic hydrocarbons, haloforms, and asbestos, which are either known or suspected to be carcinogenic. Many of these substances have been shown to be carcinogenic in the industrial setting. Thus, their newly found presence in potable water supplies is of concern to health officials. However, the concentrations of most of these materials are several orders of magnitude lower in drinking water and, as in the case of asbestos, the route of entry into the body is different. This factor presents difficulty in establishing what is an acceptable health risk. For public health reasons, it is therefore prudent to set some safe upper limit for the concentration of undesirable chemicals in drinking water.

On the other hand, certain chemicals are desirable in drinking water and have positive effects on health. There is ample evidence to support the claim that a regular intake of fluoride leads to a reduction of dental caries (World Health Organization, 1970b). Excessive quantities of fluorides result in mottling of the tooth enamel. In areas where fluoride is not naturally present in water, it is common practice to add a chemical compound to the water that will bring the fluoride concentration to about 1 mg l^{-1}. In other areas where fluoride occurrence is naturally high, it may be necessary to defluoridate the water.

Hardness in water is due to the presence of divalent metallic ions (the principal ones are calcium and magnesium), and is expressed as the equivalent quantity of calcium carbonate. Water with less than 75 mg l^{-1} of $CaCO_3$ is considered soft and above 75 mg l^{-1} of $CaCO_3$ as hard (Sawyer, 1960). Hardness in water forms scale on pipes and fittings in a water supply system, thus reducing or eliminating corrosion. Excessive hardness, however, can lead to the partial blocking of pipes, and additionally reacts with soap to form a curdy precipitate. For these reasons, water authorities have for years controlled hardness at near optimum levels—enough to restrict corrosion but not enough to restrict flow or cause consumer complaints. A vast quantity of data on hardness in municipal water supplies is therefore available—mainly for cities in Europe and North America.

These data, used in conjunction with vital statistics, have permitted numerous studies over the past 20 years on the relationship between water hardness and morbidity and mortality. Most of these studies indicate an inverse correlation between the hardness of drinking water and the incidence of cardiovascular disease (heart disease, hypertension, and stroke). Some of the studies indicate a similar inverse correlation between the hardness of water and the risk from several non-cardiovascular causes of death (Amavis *et al.*, 1976).

The specific role of individual hardness-causing components is not known. One main hypothesis is that one or more of the divalent metallic ions which cause hardness provides a protective effect. The other main hypothesis is that soft water, being more aggressive and corrosive, tends to dissolve harmful metals with which it comes in contact. Suspected harmful metals are lead, cadmium, copper, and zinc. Where water is artificially softened by the ion-exchange process, the finished water will contain relatively high concentrations of sodium and is considered by some to be unsuitable for drinking (World Health Organization, 1978).

Repeated observations in animal studies have shown that raised blood pressure results from a continued high intake of sodium. Conversely, restriction of sodium intake for hypertensive patients invariably results in lowered blood pressure levels. This evidence has led to the proposal (World Health Organization, 1979b) that, for the general adult population, sodium chloride intake should not exceed 6 g day^{-1}—equivalent to 2380 mg day^{-1} of sodium. Sodium affects the taste of water, especially when it is present as sodium chloride, and a sodium concentration in excess of 200 mg l^{-1} would be aesthetically unacceptable to most consumers. At a consumption rate of 3 l day^{-1}, water should not normally contribute more than about 25% of the amount of sodium proposed as a desirable maximum daily intake for the general adult population.

For hypertensive patients, the daily intake may be restricted to as little as 500 mg day^{-1} of sodium. Food will contribute about 450 mg—consequently the sodium concentration in water can be of critical importance. From the water treatment standpoint, it would be impractical to reduce the sodium content sufficiently to be of any use (Windle-Taylor 1978). For patients who must have a low sodium diet, the use of bottled drinking water may supply a reasonable alternative. Two brands of bottled water available in Europe—Vittel and Henniez—contain 3 mg l^{-1} and 8 mg l^{-1}, respectively.

SIGNIFICANCE OF WORLD HEALTH ORGANIZATION STANDARDS

Some countries have established national standards of water quality, and a certain degree of uniformity has been achieved both in analytical methodology and in the expression of analytical results. While this applies generally to the industrialized countries, very few of the developing countries had, as of 1970, developed their own national standards for drinking water (World Health Organization, 1973b). Of the countries which reported, 37 had either adopted the WHO International Standards for Drinking Water *in toto* or had adapted them to suit the needs of the country. For 40 other countries, standards had not yet been set, although some were planning to establish them at a future date.

The stated purpose of the WHO standards is to serve as technical guidance to the health administrations of countries wishing to establish or revise their regulations on drinking water quality control (World Health Organization, 1971). The preparation of this guidance is done in collaboration with member governments and with the assistance of experts.

WHO International Standards for Drinking Water were first published in 1958—subsequent editions came out in 1963 and 1971. The European Office of WHO published European Standards for Drinking Water in 1961 and a second edition in 1970 (World Health Organization, 1970a). The rationale provided for having two sets of standards—one 'international', the other

'European'—was that, 'In view of the different economic and technological capabilities of various countries, there will be some areas in which higher standards than those proposed for the world as a whole will be attainable—and these areas should be encouraged to attain such higher standards. It is believed that Europe is such an area and that there is, therefore, nothing illogical in setting higher standards in Europe than internationally' (World Health Organization, 1970a). Revisions to the WHO Standards are currently (1980) underway and, in the process, 'International Standards' and 'European Standards' will be merged to be replaced by a new publication on drinking water quality.

The difficulties in developing guidelines with global application are obvious. Industrialized countries suffer from few of the overwhelming constraints of technology, manpower, and financial resources which prevail in most of the developing countries. While the primary objectives contained in guidelines or standards for drinking water quality are the protection of public health, priorities are inevitably different, depending on the immediate health problems. In developing countries, communicable diseases which thrive in an unsanitary environment are serious problems. In industrialized countries, those problems are quite well under control, and the major concern is directed to contaminants which have an insidious, long-term effect on the public health.

This potential conflict in establishment of quality guidelines can be overcome by setting maximum acceptable limits (MAL) (World Health Organization, 1979a) for each quality characteristic (upper and lower limits for fluoride, pH, and hardness). In addition, 'action levels' might usefully be established. These would be some percentage of the maximum acceptable limit and, when contamination exceeded the action level for a contaminant, remedial measures would be taken. Action levels are so far (1980) undefined; selection and enforcement of action levels will be primarily the responsibility of National Ministries of Health and must, in the final analysis, be established in cognizance of the full range of prevailing local conditions.

It is interesting to note that, in the United Kingdom, a legally binding set of drinking water standards has not been adopted. The only legal requirement is that Water Authorities must supply wholesome water. The EEC has been considering the establishment of standards of drinking water quality which would be applied in all members countries. It has been argued (Kenny, 1978) that this is part of a campaign to clean up trans-frontier rivers, and that the United Kingdom therefore has no specific interest in such standards.

RE-USE OF WASTEWATER

The total amount of water on the Earth and in the Earth's atmosphere remains virtually unchanged. Consequently, the hydrological cycle—evaporation, precipitation, run-off—is a cycle of re-use. The only major change in millions of years is man's interventions in the hydrological cycle, whereby he withdraws water, uses, and contaminates it, finally returning it to a water body for further natural recycling. This was satisfactory while withdrawals from and discharges to rivers and lakes were relatively small. Increasing population and industrial development have led to increased water withdrawals so that the wastes discharged to a river by cities and industries become a portion—sometimes a substantial one—of the water supplies for downstream users. This form of re-use is known as 'indirect' re-use (World Health Organization, 1973a; International Reference Centre for Community Water Supply, 1975).

Direct or intentional re-use of wastewater occurs when the wastewater is deliberately used in a planned way for some purpose. In some arid countries, sewage—raw or primary treated—is used extensively for irrigation. In the western United States secondary treated wastewater has been used to supplement water supplies through the irrigation of agricultural land, residential greenbelts, and golf courses for many years. In these cases, any fertilizing value from constituents in the sewage is a fringe benefit; the main objective is to have sufficient water to support plant life. In other countries, which suffer from seasonal or perpetual water shortage, it is necessary to subject wastewater to complex treatment processes in order to obtain a product suitable for higher categories of use. While direct re-use usually stops short of potable water supply, there have been notable exceptions. Among these are Chanute, Kansas (Metzler *et al.*, 1958) and Windhoek, Namibia (Van Vuuren *et al.*, 1971), where, at times of critical water shortage, treated effluent was used to supplement the insufficient regular supplies.

It is evident that, as populations and the demands for water increase, there will be a corresponding increase in the re-use—both indirect and direct—of wastewater. This, coupled with increases in the number and complexity of chemicals which reach surface and groundwaters, will present new hazards to human health.

It is said, in Holland, that the water one drinks in Amsterdam has previously been a beverage for three other people. This is a rather gross exaggeration since the Rhine—Amsterdam's main water source—contains only a small quantity of wastewater relative to a rather stable average flow of about 2200 $m^3 s^{-1}$. Data on water quality in the Rhine at the German–Dutch border indicate an increase of one order of magnitude in the concentration of four chemical parameters (International Reference Centre for Community Water Supply, 1973) over the 100 year period to 1970. It is speculated that, without adequate treatment of waste discharges, these same parameters will be further increased three or more times by the year 2000. Although the reported chemicals (orthophosphate, nitrate, chloride, and iron) are not present in dangerous concentrations, their very presence is indicative of domestic and industrial wastes and signals the presence of other substances known to be harmful to health. Previous use of river water can alter its nature, thus eliminating downstream usage for reasons other than human health. The extensive use of Colorado River water for irrigation has resulted in a substantial increase in dissolved solids. The situation has so intensified in recent years that the total dissolved solids of the water entering Mexico from the United States approximates 1000 mg l^{-1} thus almost negating its further use in agriculture.

At times of low flow, certain rivers in the United Kingdom contain high proportions of sewage effluent. Notable among these are the River Lee at Chingford (47% effluent) and the Thames at Teddington (38% effluent) (Packham, 1976); figures for the sewage effluent content of the River Trent are given in Chapter 4. Data for the Great Ouse at Offord and Brownshill indicate 47% effluent, based on a flow rate which is substantially higher than minimum. Estimates of possible future concentrations of effluent in the Great Ouse forecast 80% in the year 2011.

The actual and projected increase in the indirect re-use of wastewater needs to be matched by an increase in the efficacy and reliability of water treatment processes. The problem may be diminished to some extent by the introduction of conservation measures to reduce water use. Alternatively, dual systems comprising two separate supplies—one for drinking water, the other for non-consumptive uses—might be practical in some situations. Dual systems have never achieved overwhelming popularity because of high cost and the potential hazards to health which would result from accidental cross-connection between the two supplies. Bottled drinking water is widely used in many countries, but its adoption as the drinking water component of a dual system would introduce formidable logistics problems if all residents of the area served by a water authority were to be provided with a reliable and safe supply.

The health implications of wastewater re-use, and consequently the extent of treatment required, will depend on the intended use of the treated water. Little or no treatment of sewage effluent or raw water may be necessary if the water is to be used for the irrigation of trees and certain agricultural crops. As the end use of the treated water approaches human contact and human consumption, the quality requirements become increasingly stringent and more processes must be added to the treatment train. Table 6.1 (World Health Organization, 1973a) shows the various treatment processes that are needed relative to different levels of agricultural, recreational, industrial, and municipal re-use.

While most of the research on wastewater re-use has been directed to health effects and treatment technology, a limited amount of work has been done on social factors—particularly on the attitudes of those who would be recipients of renovated wastewater. Studies have been conducted in the western United States (Bruvold, 1972; Olson and Pratte, 1978) which indicate that the public is prepared to accept expanded use of renovated wastewater within certain limits. Three psychological variables—aversion to the unclean, aversion to human

Table 6.1. Suggested treatment processes to meet the given health criteria for wastewater re-use (From World Health Organization, 1973a, *Reuse of Effluents: Methods of Wastewater Treatment and Health Safeguards*, WHO Technical Report Series no. 517. Reproduced by permission of WHO.)

	Irrigation			Recreation			Municipal re-use	
	Crops not for direct human consumption	Crops eaten cooked; fish culture	Crops eaten raw	No contact	Contact	Industrial re-use	Non-potable	Potable
Health criteria (see below for explanation of symbols)	A+F	B+F or D+F	D+F	B	D+G	C or D	C	E
Primary treatment	●●●	●●●	●●●	●●●	●●●	●●●	●●●	●●●
Secondary treatment		●●●	●●●	●●●	●●●	●●●	●●●	●●●
Sand filtration or equivalent polishing methods		●	●		●●●	●	●●●	●●
Nitrification						●		●●●
Denitrification								●●
Chemical clarification						●		●●
Carbon adsorption								●●
Ion exchange or other means of removing ions						●		●●
Disinfection		●	●●●	●	●●●	●	●●●	●●●*

Health criteria:
A Freedom from gross solids: significant removal of parasite eggs.
B As A, plus significant removal of bacteria.
C As A, plus more effective removal of bacteria, plus some removal of viruses.
D Not more than 100 coliform organisms per 100 ml in 80% of samples.
E No faecal coliform organisms in 100 ml, plus no virus particles in 1000 ml, plus no toxic effects on man, and other drinking water criteria.
F No chemicals that lead to undesirable residues in crops or fish.
G No chemicals that lead to irritation of mucous membranes and skin.

In order to meet the given health criteria, processes marked ●●● will be essential. In addition, one or more processes marked ●● will also be essential, and further processes marked ● may sometimes be required.
*Free chlorine after 1h.

waste, and overconcern for health—are attributed to be factors which set practical re-use limits somewhere short of direct personal contact and well short of human consumption.

LONG-TERM HEALTH EFFECTS

Estimates of the long-term effects on health resulting from the ingestion of contaminants in drinking water may be made in two general ways. Relatively short-term studies in the laboratory are able to provide indications of toxic effects when bacterial or animal cells are exposed to a specific contaminant, or when experimental animals are fed a lifetime diet which includes carefully measured portions of the contaminant. The results of laboratory studies are particularly useful in evaluating a 'high-risk' contaminant, and extrapolation of the findings permits an assessment of the hazard to human health. Laboratory studies can provide guidance in the design of longer-term studies of the effects of specific contaminants on human populations.

Epidemiological studies of environmental hazards to health are much longer-term and are intended to assess the risks to human health in real-life circumstances (Cantor and McCabe, 1978). The results are often unclear, however, since there is usually no effective control over exposure to the contaminant in question and the study population will probably be exposed to other contaminants in water, in food, and in the working and living environments.

Establishing upper limits for the concentration of possible contaminants in drinking water depends on data from both laboratory and epidemiological studies. Frequently, the sequence of events is that the incidence of morbidity or mortality from a single ailment is correlated with a specific environmental condition (or contaminant). This leads to the formulation of a hypothesis stating that the environmental condition is the cause of the ailment. Controlled studies in the laboratory may then confirm or disprove the cause–effect hypothesis; if confirmed, there may also be an indication of the toxic dose of contaminant for the particular species of laboratory animal used in the studies. A large-scale and long-term epidemiological study may then be undertaken to gather data on chronic exposure of humans to the contaminant and to support, supplement or contradict the laboratory data. Recognizing the inherent constraints in the control of epidemiological studies of chronic disease, especially cancer, contradiction of the results of animal experiments would require the accumulation of formidably powerful epidemiological data.

Health effects resulting from exposure to chemical contaminants may be classed as reversible, non-reversible or irreversible. Reversible effects are those that disappear quite soon after exposure ends. Non-reversible effects are those which do not disappear—totally or partially—and which may even progress at a lower rate after exposure is terminated. Irreversible effects continue to progress after cessation of exposure.

Virtually anything that man might consume would be toxic or deleterious to health if a large enough quantity was consumed. Hazardous chemical contaminants in water will have adverse effects on health at relatively small dosages, and thus it is necessary to determine the highest dose or concentration at which no adverse effects have been observed. A safety factor can be applied to the 'highest dose—no effect' concentration in order to produce a standard of water quality for the particular contaminant. The selection of a safety factor will be highly dependent on the adequacy of the data concerning the dose–effect relationship, as well as consideration of the degree of risk that is acceptable for the large human populations which may be exposed to small concentrations of the contaminant.

Such an approach, which tolerates some threshold level of a contaminant in drinking water, is valid for those contaminants whose effects are reversible and non-progressive. It is less—or perhaps not at all—valid for contaminants that cause chronic, irreversible, and progressive disease as determined by tests on laboratory animals. For example, one, or a very few, molecular event triggered by a carcinogen can initiate transformations in a cell that ultimately lead to the development of a lethal cancer. The allowable limit for known carcinogens should ideally be set at zero; this may be impractical, however, and a limit must be established based on the consideration of an acceptable level of risk.

This is well illustrated by the use of chlorine and chlorine compounds for the disinfection of drinking water. Chlorination destroys pathogenic organisms and thus is a major control measure for the prevention of water-borne communicable disease. But chlorine combines with certain organics in water to produce chloroform which has been proved carcinogenic in animal tests. Although chloroform and other chlorinated hydrocarbons must be maintained at as low a level as possible to minimize the risk of cancer, the discontinuance of chlorination, even though reducing the cancer risk, would introduce a much higher health risk from communicable disease. While the possibility exists of adopting disinfection techniques other than chlorination, they must first be proven to be effective disinfectants and to introduce no greater risks than those which result from the use of chlorine. Clearly, if the health benefits are great, it is possible to accept some small level of risk.

Knowledge of the dose–effect relationships which apply to most of the toxic inorganic chemicals has resulted from research in industrial toxicology. Since the industrial setting provides good experimental control, and many of the effects have been detected in humans, the data obtained are reliable, especially as they relate to acute toxicity. This has permitted the establishment of quite definite allowable upper limits for the concentration of toxic inorganics in drinking water. Current research is directed to the low-dosage long-term effects of toxic inorganics, a subject which still contains uncertainties. It is generally considered, however, that the upper limit of concentration of toxic inorganics in drinking water recommended by WHO (World Health Organization, 1971) provides an adequate margin of safety against adverse effects to human health.

The situation concerning organics in drinking water is less straightforward. There are literally thousands of synthetic organic compounds which are discharged to rivers in sewage effluent and ultimately appear in drinking water supplies. Some are known to be carcinogenic, mutagenic or teratogenic in man; others are suspect. Many are difficult to detect and are present in trace concentrations detectable at the microgram per litre level. Major concern is directed toward the trihalomethanes which result from the discharges of wastes and the chlorination of water. Much more research is needed to determine accurately the long-term effects on health. Such information is simply not available for all organics, and with the vast number of potentially hazardous chemicals present in the environment, it will require a long time to generate the comprehensive and reliable data that is required before positive limits can be established. In the meantime, it is prudent to be cautious where chemical contaminants in water are concerned, and to make all reasonable efforts to keep their concentration at as low a level as is practicably possible.

The National Academy of Sciences, National Research Committee, Committee on Drinking Water (1977) has proposed four principles which provide useful guidance to the assessment of safety and risk to human health from the presence of carcinogenic chemical contaminants in water. These principles are:

(1) effects in animals, properly qualified, are applicable to man;
(2) methods do not now exist to establish a threshold for long-term health effects of toxic agents;
(3) The exposure of experimental animals to toxic agents in high doses is a necessary and valid method of discovering possible carcinogenic hazards in man; and
(4) material should be assessed in terms of human risk, rather than as 'safe' or 'unsafe'.

CHAPTER 7

Water Quality Control

L. B. Wood and D. Sheldon

The water cycle within a catchment ecosystem may be represented diagrammatically as shown in Figure 7.1. To the right is the natural cycle (excluding rainfall), and on the left man's interaction with it, involving abstraction of water, its treatment, use, and recycle of the treated wastewater. In order to ensure that water is of a desirable quality and can meet the variety of demands made upon it, there must be continual monitoring, and the appropriate water quality standards must be set and enforced by the controlling authority.

The focal points of such systems are the rivers and, in England and Wales, Regional Water Authorities now have complete responsibility for the whole of the water cycle in the area of one or more river catchments. The following is a description of the purification and control of potable water sources and waterborne wastes.

WATER TREATMENT PROCESSES

Water supplies are derived from two main sources—groundwater (wells and boreholes) and surface water (rivers and lakes). The demand for water is predicted to increase by 42% in the United Kingdom by the year 2001 (National Water Council, 1978a) and, to meet this, supplies may be augmented by inter-regional transfer from an upland source to a lowland river (e.g. the Severn–Thames link), the use of treated surface water to recharge aquifers during high winter flows (e.g. the Thames Water Authority River Lee recharge scheme), or supplementing river flows in dry weather by pumping from augmentation boreholes (e.g. the Thames Water Authority River Lambourn scheme).

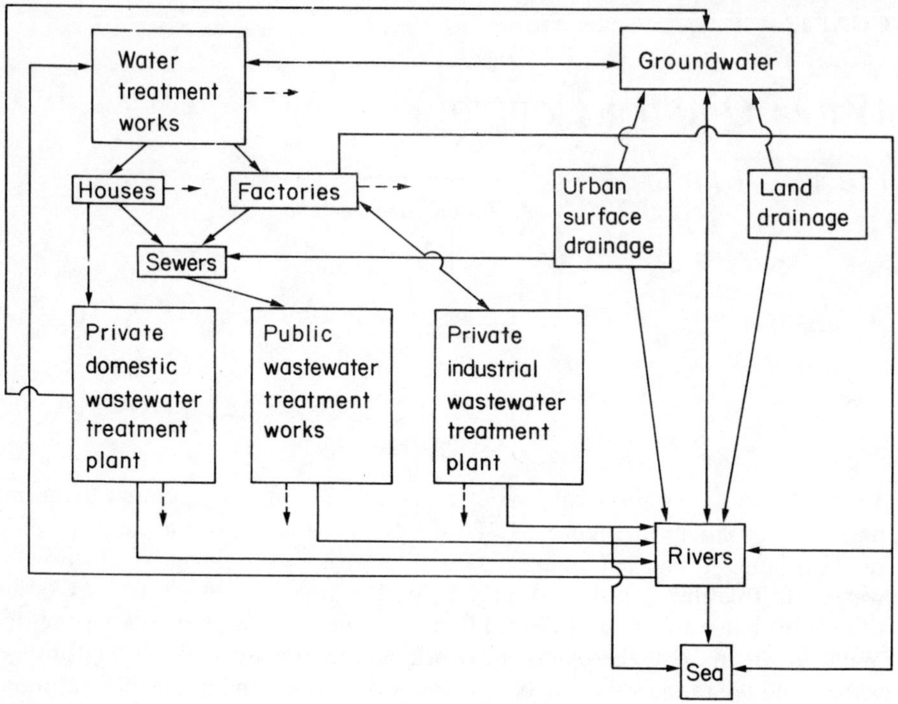

Figure 7.1. The water cycle in a catchment ecosystem: - - → indicates solid waste disposal

Treatment of Groundwater Supplies

The treatment necessary for raw groundwater (Figure 7.2) will depend on its quality; generally, this is better than that of surface water, and little treatment other than disinfection is required. Aeration may be necessary to remove carbon dioxide (which is corrosive to the distribution system), and dissolved manganous and ferrous salts by oxidation, when the insoluble products are removed on a sand filter.

Treatment of Surface Water Supplies

This is outlined in Figure 7.3.

Reservoir Storage

The storage of raw water in reservoirs, apart from providing resources in periods of drought and at times of accidental contamination of rivers, has, in addition, the advantages of effecting (1) sedimentation of suspended solids, and

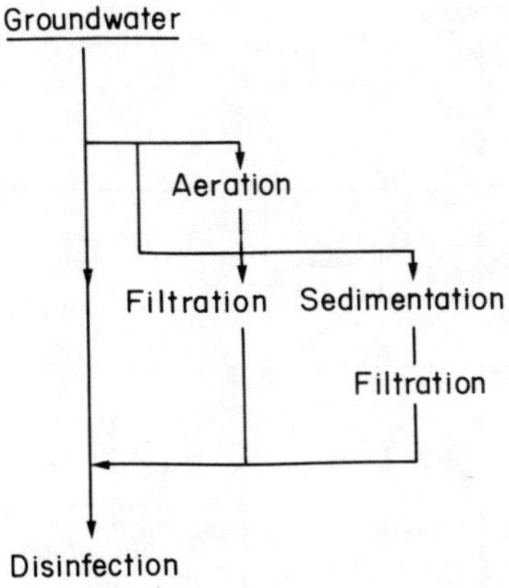

Figure 7.2. Water treatment processes for groundwater

(2) removal of pathogenic organisms. However, prolonged water storage sometimes provides opportunity for algal growth which can impart undesirable taints ('grassy' or 'mouldy' tastes) and reduce filtration efficiency. The problem can be mitigated by jet circulation of the water in the reservoir and careful biological control.

Primary Filtration

Rapid gravity sand filters or microstrainers can be used to remove suspended solids. The former are beds of sand, 1.0–1.3 m deep, resting on gravel and contained in a concrete tank supplied with underdrainage. Raw water under a head of 2 m flows downwards at about 6 m hr^{-1}, the sand being backwashed at intervals by water and compressed air. A microstrainer (Figure 7.4) is a rotating drum covered with fine mesh material through which water passes, the filtered solids being washed by jets into a trough at the top of the drum. Microstrainers are suitable for removal of coarse solids, such as algae, but not finely divided solids.

Slow Sand Filtration

In this process, suspended solids are filtered out and a film of biological slime which accumulates on the surface of sand particles oxidizes colloidal and dissolved organic impurities and ammonia, and effects some removal of pathogens and viruses. The increasing build-up of film reduces the permeability of the filter, and it becomes necessary periodically to drain the bed and remove and replace the top 20 mm of sand.

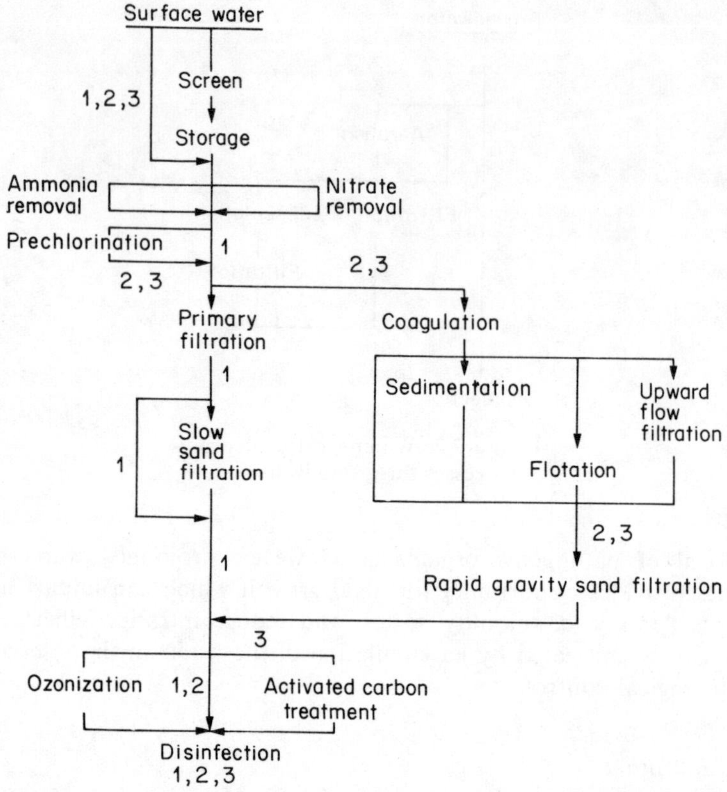

Figure 7.3. Water treatment processes for surface water. Numbers 1, 2, and 3 refer to EEC prescribed treatment processes for raw water qualities A1, A2, and A3 (of decreasing purity), respectively

Coagulation and Flocculation

This is a commonly used alternative process to slow sand filtration. It is a superior method for the removal of colour from water, but less effective in removing unwanted tastes. Aluminium or iron salts are the commonly used flocculating agents, sometimes with the addition of polyelectrolytes. Colloidal and coloured matter is adsorbed on the hydroxide flocs under controlled conditions of pH, stirring, and residence time. The flocculated solids are separated from the water in several ways, as shown in Figure 7.3.

Disinfection

This is the most important single stage in the treatment process (Fish and West, 1978). The usual method is by chlorination, but ozonisation is used

Figure 7.4. A microstrainer in operation at a water treatment works. (From Fish and West, 1978, *Water*. Reproduced by permission of Thames Water Authority and National Water Council.)

largely in France. In the London region, a solution of chlorine gas is injected continuously as a final treatment stage and allowed to remain in contact with the water for up to 2 h. The dose is adjusted to give disinfection without imparting a chlorinous taste to the water. 'Free chlorine' is a much more powerful bactericide and virucide than 'combined chlorine' (i.e. chlorine present as chloramines), and for the chlorination of groundwater considered to be at risk, breakpoint chlorination with a dosage of 0.5 mg l^{-1} of free chlorine maintained for 30 min followed by dechlorination with sulphur dioxide is favoured. Breakpoint chlorination is the destruction of chloramines by oxidation using excess chlorine.

Advanced Processes

With the increasing sewage effluent content of surface waters, the importance of ammonia and nitrate removal at the treatment stage is also increasing. A biological fluidized-bed process for the nitrification of ammonia is in use in a few sites in the UK using upflow clarifiers. Fine sand grains are used as a support for nitrifying bacteria which function at upflow velocities up to 25 m hr^{-1} and at low temperatures. This process is more economical than break-point chlorination at ammonia levels exceeding 0.2 mg l^{-1} (Short, 1975).

Nitrate removal processes may occur naturally in the river and storage reservoir, especially in warm weather. If abstracted water for potable supplies contains more than the proposed EEC maximum admissible concentration of 50 mg l^{-1} NO_3, then it is necessary to stop abstraction or blend the supply with low-nitrate water. Very few artificial denitrification processes are operating in EEC countries at present, but they are likely to show a large increase in number in the near future (Water Research Centre, 1977a). Anglian Water Authority has estimated that a £10–15 million expenditure on nitrate removal plant would be required in their region if the proposed EEC limit is to be met (Anglian Water Authority, 1978). Ion-exchange processes are in use in the USA, whilst a fluidized-bed biological denitrification plant is being installed to treat surface water by the Anglian Water Authority (Water Research Centre, 1978).

For removal of taste and odour, activated-carbon treatment has been used extensively in Europe as an emergency measure, a dose of 3–10 mg l^{-1} of powdered carbon being applied (Water Research Centre, 1977a). Fixed-bed granular activated-carbon filters are used in Europe and the USA for the removal of trace organic pollutants. In the USA, mixed sand/activated-carbon filters are used after the sedimentation stage in place of the conventional rapid gravity sand filters.

The Water Research Centre (WRC) have published a useful summary (Water Research Centre, 1976) of the costs of water (and wastewater) treatment unit processes so that a cost/benefit analysis of the water cycle can now be made.

WASTEWATER TREATMENT PROCESSES

Private Domestic Plants

Septic tanks are commonly used for the treatment of sewage from houses not connected to a public sewerage system. These effect sedimentation of suspended solids in the sewage and anaerobic digestion of the deposited sludge whereby fat, carbohydrate, and protein are converted to gaseous methane, ammonia, and carbon dioxide by bacterial action in the absence of air. The volume of sludge is thereby reduced and such tanks require infrequent de-sludging (about every 12 months). The effluent from a septic tank has a high biochemical oxygen demand (BOD) and must not be discharged direct to a river or aquifer without further treatment, such as passage through an unsaturated soil zone. The number of septic tanks must be controlled at the planning stage to prevent pollution of aquifers, e.g. by nitrates.

Public Sewage Works

The basic processes of sewage treatment are shown in Figure 7.5, which indicates alternative methods and standards commonly achieved. Cumulative treatment costs are shown (Truesdale and Taylor, 1975) and it is seen that the final stages required for water reclamation double the cost of treatment, adding more than 7.5 p m^{-3} to the total cost. This is greater than the cost of conventional treatment of river water (3.5–3.8 p m^{-3}); hence, at present, the use of sewage effluent as a source of potable water is uneconomic. Partially treated effluents have been used commercially, e.g. for cooling purposes, but care is then required in their distribution to prevent cross-connection with domestic water supplies.

Sewage works are typically designed to treat a maximum volume of three times the dry weather flow (DWF). Any flow exceeding this is diverted after screening to stormwater tanks for temporary storage. Flow in excess of $6 \times$ DWF, which may occur during very heavy rainfall and is usually beyond the capacity of a works, is passed direct to the river.

Preliminary Treatment

Sewage is first passed through bar-screens to remove rags and large objects which would interfere with subsequent pumping operations, and then grit is removed in grit channels or detritors. Occasionally, the sequence of these operations is reversed. Screenings are either macerated and returned to sewage flow or preferably, removed, dried, and incinerated. The use of comminutors for the combined screening and maceration of rags is now falling into disfavour, due to the rapid wear and high cost of replacement of cutting blades. Grit is removed from the sewage by reducing the rate of flow from 1 m s^{-1}, the minimum velocity for self-cleaning sewers, to 0.3 m s^{-1} when grit will settle. This is effected in stirred sedimentation vessels (detritors) or in constant-velocity grit channels.

Primary Sedimentation

Sewage is passed slowly and continuously through a rectangular or circular sedimentation tank, when much of the suspended solid matter settles under

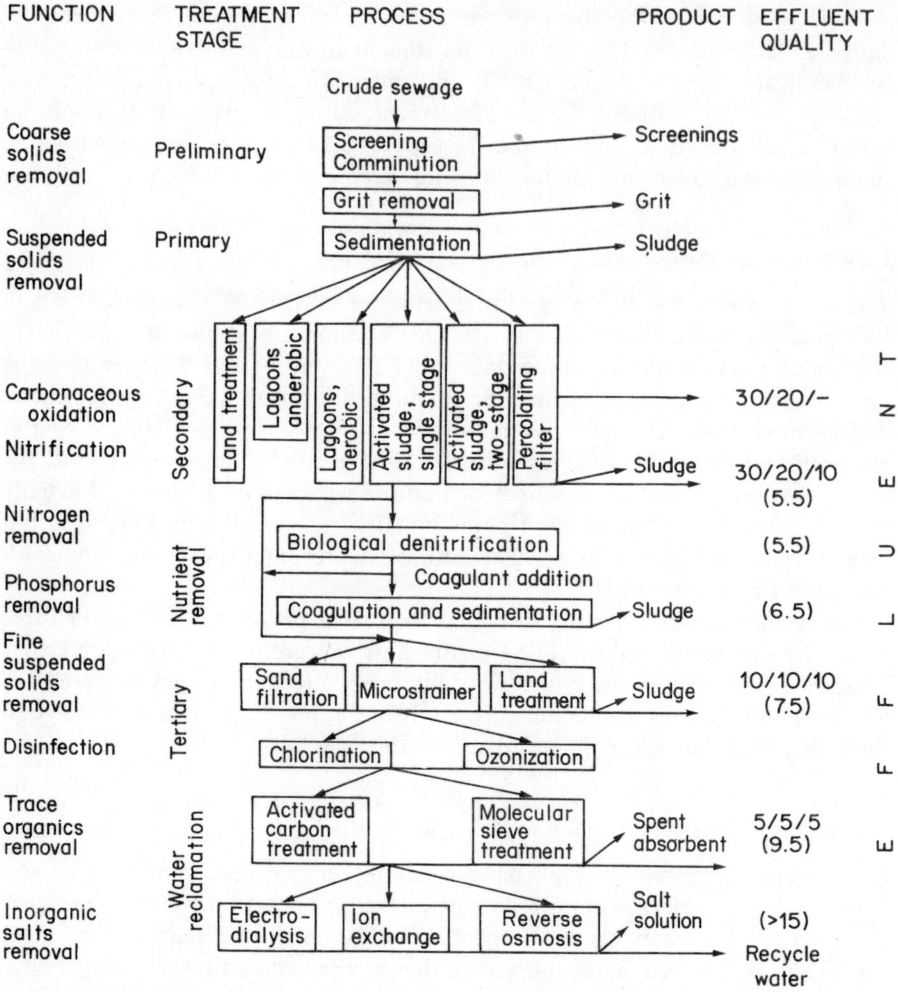

Figure 7.5. Wastewater treatment processes, effluent qualities, and treatment costs. Effluent quality is given as: suspended solids/biochemical oxygen demand (ATU)/NH_3-N all in mg l^{-1}; below (in parentheses) is given the cost in pence m^{-3} in 1976

gravity to form a sludge on the tank floor. This is conveyed by mechanical scrapers or suction devices to a sludge hopper for further treatment. A skimming mechanism removes oil and scum from the tank surface.

The tank efficiency is measured in terms of the proportion of solids removed from the sewage and the solids content of the sludge produced. In both respects, the efficiency is reduced when the incoming sewage contains surplus activated sludge due to the low sludge density of this material. Typical removal efficiencies are 60–75% for suspended solids and 45–60% for BOD. The dry matter content of deposited sludge is typically 3% (w/v) for primary mixed sludge. At low temperature, removal efficiencies are reduced, and this can cause overloading of the secondary treatment stage in winter.

Secondary Treatment

Biological oxidation is designed to accelerate the natural processes of biodegradation, whereby polluting matter in settled sewage is partly oxidized and partly converted to new cells by a culture of micro-organisms maintained in the sewage treatment plant. The new cells themselves undergo progressive auto-oxidation. In land treatment, the micro-organisms (bacteria, protozoa, etc.) are held on the surface of vegetation and in the soil. In percolating filters, a zoogloeal slime of organisms covers the porous medium of the filter, and in the activated-sludge process the organisms are suspended in flocs floating in the sewage medium. Carbonaceous compounds are oxidized to carbon dioxide and water, and nitrogenous compounds to ammonia, nitrites, and nitrates.

From the oxidation plant, the flow may pass to secondary settling (or humus) tanks, where the final sludge is separated, allowing a clarified effluent to be discharged from the works. The efficiency of the biological oxidation process is determined by the biodegradability or ease of breakdown of the components of the wastewater, the rate of loading of the plant (the food/micro-organism ratio), and the temperature. All these factors are subject to diurnal and seasonal fluctuations, causing variations in effluent quality. Typical frequency distributions of effluent quality from two works, one of which produces an average effluent quality of 20 mg l^{-1} for a parameter such as BOD, the other an average of 35 mg l^{-1}, are illustrated in Figure 7.6 (A. G. Boon, private communication). It is seen that in each example the 95 percentile is a much higher value of the parameter than the mean value. Works design is based on average figures, whereas effluent standards are expected to be achieved by at least 95% of snap samples taken (the 95 percentile basis; National Water Council, 1978b). As a result of the examination of effluents from 14 sewage treatment works, Porter and Boon (1971) showed that a linear relationship existed between the mean and the 95 percentile concentrations of BOD (Figure 7.7) and suspended solids. This is a useful guide when relating plant control data to the required effluent standards.

From Figure 7.5, it can be seen that there are a number of methods of secondary treatment available:

(1) Land treatment of sewage is still a common method of disposal in some countries. Thus 46% of all sewage is discharged directly to land in the Republic of Ireland (Water Research Centre, 1977a). Disadvantages of the method are the large area of land required (at least 0.4 h per 1000 persons) and

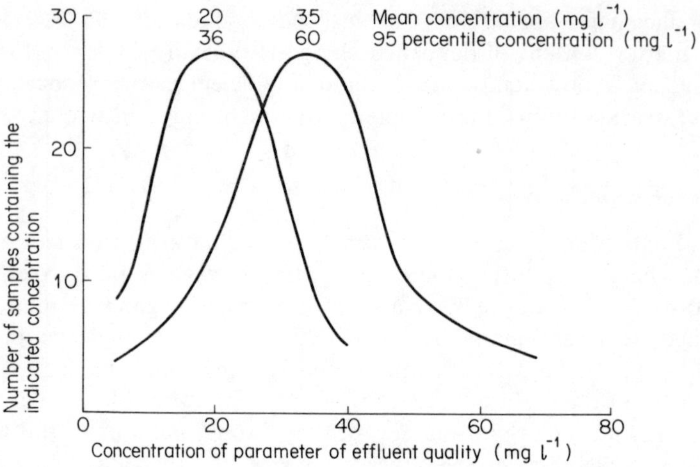

Figure 7.6. Typical frequency distributions of effluent quality (A. G. Boon, private communication)

the health hazards. It is used for the treatment of some industrial effluents, e.g. from food processing industries.

(2) Lagoons or oxidation ponds are widely used for the oxidation of sewage in semi-tropical countries and where land is plentiful, the oxygen required for the biodegradation process being supplied by dissolution of air at the pond surface and also by the photosynthetic activity of pond algae. The ponds are usually 1.0–1.5 m deep and may or may not have an outlet. Typically, pond loadings (which will be limited by the rate of BOD removal) are low, but these may be increased by connecting a lagoon which is anaerobic (devoid of oxygen) in series with an aerobic lagoon or by installation of mechanical aeration devices (Eckenfelder, 1966). Table 7.1 shows typical BOD loadings and percentage BOD removals for lagoons and other sewage treatment processes.

(3) Percolating filters consist of beds packed with pieces of hard material such as coke or granite, preferably having a large surface area. Each bed may be circular or rectangular, about 2 m deep and enclosed by walls. Settled sewage is sprayed onto the surface of the bed from moving distributor jets, whilst air passes upward through the bed from ventilating and drainage channels in the base. The medium of the bed becomes covered with a biological film into which oxygen and organic substrate from the wastewater diffuse, and where oxidation processes and synthesis of new cells occur. Oxidation products diffuse back into the liquid stream, and are carried away with surplus film (humus) to humus tanks where suspended solids are deposited and largely removed from the effluent. The biological film is scavenged by insect larvae and worms, which helps to prevent excess film blocking the filter medium thereby causing 'ponding' or flooding of the filter.

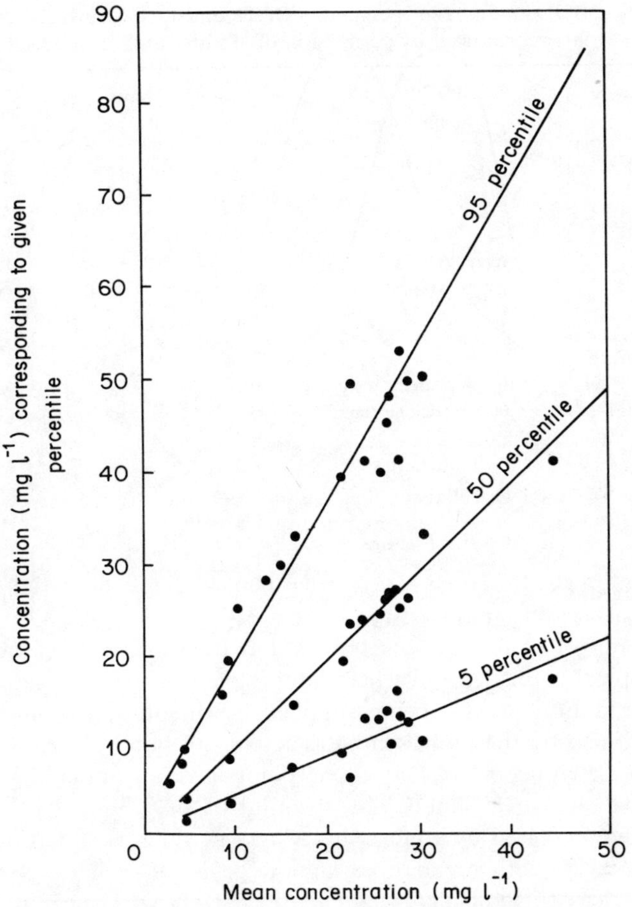

Figure 7.7. Relation between mean BOD and the corresponding 95 percentile BOD for sewage effluents. (From Porter and Boon, 1971. Reproduced by permission of Her Majesty's Stationary Office)

Alternative arrangements of filters are shown in Figure 7.8. Recirculation of effluent and the system of alternating double filtration (ADF) enable higher BOD loadings of the filters to be achieved without ponding (see Table 7.1). The ADF system requires the direction of sewage flow to be reversed periodically between two filters in series, so that surplus film which accumulates on one filter in the first cycle is removed in the second cycle. These systems, while increasing the rate of carbonaceous oxidation, reduce the rate of oxidation of ammonia (nitrification) due to the wash-out of nitrifying organisms. For complete nitrification, the two-stage system (Figure 7.8(c)) is preferred, nitrifying organisms being retained on the second filter.

Table 7.1. Typical BOD loadings and percentage BOD removals for sewage treatment processes (part reproduced by permission of Hodder and Stoughton)

Treatment	BOD loading ($kg\ m^{-3}\ day^{-1}$)	BOD removal (%)	Reference
Land	0.04*	55–60	Truesdale and Taylor (1975)
Lagoon			
aerobic	0.009–0.018†	80–95	Eckenfelder (1966)
anaerobic	0.027–0.045†	50–70	Eckenfelder (1966)
Percolating filter			
simple	0.064 settled sewage	92.5	Mills (1945)
recirculation, 1/1	0.18 settled sewage	92	Greater London Council (1973)
ADF	0.19 settled sewage	92	Mills (1945)
plastics media	0.5–6.0	90–30	Callely *et al.* (1977)
disc filter	0.6 crude sewage	97	Bruce *et al.* (1973)
aerated submerged plastics medium	6.5 settled sewage	94	Bruce *et al.* (1973)
Activated sludge process			
extended aeration (carousel)	0.25 crude sewage	98	Johnstone *et al.* (1978)
conventional	0.4–0.6 settled sewage	95	Greater London Council (1973)
deep shaft	3.6 crude sewage	86 crude sewage 82 settled sewage	Bolton and Ousby (1977)
Unox oxygen (Newton Creek)	4.3–6.1 crude sewage	90–89	Brenner (1974)

*Equivalent to 140 $kg\ ha^{-1}\ day^{-1}$.
†Assumes lagoon is 1 m deep.

The rate of biological oxidation is increased by increasing the concentration of organisms in the system, and on a percolating filter this is achieved by increasing the specific surface area *S* of the filter medium, where *S* is the surface area of pieces of media per unit volume of packed bed. *S* is increased as the size of pieces of media is decreased, but the tendency for the filter to block is thereby increased. To overcome this, plastics media such as 'Flocor' and 'Biopac' have been developed which have a high *S* value combined with high voidage space. These are used chiefly in the treatment of high-strength industrial wastes.

(4) A rotary bio-disc filter consists of a row of closely spaced vertical metal or plastic discs on a horizontal shaft, mounted on bearings over a tank of sewage. The discs are equivalent to the medium in a percolating filter and

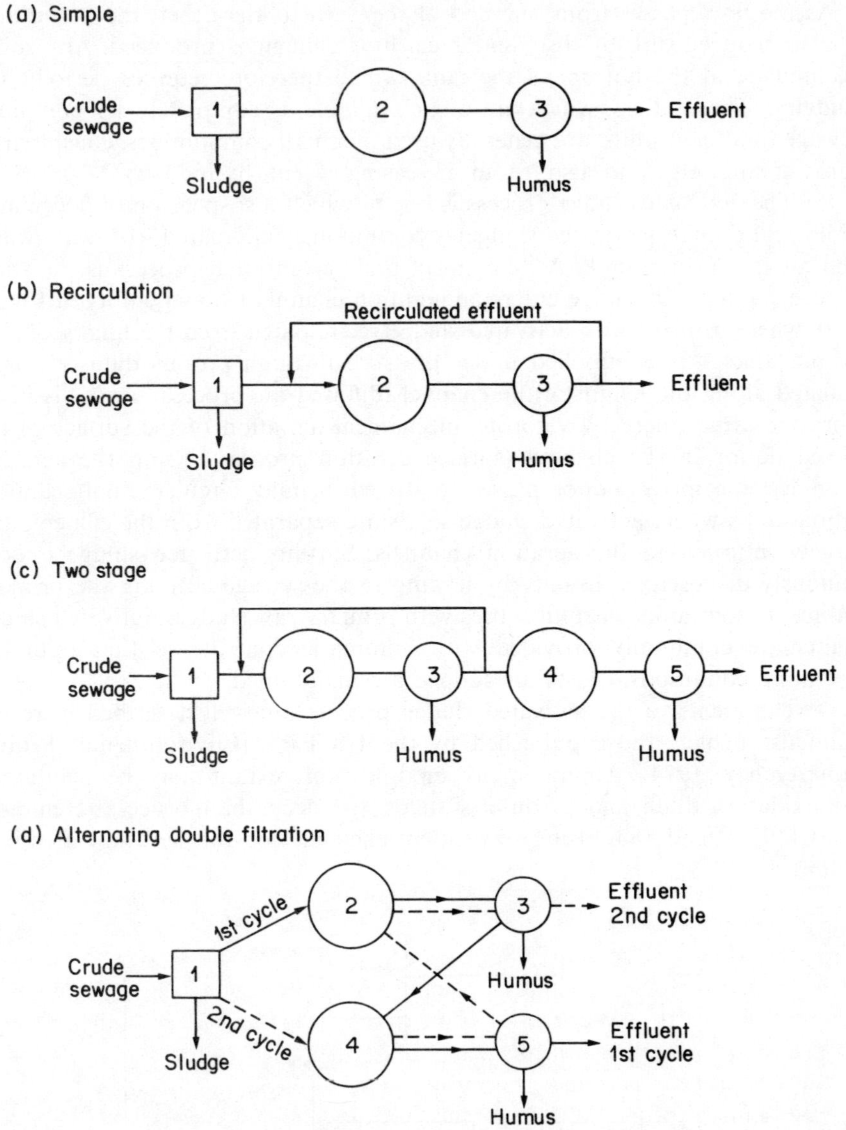

Figure 7.8. Arrangements of biological filters: 1, primary sedimentation tank; 2 and 4, percolating filters; 3 and 5, humus tanks.

provide an enlarged surface area for attachment of the microbial film and the grazing fauna. As the discs rotate (1–3 revolutions per minute), the organisms are alternately immersed in the sewage bath and exposed to the air for oxygenation.

As the flow passes from one end of the bath to the other, the pollutional load is reduced disc by disc, and a clarified effluent is produced. Any solids accumulate at the bottom of the tank, which therefore requires periodic desludging. These filters, which are often available as completely self-contained sewage treatment units, are generally used for small communities, coach parks, caravan sites, etc., and also, to an increasing extent, by industry.

(5) The activated-sludge process is one in which a suspension of flocculated biological growth (activated sludge) is continuously circulated with wastewater and air in an open tank. A diagram of the conventional process is shown in Figure 7.9. Settled sewage enters an aeration channel (shown empty in Figure 7.10) where it mixes with activated sludge recirculated from the final sedimentation tanks. Air is supplied under pressure through porous diffuser domes situated along the length of the channel (diffused air process), or is dissolved from the atmosphere by vigorous mechanical agitation of the surface of the mixed liquor in the channel (surface aeration process). From the aeration channels, the mixed liquor passes to stirred, usually circular, final sedimentation tanks where activated sludge solids are separated from the effluent, and largely returned to the aeration channels. Surplus activated sludge is continuously disposed of, usually by feeding to the sewage entering the primary sedimentation tanks where it settles with primary raw sludge. Activated sludge plants are commonly provided with automatic controls so that optimum operating conditions, e.g. of air supply, are maintained.

Developments of the activated sludge process are well described in recent manuals such as those published by the US EPA (Environmental Protection Agency, 1974). Among many modifications extant may be mentioned the oxidation ditch and carousel systems, the deep shaft process (Hemming *et al.*, 1977), and that using oxygen-enriched air for the aeration of mixed liquor.

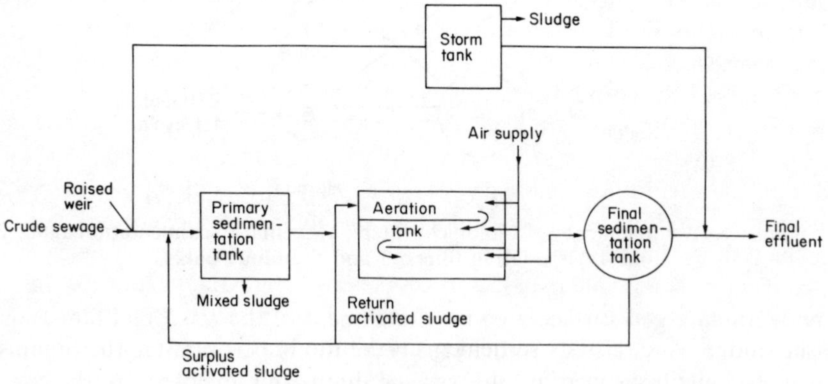

Figure 7.9. Conventional activated-sludge process

Figure 7.10. Empty aeration tank at Deephams Sewage Treatment Works, Thames Water Authority. (Reproduced by permission of Thames Water Authority.)

The oxidation ditch and carousel (Figure 7.11) plants utilize a compressed-ring-shaped aeration channel in which mixed liquor is circulated continuously by the rotation of a mechanical aeration device. This may be in the form of a horizontal spiked metal cylinder spanning the oxidation ditch, or in carousel plants, cone-shaped aerators on a vertical axis. Both systems are examples of extended aeration plants in which the residence time of sewage in the aeration channel is about 24 h, which allows bacteria to operate in the endogenous growth phase when considerable auto-oxidation of cells takes place and sludge production is low. These are very versatile plants which treat crude sewage of industrial or domestic origin and are adapted to handling shock loads.

The deep shaft process employs a vertical borehole as the aeration channel divided into two concentric cylinders (Figure 7.12). Air injected into the downcomer tube causes circulation of mixed liquor through the system, while the high hydrostatic pressure at the base of the shaft enhances the solubility and rate of dissolution of oxygen. This enables the activity of a high concentration of organisms to be supported, and high rates of biodegradation to be achieved. A similar result is obtained by the use of oxygen-enriched air in the conventional activated sludge process (Water Research Centre, 1975a), and some workers claim for this process a lower rate of sludge production and a denser sludge which assists settlement in the final tanks. However, it has been shown that nitrification rates are reduced due to a lowering of the pH of the mixed liquor (Wood *et al.*, 1976).

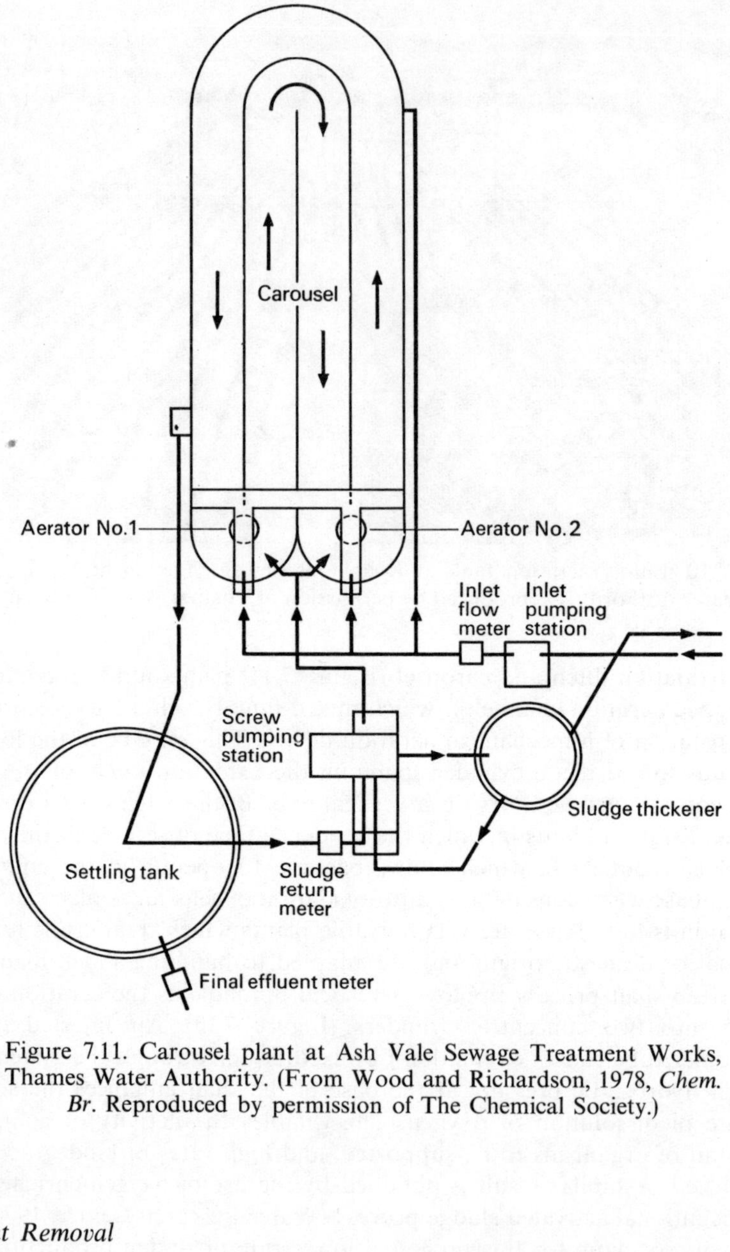

Figure 7.11. Carousel plant at Ash Vale Sewage Treatment Works, Thames Water Authority. (From Wood and Richardson, 1978, *Chem. Br.* Reproduced by permission of The Chemical Society.)

Nutrient Removal

Following secondary treatment, a well nitrified domestic sewage effluent will usually contain 30–40 mg l^{-1} of nitrate nitrogen and 10–12 mg l^{-1} of phosphate as P. It is becoming increasingly necessary in the UK to remove

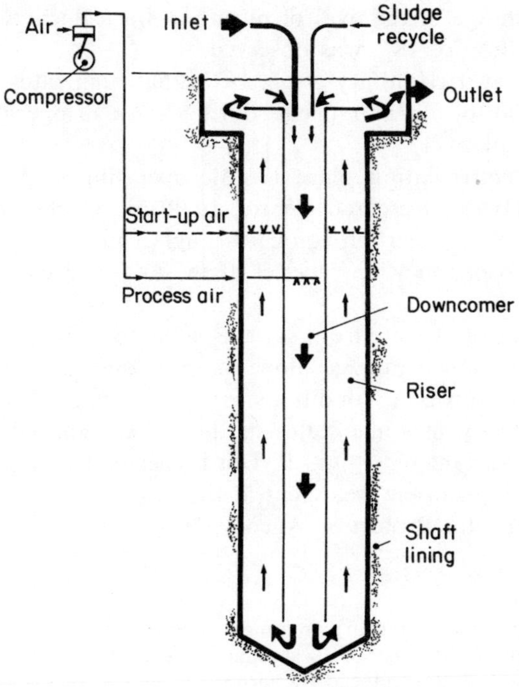

Figure 7.12. ICI deep shaft process. (From Bolton and Ousby, 1977, *Prog. Water Technol.* Reproduced by permission of Pergamon Press.)

some of the nitrogen from the effluent before discharge to a river used as a source of potable water, and/or from the river water after abstraction, in order that the WHO maximum recommended level of nitrate in drinking water (50 mg l^{-1}) shall not be exceeded. In order to eliminate eutrophic conditions in bodies of impounded water, it is necessary to reduce concentrations of NO_3-N in the receiving water to below 0.2–0.3 mg l^{-1} and of PO_4-P to below 0.01 mg l^{-1} (Truesdale and Taylor, 1975). For this reason, certain countries take steps to reduce the concentration of these nutrients in sewage effluent. Biological denitrification has been effected at the Rye Meads Sewage Works of Thames Water Authority (Cooper *et al.*, 1977) by conversion of the inlet bay of each four-bay aeration channel of the activated sludge plant to an anoxic zone, i.e. one containing no dissolved oxygen but having suitable available sources of combined oxygen such as nitrates. This has been achieved by removing the air diffuser domes from these bays and fitting stirrers to maintain the activated sludge in suspension. Nitrate recirculated with the return sludge is converted to nitrogen gas by facultative bacteria which proliferate in the anoxic activated sludge. With a return sludge/sewage ratio of 1/1 in this arrangement, the maximum percentage of removal of nitrate in the effluent would be 50%; in practice, a reduction of 48% was observed.

It has been shown (Hanbury *et al.*, 1978) that high rates of denitrification (80–90% removal of nitrate) occur during normal operation of carousel activated sludge plants.

A full-scale denitrification plant is now operating at East Hyde Sewage Works, Luton (Water Research Centre, 1978), in which the final effluent is fed with methanol as a carbon source through a modified upward-flow sand filter under anoxic conditions, better than 70% removal of nitrate being achieved.

The removal of phosphate from sewage effluent has received little attention in the UK, since land drainage alone supplies sufficient phosphate for the establishment of eutrophic conditions in surface waters. Removal of phosphates is mandatory in some states in the USA, required maximum limits ranging from 0.1–2.0 mg l^{-1} PO_4-P. This is effected by addition of lime, alum or iron salts during sewage treatment at the primary or secondary treatment stages (Environmental Protection Agency, 1976).

Table 7.2. Comparison of tertiary treatment methods when treating well oxidized secondary effluents of at least 20/30 standard. (From Truesdale and Taylor, 1975, *Water Pollut. Control*. Reproduced by permission of The Institute of Water Pollution Control.)

Method	Rate of treatment ($m^3\ m^{-2}\ day^{-1}$)	Performance (% removal)		
		SS	BOD	Bacteria coli-aerogenes
Grass plots	up to 0.85	60–75	55–60	90
Lagoons				
short retention	0.25–0.45	35–55	35–50	70
long retention	0.28	80	65	99
Slow sand filters	up to 3	60–65	35–45	50
Rapid gravity sand filters	115–230	70–90	50–70	30
High rate upward flow sand filters	230–450	50–75	45–55	25
Microstrainers, 1500 apertures per mm	350*	55–75	25–50	15
Upward flow gravel bed clarified	14–23	50–60	25–40	25

* Flow per unit area of fabric.

Tertiary Treatment

Owing to the increased re-use of surface waters, sewage effluent standards better than 30 mg l^{-1} of suspended solids and 20 mg l^{-1} for BOD are required for many works, and, to comply with this, tertiary treatment of the effluent is necessary. This involves removal of suspended solids by sedimentation or filtration methods similar to those described for the treatment of potable water sources. Truesdale and Taylor (1975) have compared the efficiency of these processes and their data for effluents from biological filters are reproduced in Table 7.2.

In 1973, 48% of all municipal wastewater treatment plants in the USA were equipped with chlorination facilities for disinfection of sewage effluents. The adoption of disinfection procedures for sewage works effluent will no doubt become general throughout the USA as a result of the enforcement of bacteriological effluent standards. In the UK, disinfection of effluents from sewage works is not usually practised. The recommended dosage for biological filter effluents is 3–10 mg l^{-1}, for activated sludge effluents 2–8 mg l^{-1}, and for sand filter effluents 1–5 mg l^{-1} (Environmental Protection Agency, 1974). Disinfection using chlorine is known to produce trace amounts of halogenated hydrocarbons from carbonaceous impurities in the effluent which could present toxic hazards if the effluent was then abstracted for potable water use (Arrhenius, 1977). Other methods of disinfection being evaluated are the use of ozone and ultraviolet light.

Water Reclamation

The final stages in the conversion of sewage effluent to a high-purity water are the removal of trace non-biodegradable organic compounds and inorganic salts. Trace organics may be partly removed by adsorption on activated carbon and/or use of a recently discovered type of molecular sieve (Flanigen *et al.*, 1978). Inorganic salts may be partly eliminated by the process of reverse osmosis, ion exchange and, less effectively, by electrodialysis. Reverse osmosis removes dissolved salts by passage of water under pressure through a semi-permeable membrane. Electrodialysis separates ions from water by passing it through a membrane under the influence of an applied electric potential.

Until improvements are made in the efficiency of adsorbents and osmotic membranes, additional water storage schemes would appear to be a more economic and satisfactory method of coping with occasional droughts than the reclamation of water from sewage effluents. However, the indirect re-use of sewage effluents is increasing. Sewage effluent now forms 20% of abstracted water in the River Lee under normal conditions and 60–70% in times of drought. Hence, advanced effluent and water treatment processes will find their major use in the treatment of industrial discharges and potable water supplies, by eliminating refractory compounds which represent an increasing toxic hazard to the ecosystem, including man.

Industrial Wastewater Treatment

About half of the 14 million cubic metres (14000 Ml) of sewage discharged daily in the UK comes from factories (Boon and Borne, 1976). In the UK, traders are legally entitled to discharge trade effluents to sewers and controlled waters subject to consent conditions and charges, where appropriate, determined by the Water Authority. In order to meet these conditions, which are summarized for the Thames catchment in Table 7.3, manufacturers commonly apply some form of pretreatment to their wastewaters. These comprise physicochemical and/or biological processes which have been summarized for particular industries by Tomlinson (1976), from whose review Table 7.4 is derived. Detailed discussions of these processes are given by Southgate (1948), Eckenfelder (1966), and Callely *et al.* (1977). Of importance for the elimination of toxic metals from sludge is the direct recycle of metals from wastes in the electroplating industry (Swanwick and O'Gorman, 1973) by the application of advanced technologies. Valuable by-products can also be recovered from wastes from the food industries or they can be used as substrate for single-cell protein production (Wheatland *et al.* 1976).

Sludge Treatment and Disposal

The cost of disposing of the 1.1×10^6 tonnes dry weight of sewage solids produced in the UK was £100 million in 1975/6 or 13% of the revenue expenditure of Regional Water Authorities (Water Research Centre, 1977b). The bulk of the sludge comprises a mixture of primary and secondary sludges containing 3–5%(w/v) dry matter of domestic and industrial origin. In the UK, 46% is disposed of to farm land, 31% to landfill sites, 20% to sea, and 3% by incineration (National Water Council, 1978a). The method of disposal of the sludge determines its prior treatment, about half of the raw sludge being anaerobically digested in closed tanks before disposal (Swanwick *et al.*, 1972). Of course, disposal of sludge, for example, on land, is not without its problems; guidelines have been published in the UK (Department of the Environment/National Water Council, 1977) which recommended the maximum permissible quantities of toxic elements (e.g. copper, nickel, zinc, and other metals largely of industrial origin) which may be added to land in sludge over 30 years.

Agricultural Waste Treatment

The major components of agricultural waste are the excreta from livestock which, in the UK, exert a BOD load to the environment equivalent to a

Table 7.3. Typical ranges of consent limits for trade effluents discharging to sewers and to rivers. (Reproduced by permission of Thames Water Authority.)

Substance or parameter	Maximum value of parameter for effluents discharging to	
	Some Metropolitan sewers	Rivers
Settleable solids ($mg\ l^{-1}$)	200–1000	—
Suspended solids ($mg\ l^{-1}$)	—	5–30
Cyanide as CN ($mg\ l^{-1}$)	10	0–1
Sulphide as S ($mg\ l^{-1}$)	5–10	No offensive matter
BOD ($mg\ l^{-1}$)	—	5–20
Free ammonia ($mg\ l^{-1}$)	50–100 as NH_3	0.1–10 as N
Sulphate as SO_3 ($mg\ l^{-1}$)	1500	—
Available chlorine as Cl ($mg\ l^{-1}$)	50	0–0.5
Available sulphur dioxide as SO_2 ($mg\ l^{-1}$)	5–10	—
Formaldehyde as HCHO ($mg\ l^{-1}$)	200	Nil
Grease and/or oil (saponifiable) ($mg\ l^{-1}$)	100–400	0–2
Grease and/or oil (unsaponifiable) ($mg\ l^{-1}$)	50	0–2
Cadmium as Cd ($mg\ l^{-1}$)	1 or <0.25 kg day^{-1}	0.1–<1
Chromium as Cr ($mg\ l^{-1}$)	10	0–1
Copper as Cu ($mg\ l^{-1}$)	5	0.2–1
Lead as Pb ($mg\ l^{-1}$)	10	3
Nickel as Ni ($mg\ l^{-1}$)	5	<1
Zinc as Zn ($mg\ l^{-1}$)	5	<1–20
Total toxic metal ($mg\ l^{-1}$)	—	0–0.5
pH	6–11	6–9
Temperature (°C)	43.3	Ambient–31
Halogenated hydrocarbons	Prohibited	Nil
Petroleum spirit	Prohibited	—
Calcium carbide	Prohibited	—
Toxic chemicals	—	Nil

human population of approximately 150 million. Much of this is distributed to grazing land by free-ranging animals, where it acts as a valuable fertilizer. At present, there is sufficient land to accommodate the load, provided it can be uniformly dispersed. However, the continuing trend towards the intensification of farming methods—the factory farming of cattle, pigs, and poultry—requires cattle and pig slurries of urine and faeces to be disposed of to land artifically, e.g. by means of rain guns. Problems with pig slurry may arise due to the presence of copper (up to 400 $mg\ l^{-1}$) and arsenic (200–400 $mg\ l^{-1}$) derived from feed additives,

Table 7.4. Effluent treatment processes and industries using them. (From Tomlinson, 1976, *Industrial Pollution Control Yearbook*. Reproduced by permission of the author and Industrial Newspapers Ltd.)

(1) Primary treatment (screening, settlement)
(2) Biological filtration
(3) Activated sludge
(4) Anaerobic digestion
(5) Chemical treatment
(6) Chemical flocculation
(7) Electrolysis
(8) Evaporation
(9) Irrigation
(10) Activated-carbon absorption
(11) Ion exchange
(12) Reverse osmosis, electrodialysis, ultrafiltration

Industries	Processes used	Special components of wastewater
Iron and steel	5,6,7	Acids, iron salts
Coke ovens	2,3	Phenols, thiocyanate, thiosulphate
Non-ferrous metals	5,6,7	Toxic metals
Electroplating	5,6,7,11	Toxic metals, cyanide
Paper	1,2,3,4,5,6,8,	Cellulose fibre
Textiles	2,3,4,5,6,9	Dyes, bleach
Leather	2,3,5,6	Sulphide, chromium
Chemicals	2,3,5,6,10,12	Inorganic and organic chemicals
Electric power	11,13	—
Petroleum	1,5,6	Oils
Automotive	1,5,6	Oil, detergent
Paint and ink	1,5,6	Solvents
Construction	1	Suspended solids
Ore mining	1	Suspended solids
Soap and detergents	2,3,5,6	Detergents
Plastics	2,3,5,6	Organic chemicals
Rubber	2,3,5,6	Toxic additives
Farming	1,2,3,4,9	High BOD, NH_3
Food and drink	1,2,3,4,8,9,12	High BOD, fats

and also due to pathogens, necessitating the careful choice of disposal area and the prevention of leakage from storage lagoons, so that water resources will not be adversely affected.

Simpson (1971) estimates that, of all farm waste produced in England and Wales, only that equivalent to sewage from a population of 3 million people is subject to treatment by methods other than land disposal. Land disposal is the most satisfactory way of treating drainage from silage vats, which is very strongly polluting with a BOD ranging from 12000–66000 mg l^{-1}. Besides animal excreta, which are 2–3 times more difficult to treat by biological oxidation than human wastes, farm wastes include washings from dairies and milking parlours, vegetable wash liquors, and spent sheep dips.

Where facilities exist, disposal to the public sewers is commonly practised, and since 1961 such discharges have been legally recognized as trade effluents and charged for on this basis. Unfortunately, the choice of on-site treatment methods listed in Table 7.4 is severely limited by economic restraints. Gowan (1972) quotes the total cost of high-rate biological filtration, and activated-sludge treatment using an oxidation ditch and surface aerators as £13, £5.83, and £5.06 per kg BOD treated per day, respectively.

PROTECTION OF WATER QUALITY FROM THE EFFECTS OF LANDFILL

In the past years, industry has been attracted to and sited near mineral resources. The worked-out excavations from mineral-winning provide convenient sites for disposal of industrial and other wastes as landfill. In some cases, the mineral deposits are adjacent to watercourses (for example, terrace gravels) and the excavated site is in hydraulic continuity with the river. Other sites are located in water-bearing strata, for example, chalk, which is a major aquifer and used as a source of water supply.

Seventeen million tonnes of domestic and commercial waste and 23 million tonnes of industrial waste were estimated to be disposed of in England in 1975 (Department of the Environment, 1976a). Of all industrial solid waste, 88% is disposed of to landfill sites, its composition being (Department of the Environment, 1970): general factory rubbish, i.e. uncontaminated by process wastes (10.4%), relatively inert process wastes (82.8%), flammable process wastes (1.1%), acid or caustic wastes (3.9%), and indisputably toxic wastes (1.8%).

Since about one-third of the nation's potable water supplies are drawn from underground resources, and the demand is expected to increase, it is essential that our aquifers are protected. If surface water becomes polluted, the remedy lies in removing as much pollutant as possible and allowing the remainder to be diluted as it is washed downstream; the resource is soon restored. By comparison, the movement of groundwater is usually very slow and, if it becomes contaminated, remedial action may be impracticable. However, in time, natural dilution may be available to reduce contaminant concentrations to acceptable levels. Alternatively, a scavenging system may be installed utilizing a production borehole to remove and pump to waste as much contaminated water as possible.

The Water Resources Act 1963, Section 72, prohibited the discharge of polluting matter to underground strata by means of a 'well, borehole or pipe' without the Regional Water Authority's consent, but did not specify seepage from diffuse sources such as leachate generated from a waste disposal site.

The Deposit of Poisonous Waste Act 1972 was designed more to protect the general public from hazards associated with direct contact with indiscriminately dumped toxic wastes than to protect water resources, and gave the Waste Disposal Authority (WDA) the right only to give or withhold permission to tip waste, but not to impose conditions for acceptance, or for sealing the site to prevent underground contamination.

The Control of Pollution Act 1974, Part 1, introduced regulations for the disposal of controlled waste, aimed, amongst other objectives, at protecting surface and groundwater resources. Two major implications of the Act were:

(1) the WDA is required to prepare and revise waste disposal plans for its area and in so doing to consult the relevant Regional Water Authority; and
(2) it is made an offence to dispose of controlled waste to land without a licence, to be issued by the WDA after consulation with the Regional Water Authority.

In 1973 the Department of the Environment set up a programme of research into the behaviour of hazardous wastes in landfill sites. Their report (Department of the Environment, 1975) emphasizes that when drainage (leachates) from landfill comes into contact with rock minerals and groundwaters, the composition of the liquid may be modified by many possible processes. These include physical processes such as dilution, dispersion, filtration, and gas exchange, and biochemical and chemical processes such as oxidation, formation of complexes, acid–base reactions, precipitation, ion exchange, and adsorption (Mather and Bromley, 1977).

Regional Water Authorities are reviewing possible waste disposal sites to determine how far the natural attenuative ability of an unsaturated zone can protect groundwater from the results of landfill operations.

Landfill sites fall into three types (Department of the Environment, 1976b):

(1) those providing a significant element of containment for wastes and leachates:
(2) those allowing slow leachate migration and significant attenuation; and
(3) those allowing rapid leachate migration and insignificant attenuation.

Sites of the first category may derive their capacity for containment either from existing geological conditions, or it may be necessary to increase the impermeability by processes such as puddling of chalk or the provision of an impervious lining. A site of this kind retains leachate which must be removed and treated, and there are sometimes problems with gas generation.

From the WDA point of view, the sites allowing slow migration of leachate are probably most suitable, but to protect the aquifer there must be an unsaturated zone at least two metres thick. In this zone, the movement of leachate is slow and will be attenuated by aerobic biodegradation (using the oxygen present in the gas phase) and other chemical reactions. Sites allowing rapid leachate migration, such as those on fissured formations or gravels, are the least suitable for disposal of hazardous wastes. Many Regional Water Authorities are now formulating aquifer protection policies to identify activities which are considered to present risks to water quality in certain defined protection zones.

WATER QUALITY MONITORING

Water Authorities are obliged to classify streams in their catchment according to the use and amenities required of them. A scheme proposed by the National Water Council (1978b) is shown in Table 7.5. Associated with each classification are 95 percentile quality criteria which must be met if the potential uses of the streams are to be achieved. These criteria are expressed in terms of dissolved oxygen concentration, BOD, and ammonia, and European Inland Fisheries Advisory Commission (EIFAC) and EEC quality requirements. For effluents discharging to the streams, consent standards are set which must be of sufficient severity to enable the river to achieve its quality objectives. Such consents are usually stated in terms of BOD, ammonia, and suspended solids.

Quality Monitoring at Water Treatment Works

Surface water abstracted for potable supply will be required to meet conditions of an EEC Directive (EEC, 1975a). A (proposed) Directive (EEC, 1975b) will eventually replace the WHO recommended limits now observed for water in the distribution system (World Health Organization, 1970). The proposed EEC Directive specifies chemical and bacteriological standards and the frequency of sampling required of water from distribution networks. In the London region, potable water samples at all stages of treatment are subject to daily bacteriological examination and regular chemical assessment.

Bacteriological requirements of water supply are:

(1) Throughout any year, 95% of samples should not contain any coliform organisms or *Escherichia coli* in 100 ml.
(2) No sample should contain more than 10 coliform organisms per 100 ml.
(3) No sample should contain more than 2 *E. coli* per 100 ml.
(4) No sample should contain 1 or more *E. coli* per 100 ml in conjunction with a total coliform count of 3 or more per 100 ml.
(5) Coliform organisms should not be detectable in 100 ml of any two consecutive samples.

Quality Monitoring at Wastewater Treatment Works

The purposes of monitoring in this field are:

(1) to ensure that plants operate at maximum efficiency; and
(2) to ensure compliance with consent conditions.

Table 7.5. Classification of river quality. (From National Water Council, 1978b, *River Water Quality, the Next Stage. Review of Discharge Consent Conditions.* Reproduced by permission of NWC.)

River class	Quality criteria—Class limiting criteria (95 percentile)	Remarks	Current and potential uses
1A	(i) Dissolved oxygen (DO) saturation greater than 80% (ii) Biochemical oxygen demand (BOD) not greater than 3 mg l^{-1} (iii) Ammonia not greater than 0.4 mg l^{-1} (iv) Where the water is abstracted for drinking water, it complies with requirements for A2† water (v) Non-toxic to fish in EIFAC terms (or best estimates if EIFAC figures not available)	(i) Average BOD probably not greater than 1.5 mg l^{-1} (ii) Visible evidence of pollution should be absent	(i) Water of high quality suitable for potable supply abstractions and for all other abstractions (ii) Game or other high-class fisheries (iii) High amenity value
1B	(i) DO greater than 60% saturation (ii) BOD not greater than 5 mg l^{-1} (iii) Ammonia not greater than 0.9 mg l^{-1} (iv) Where water is abstracted for drinking water, it complies with the requirements for A2† water (v) Non-toxic to fish in EIFAC terms (or best estimates if EIFAC figures not available)	(i) Average BOD probably not greater than 2 mg l^{-1} (ii) Average ammonia probably not greater than 0.5 mg l^{-1} (iii) Visible evidence of pollution should be absent (iv) Waters of high quality which cannot be placed in Class 1A because of high proportion of high-quality effluent present or because of the effect of physical factors such as canalization, low gradient, or eutrophication (v) Class 1A and Class 1B together are essentially the Class 1 of the River Pollution Survey	Water of less high quality than Class 1A but usable for substantially the same purposes

2	(i) DO greater than 40% saturation (ii) BOD not greater than 9 mg l^{-1} (iii) Where water is abstracted for drinking water, it complies with the requirements for A3† water (iv) Non-toxic to fish in EIFAC terms (or best estimates if EIFAC figures not available)	(i) Average BOD probably not greater than 5 mg l^{-1} (ii) Similar to Class 2 of RPS (iii) Water not showing physical signs of pollution other than humic colouration and a little foaming below weirs	(i) Waters suitable for potable supply after advanced treatment (ii) Supporting reasonably good coarse fisheries (iii) Moderate amenity value
3	(i) DO greater than 10% saturation (ii) Not likely to be anaerobic (iii) BOD not greater than 17 mg l^{-1} *	Similar to Class 3 of RPS	(i) Waters which are polluted to an extent that fish are absent or only sporadically present (ii) May be used for low grade industrial abstraction purposes (iii) Considerable potential for further use if cleaned up
4	Inferior to Class 3	Similar to Class 4 of RPS	Waters which are grossly polluted and are likely to cause nuisance
X	DO greater than 10% saturation		Insignificant watercourses and ditches not usable, where objective is simply to prevent nuisance developing

Notes

(a) Under extreme weather conditions (e.g. flood, drought, freeze-up), or when dominated by plant growth, or by aquatic plant decay, rivers usually in Classes 1, 2, and 3 may have BODs and dissolved oxygen levels or ammonia content outside the stated levels for those Classes. When this occurs, the cause should be stated along with analytical results.

(b) The BOD determinations refer to five-day carbonaceous BOD(ATU). Ammonia figures are expressed as NH_4.

(c) In most instances, the chemical classification given above will be suitable. However, the basis of the classification is restricted to a finite number of chemical determinands, and there may be a few cases where the presence of a chemical substance other than those used in the classification markedly reduces the quality of the water. In such cases, the quality classification of the water should be downgraded on the basis of the biota actually present, and the reasons stated.

(d) EIFAC (European Inland Fisheries Advisory Commission) limits should be expressed as 95 percentile limits.

(e) RPS refers to the 'River Pollution Survey of England and Wales: Updated 1973', published by Her Majesty's Stationery Office, London, for the Department of the Environment and the Welsh Office.

*This may not apply if there is a high degree of re-aeration.

† EEC category A2 and A3 requirements are those specified in the EEC Council Directive of 16 June 1975 concerning the Quality of Surface Water intended for Abstraction of Drinking Water in the Member States.

To control plant operations, samples are taken after successive stages of treatment, usually manually at smaller works and automatically at larger works, and presented for analysis as a flow-weighted composite of individual samples (Wood and Stanbridge, 1968). The 'control' sample of final effluent so obtained will indicate average quality over a period of time, and whilst this is satisfactory for plant control, it is not necessarily suitable as an 'audit' sample to check that the effluent meets its consent standard at all times. For a Water Authority to be seen to be accountable to the general public, it is desirable that audit samples should be taken at unspecified times by officers without involvement in wastewater treatment, such as those from a unifunctional River Division (Nicolson and Wood, 1974).

Quality Monitoring of Rivers

Pollution control in the catchment hinges upon river water quality for the control of which, in addition to regular manual sampling, automatic river side monitoring stations have been set up throughout the world (Department of Industry, 1975). In Lea Division of Thames Water Authority, five such stations have been established to monitor the river automatically for five parameters, namely: dissolved oxygen concentration, temperature, suspended solids, pH, and electrical conductivity (Hinge, 1977). A computerized telemetry system interrogates the stations at hourly intervals using Post Office land lines. The system has provided early warning of river pollution; for example, after an accidental escape of cyanide to a sewerage system, the receiving sewage works discharged an effluent with an increased ammonia content. The progress of this downstream was monitored, thus enabling abstraction intakes to be closed for the minimum of time. Biological monitoring (discussed more fully in Chapter 5) serves as an adjunct to chemical monitoring for the detection of long-term quality trends and of sporadic pollution incidents.

PROBLEM OF DETECTION AND REMOVAL OF SUBSTANCES OF POTENTIAL HEALTH HAZARD FROM RAW AND POTABLE WATERS

The assumption that a body of water is of an acceptable quality and, for example, that a water supply is wholesome provided that it complies with the currently accepted national and international standards, is valid only if the supply authority takes all practical steps to ensure that the degree of pollution is firmly controlled both in general terms as well as regards specific chemicals. A technique for the latter has been developed (Fish and Torrance 1977, 1978; Wood and Richardson, 1978), which consists basically of assessing and controlling, where necessary, the nature and extent of persistent low-level chemical contamination passing to rivers from industrial, domestic, and agricultural sources. It is therefore designated 'Catchment Quality Control' and can be applied to both point and diffuse types of discharge.

Industrial contaminants are assessed by preparing a mass balance of all potentially toxic chemicals discharged to sewers or directly to streams, and which will be present at a potable water abstraction point in concentrations in excess of 1 part in 10^{10} (0.1 $\mu g\ l^{-1}$). Dischargers of trade effluents are consulted to obtain the quantities of such substances which they discharge annually. If the extent to which these contaminants are removed at sewage treatment works is known, due account is taken, otherwise they are assumed to pass through unchanged and this provides a margin of safety in the estimation.

Substances present at concentrations below 0.1 $\mu g\ l^{-1}$ are disregarded because a study of worldwide standards of quality for potable water showed no quality limit more stringent than this.

It is probable that, of the pharmaceutical products, detergents, cosmetics, and other similar domestic products and their metabolites, greater quantities are discharged to sewers from domestic premises than from premises manufacturing and packaging them. To determine the levels of such substances present in rivers at water abstraction points, it is necessary to establish an average per capita discharge. This can be evaluated from data obtained from such sources as manufacturers' sales and records of the Department of Health and Social Security of prescriptions issued.

It is much more difficult to assess the micropollutants which are likely to arise from highway drainage, and more investigational work is necessary before a satisfactory assessment can be made.

Pesticides, fungicides, crop conditioners, sheep dips, and other similar chemicals used agriculturally may be present in land run-off to give rise to potentially hazardous concentrations at water abstraction points. Consultation with central Government and other bodies with agricultural interests can assist in assessing distribution of these substances in watercourses.

From the above, the predicted concentrations of all micropollutants at abstraction points can be evaluated. The next stage is to sample at these sites for the listed chemicals to confirm or allow modification of the results. The final procedure is to refer the questionable chemicals present in excess of 0.1 $\mu g\ l^{-1}$ to medical and toxicological experts for assessment of public health significance.

The whole procedure of Catchment Quality Control must be one of frequent review, not only of the levels of chemicals present but also of their significance. With greater knowledge, it may be found that substances previously regarded as insignificant are indeed hazardous. The concentrations of such substances will be known and the appropriate action can be taken. This procedure supplements information obtained from direct analysis of water samples. These micropollutants cannot, however, be identified merely by chemical analysis. Until a mass balance is made, it is not known which substances are being sought. Even when they are identified, their analysis involves sophisticated instrumentation using gas chromatography, high-pressure liquid chromatography, and mass spectroscopy (Water Research Centre, 1975b), and analytical methods have to be developed for some constituents. These methods are time-consuming and expensive and not suited to frequent routine monitoring. For these reasons, the application of Catchment Quality Control techniques as outlined is the only practicable approach.

RECENT LEGISLATION OF SIGNIFICANCE FOR TREATMENT PROCESSES

Probably the two most significant pieces of recent legislation related to control of pollution in the UK are the Water Act of 1973 and the Control of Pollution Act 1974. A national policy introduced by the former involved the establishment of a number of bodies with interrelated functions as shown in Figure 7.13. The most important impact upon water treatment was undoubtedly the setting up, in April 1974, of the ten Regional Water Authorities each with complete responsibility for the water cycle in the area of a watershed. The unified system of control has been severely tested, especially during the 1976 drought and on other occasions, and its advantages were apparent. Some concern was felt about the Water Authorities being the largest dischargers to streams, and at the same time having regulatory powers for pollution control, but it has been shown that Water Authorities can preserve unbiased accountability (Nicolson and Wood, 1974).

The Control of Pollution Act 1974 received the Royal Assent on 31 July 1974 but the date of its full implementation is not yet decided. It dealt in six parts with waste on land, pollution of water, noise, and pollution of the atmosphere; reference has already been made in this chapter to the legislation controlling waste on land.

The Act repealed and updated many of the provisions of previous Acts. The Control of Pollution Act, Section 31 (1), reinforces the Rivers (Prevention of Pollution) Acts 1951 and 1961, and makes it an offence knowingly to permit poisonous or noxious matter to enter any 'relevant waters', or to enter a stream whereby pollution is substantially aggravated or the flow is impeded, or solid wastes to enter a stream or 'restricted waters'. Since 'relevant waters' include specified underground waters, the Act will remove the limitation implied in the Water Resources Act to discharge from a 'well, borehole or pipe,' and give effective legislation for control of pollution of all water resources.

The Rivers (Prevention of Pollution) Acts gave the regulatory authorities powers to issue consents for discharges to watercourses and required them to keep registers of such consents with the conditions imposed, to be available for

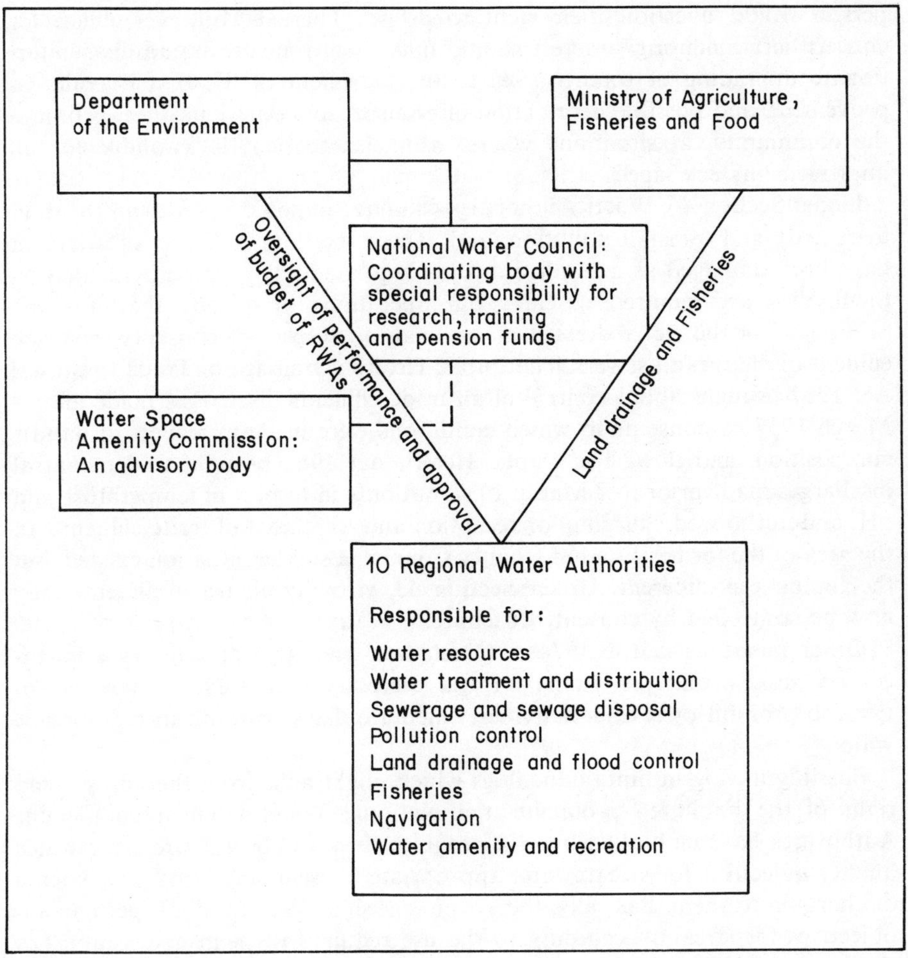

Figure 7.13. Organization of water management in England and Wales. (From Wood and Richardson, 1978, *Chem. Br.* Reproduced by permission of The Chemical Society.)

inspection by the public at large. The new Act goes further and requires (Section 41) similar registration of details of samples of water and effluent taken by Water Authorities in pursuit of their obligations. An important difference in the new Act is that, whereas the former legislation stipulated that prosecutions for offences could be brought only by the regulatory authority or the Attorney General, this restriction has been omitted from the Control of Pollution Act with the result that it will allow members of the public to institute legal proceedings for alleged pollution. At first sight, no reasonable person would question their right to do so. There is, however, cause for concern lest a minority interest should make use of this to obtain disproportionate upgrading of watercourses to the detriment of the best overall improvement, and possible deprivation of essential services to another section of the community, in situations where insufficient capital is available for all improvements envisaged.

Under Section 46, Water Authorities will have imposed upon them the duty to remedy and forestall pollution, and, where the flora or fauna of a stream have been damaged as a result of a discharge made by a consent granted by them, they are required to carry out operations to remedy the situation. Section 43 of the Act redressed an anomaly in respect of consents for trade effluent discharges to sewers. The Public Health (Drainage of Trade Premises) Act 1937 brought about control of all trade effluents discharges made after 3 March 1937 by consents to which conditions were made in respect of nature, composition, and flow. The Public Health Act 1961 brought under control discharges made prior to 3 March 1937, but only in respect of temperature and pH, and authorized charging for reception and treatment of trade effluents. In the area of the former London County Council, a similar situation existed, but the timing was different. Under Section 43, virtually all trade effluents may now be controlled by consent procedure.

Other measures in the 1974 Act include the control of sanitary appliances on vessels, and provisions for the Secretary of State to make regulations to prohibit or restrict activities which are likely to result in pollution of water.

Possibly having in mind difficulties which might arise from the unrestricted right of the public to prosecute, the National Water Council and Water Authorities have embarked upon a programme of establishing realistic water quality objectives for streams and appropriate consent standards for effluents discharging to them. Basically, the system classifies 'long-term' or 'permanent' objectives for streams according to the use required of them and establishes the relevant water quality parameters. Then, by mathematical modelling, the necessary consent standards for effluents can be evaluated on a reasonable and scientific basis (Fish, 1978a, 1978b). Since it is illogical to set stream objectives that cannot be met because capital is not available to upgrade works which cannot produce other than unsatisfactory discharges, such streams are given 'short-term' or 'interim' objectives, and discharge consent standards are based upon what can be expected by good management from the existing plant. It has been suggested that this will lead to deterioration of quality of watercourses but most Water Authorities have stipulated that there must be no deterioration from present stream quality, and at least one Authority (for the purpose or establishing priorities for improvements) has introduced a 'score' system based upon environmental needs which will ensure that the best cost/benefit is achieved.

Legislation in Other Countries

Water pollution control legislation varies considerably from one country to another and in some, France in particular, it can be very complex. The legislative provisions of EEC countries for inland waters are set out in Table 7.6.

International legislation on water pollution control, especially in the case of rivers or water-bearing substrata crossing national frontiers, is not well developed. Early attitudes in the United States supported the Harman Doctrine that a state was free to use rivers without restriction as they flowed through its territory. It is doubtful whether many legal authorities would now subscribe to this: rather the principle, *sic utere tuo ut alienum non laedas**, would be more acceptable. The Helsinki Rules on the Uses of International Rivers 1966, Article X, states:

> 'Consistent with the principle of equitable utilisation of the waters of an international drainage basin, a State:
> (a) must prevent any new form of water pollution or any increase in the degree if existing water pollution in an international drainage basin would cause substantial injury in a co-basin State and
> (b) should take all reasonable measures to abate existing water pollution in an international drainage basin to such an extent that no substantial damage is caused in the territory of a co-basin State'.

Unfortunately, there are currently no institutions which can oversee the implementation of these principles, and their application is usually determined by treaties between the states concerned.

On 22 January 1972 the Treaty of Accession was signed on behalf of the United Kingdom when it became a member of the enlarged European Community. The EEC fulfils its administrative duty (Waters and Cockburn, 1978) by the use of:

(1) *Regulations*, which are mandatory requirements applicable to all Member States.
(2) *Directives*, i.e. mandatory instructions to all Member States to achieve specified objectives.
(3) *Decisions*, which are mandatory instructions to designated individuals or bodies within the Community.
(4) *Recommendations* and *Opinions*, which are non-mandatory statements which may or may not be acted upon.

The Commission of the EEC drafts proposals which may be commented

*'So use your own property as not to injure your neighbours'.

Table 7.6. Controls in EEC countries. (From McLoughlin, 1976, *The Law and Practice Relating to Pollution Control in the Member States of the EEC: A Comparative Study*. Reproduced by permission of the ECSC, EEC, and EAEC, and Graham and Trotman Ltd.)

Legislative provision	Belgium	Denmark	France	German Fed. Rep.	Ireland	Italy	Luxembourg	Netherlands	United Kingdom
Control of discharge siting	Yes	Yes	Partial	Yes	Yes	Yes	Yes	Yes	Yes
Control of industrial processes	Yes	No	Yes	Yes	Yes	Yes	No	Yes	Yes
Control of siting of storage tanks	Yes	Yes	No	Yes	Yes	Yes	No	Yes	Yes
Provisions for pretreatment of effluent	Yes	Yes	Yes	Yes	No	Yes	Yes	Yes	No
Limitations on discharges	Yes	Yes	Yes	Yes	Yes	Yes	Yes	Yes	Yes
Compulsory monitoring by discharger	Yes	Yes	Yes	Yes	Yes	Partial	No	Yes	Yes
Control of dumping in or near inland water	Yes	Yes	Yes	Yes	Yes	Yes	No	Yes	Yes
Restrictions on waterborne craft	No	Yes	No	Yes	No	Yes	No	Partial	Yes

Table 7.7. EEC Directives relating to environmental pollution and water quality. (Reproduced by permission of Thames Water Authority.)

Reference	Title	Description	Effective dates
73/404/EEC OJ No. L347	On the Approximation of the Laws of the Member States Relating to Detergents	Requires detergents to be at least 90% biodegradable	Notification 22 Nov. 1973
75/439/EEC OJ No. L194	On the Disposal of Waste Oils	Requires safe collection and disposal of waste oils, and *inter alia* prohibits discharge of oil to the aquatic environment	Notification 16 June 1975
75/440/EEC OJ No. L194	Concerning the Quality Required of Surface Water Intended for the Abstraction of Drinking Water	Characterizes raw surface water by parameters the values of which are determined by the type of treatment the water is to receive	Notification 16 June 1975
76/160/EEC OJ No. L31	Concerning the Quality of Bathing Water	Requires that within ten years the quality of fresh or sea water for bathing should meet the parametric values laid down in the Directive	Notification 8 Dec. 1975
76/403/EEC OJ No. L108	On the Disposal of Polychlorinated Biphenyls and Polychlorinated Terphenyls	Controls general disposal of these substances to minimize the risk of their dispersion into the environment	Notification 6 Apr. 1976
76/464/EEC OJ No. L129	On Pollution caused by Certain Dangerous Substances Discharged to the Aquatic Environment of the Community	Seeks to control emission of specified substances into the aquatic environment by emission standards and environmental quality objectives	Notification 4 May 1976
78/176/EEC OJ No. L54	On Waste from the Titanium Dioxide Industry	Seeks to control dispersion of wastes from this industry and to encourage recycling	Notification 20 Feb. 1978
78/659/EEC OJ No. L222	On the Quality of Fresh Waters Needing Protection or Improvement in Order to Support Fish Life	Requires Member States to designate stretches of fresh water which have to be brought into conformity with quality standards	Notification 18 July 1978

upon by the Economic and Social Committee and the European Parliament. The Council, comprising representatives from Member States, is the legislative body, though the European Commission is responsible for implementation of Directives and Regulations, and Member States become responsible for their implementation in their own countries. They usually have powers to delegate this to a competent body, such as a Water Authority, but the Government of any State would have to account to the European Court of Justice for any infringement. The Official Journal of the Community publishes proposals and notices as 'C' code entries, and actual legislation, e.g. Directives and Regulations, are 'L' coded.

Table 7.7 lists Directives currently notified; some of these are discussed further in Chapter 8. In addition, Decisions have been notified on:

(1) prevention of marine pollution from land-based sources;
(2) a common procedure for an inventory of information on the environment; and
(3) a common procedure for exchange of information on surface water quality.

Proposed Directives have been published on:

(1) reduction of water pollution from wood pulp mills;
(2) quality of water for human consumption;
(3) dumping of wastes at sea; and
(4) quality requirements for waters favourable to shellfish.

CHAPTER 8

Water Quality Objectives and Standards

D. C. Renshaw

The aim of this chapter is to disentangle and consider briefly the major influences which operate on the setting and enforcement of water quality objectives and standards. The subject is of interest in itself, but some understanding of it is also necessary in order to give perspective and context to the questions considered in other parts of this book.

The last five years has seen the beginning of two major changes in the handling of water quality objectives and standards in the UK. First, we are beginning to move from the widespread use of effluent standards related mainly to generalized or implied objectives towards more coherent sets of standards explicitly related to rationally predetermined objectives. Secondly, there is an increasing tendency for standards and objectives themselves to be the subject of legal instruments of some detail and complexity (this being an especial feature of EEC legislation affecting water matters), in contrast with the long-standing reliance of the UK upon a fragmented pragmatic approach based upon broad general legal requirements. At the same time, public awareness of, and opportunity to participate in, the determination of objectives and standards is growing and is being reinforced by significant legislative changes. Add to this the fundamental reform (in England and Wales) of administrative responsibility for water management and sharply increased activity in all aspects of the subject by the EEC, and the complexity of the current picture can be seen.

As a basis for understanding some of these changes, it is convenient first to examine some of the important institutional and legislative developments since 1973; the significance of these is not always recognized, but it is, in my view, considerable.

LAWS AND INSTITUTIONS

The Water Act of 1973, the Control of Pollution Act 1974, and the United Kingdom's accession to the European Economic Community in 1973 are significant events in the recent development of water quality policy in this country. The 1973 Act gave concrete administrative expression to the concept of integrated management of the hydrological cycle, on a catchment basis, by setting up the ten Regional Water Authorities now responsible for this task in England and Wales. For the first time, one institution was to be in a position to control the major factors influencing water quality in a catchment. Not only did all the legal powers and duties relating to the supply of water pass to the new bodies (with the exception, not significant in the present context, of the areas where private water companies continued to undertake the task), but the Authorities also took over the powers to control effluent standards contained in the Rivers (Prevention of Pollution) Acts 1951 and 1961 and allied legislation, and—perhaps most importantly—the management and operation of sewerage systems and sewage treatment works. (Again, the sewerage agency functions exercised by some District Councils are not significant in this context.) The previous fragmentation of responsibility was at an end, and those involved with questions of water objectives and standards were reduced in effect to the Regional Water Authorities as managers and suppliers, on the one hand, and the users—such as industry and agriculture—on the other. But, at the same time, the Water Authorities themselves became users in one significant way, since a majority of discharges to inland waters originate from the sewage treatment works which came under their control. So the new advantage of unified hydrological control within a catchment also placed the Water Authorities in a potentially difficult position as possible polluters of the waters whose quality they were charged with controlling.

The advent of the Regional Water Authorities also brought about some changes in the role of central Government in relation to water quality matters. Section 1 of the 1973 Act lays upon the Secretary of State for the Environment duties not only to promote a national policy for water in England and Wales but also to secure the effective execution of that policy in so far as it relates *inter alia* to the treatment and disposal of sewage and other effluents, the restoration and maintenance of the wholesomeness of rivers and other inland waters, the use of inland waters for recreation, and the enhancement and preservation of amenity in connection with inland waters. This is not, of course, an executive role, but it does give the Secretary of State a general locus in the scheme of things. At the same time, the more compact organization of the water industry, and particularly the setting up of the National Water Council as a sounding board and clearing house for the Regional Water Authorities, created the possibility of new channels for informal contact and discussion between Government and Water Authorities. In the water quality context, the two major areas of interaction between Water Authorities and

Government have so far been in relation to EEC legislation and the implementation of Part II of the Control of Pollution Act 1974.

At the same time as the new Regional Water Authorities were beginning operations, the Control of Pollution Act 1974 received the Royal Assent. Part II of the Act deals with water pollution. It consolidates and updates previous legislation relating to the control of individual discharges to water through the system of consents to be administered by the Water Authorities. But, like the previous legislation which it supersedes, it has little to say about discharge standards. Indeed, the only specific test to be applied in deciding an individual consent application is that of 'reasonableness'. Consents may not be unreasonably refused by the Water Authority, but they may be granted subject to reasonable conditions. Nonetheless, the Act is significant in several respects for the whole process of setting standards.

First, it extends physically the areas of control and the types of discharge to be controlled. Hitherto, some significant estuaries and most areas of coastal water have been outside the system of control, and discharges to them could be legally made without consent. The new Act changes that by bringing under control all tidal waters up to, roughly, the three-mile limit. And it will control, for the first time, some types of discharge inland also, particularly those which could affect the quality of groundwater.

Secondly, the Act provides for full public information about water quality and discharges, and for public participation in the control process. Water Authorities will have to set up public registers which will give full details of discharge consents applied for and granted, and the results of the analyses of water samples taken by the Authorities for pollution control purposes. Moreover, an application for a discharge consent which could have an appreciable effect on the receiving water will have to be advertized, and representations by the public must be taken into account.

Finally, the Act confirms the Secretary of State's formal functions in the control of standards. He is (as previously) to decide appeals by individual dischargers against decisions of the Regional Water Authorities on their consent applications, and is also to be the final judge of whether the consent conditions which Water Authorities propose to apply to their own discharges are acceptable, or whether they should be amended.

At the beginning of 1979, Part II of the Act was not actually in force. But the Government had announced its intention of implementing the new provisions by the end of 1979, and a great deal of preparatory work was in hand, both by Government and by the Regional Water Authorities. Implementation was to be preceded by a review of the conditions of consent for all existing discharges, and a comprehensive exercise by the Water Authorities to state formally river quality objectives, often in great detail, for all significant inland waters in England and Wales. I shall return to this subject more fully later in this chapter.

THE EUROPEAN ECONOMIC COMMUNITY

Britain's membership of the EEC has introduced a wholly new factor on the water quality scene. The Council of Ministers instituted in 1973, and re-affirmed in 1977, an Environment Action Programme with the aim of improving the quality of life in all Member States. Water management and water quality occupy a significant place in the aims of this programme, and both the Commission and the Council have devoted considerable effort to carrying the programme into effect over the last five years. The main instruments have been a series of Directives adopted by the Council touching many aspects of water quality and standards. These Directives are, of course, intended for Community-wide application and cannot, therefore, be built around quality requirements in particular areas or catchments. It is this feature which has given rise to one of the most fundamental and important discussions in recent years in the UK about the whole nature and purpose of water quality control—the debate between the advocates of control by uniform emission standards and those who prefer environmental quality objectives as a basis. I shall refer to this in more detail later.

It is worth noting, at this stage, the areas which have been of particular concern to the EEC Council, as reflected in the Directives which have so far been adopted for implementation in the Member States (as at the beginning of 1979). The most significant is probably the Directive on Pollution Caused by Certain Dangerous Substances Discharged into the Aquatic Environment (EEC, 1976a), which sets up a framework for the control of discharges of particular types of potential pollutant. This Directive introduced the concept of the so-called 'black list' and 'grey list' substances—the former being pollutants regarded as particularly dangerous because of their toxicity, persistence, and bio-accumulation, the latter being substances whose control was desirable in the interests of diminishing or avoiding pollution. The Directive provides for Member States to set up their own machinery for controlling pollution by grey list substances, based on nationally determined quality objectives. For black list substances, the picture is somewhat different. The Directive provides for these substances to be controlled either by the application of uniform emission standards or of community-level quality objectives, but both the standards and the objectives will be determined for each individual substance or family of substances in a further series of Directives to be proposed by the Commission to the Council; at the beginning of 1979, no such further Directives had appeared.

The Council has adopted three Directives based upon the notion of controlling the quality of water according to the use to which it is to be put. The Directive on the Quality of Bathing Water (EEC, 1976b) lays down standards to be achieved by all inland and coastal waters defined as bathing waters under the Directive; like many Brussels texts, this one is not as straightforward as it seems on the surface, largely because the rather obscure definition of 'bathing waters' makes it difficult to interpret. The Directive on the quality of water intended for the abstraction of drinking water (EEC, 1975) and the Directive on the quality of water for freshwater fish (EEC, 1978b) similarly lay down standards for inland fresh water intended for these two uses.

The Council also agreed, at the end of 1978, upon the Directive on the quality of water intended for human consumption, though at the beginning of 1979 it had not been formally adopted and no official text had been published. This Directive is of great potential significance, since it lays down Community-wide standards for a wide range of parameters fixing the quality of water supplied for consumption (including food preparation).

The Directive on Waste from the Titanium Dioxide Industry (EEC, 1978a) is of a rather different kind. It is an attempt to deal with the pollution problems which could be caused by a particular industry, but, unlike other Directives, does not lay down quantified standards to be achieved either by the effluent from the industry or in the water which receives such effluent. It is based rather on the notion of reducing or eliminating pollution caused by the industry, and operates through requiring the preparation of programmes in the Member States for this purpose.

From this brief mention of existing Directives, it can be seen that the EEC's interest in water quality is potentially of very wide range, and other texts under discussion at the moment confirm this: the Commission has made proposals to the Council for Directives on the protection of groundwater quality, the prevention of pollution by paper pulp mills, and the quality of water required for shellfish breeding.

WATER QUALITY OBJECTIVES

Against the background of the previous sections, it is now possible to consider the setting of water quality objectives. For this purpose, I shall regard an objective as a general statement of the aim to be achieved by water management policies, and I shall distinguish three such aims:

(1) the protection of public health, largely in relation to the consumption of water;
(2) the protection of the aquatic environment, in order to make it fit for the various uses to which it can be put (for example, industrial and agricultural abstraction, commercial and recreational fisheries, and other forms of recreation); and
(3) the protection of aquatic life and ecosystems.

Some or all of these aims are recognized, implicitly or explicitly, by most of those with the responsibility for water quality management. In the UK, no doubt because of our dependence upon surface water abstractions for the supply of drinking water in many parts of the country, the protection of public health has tended to predominate. And it does, in practice, come top of the hierarchy, since the achievement of that objective through the setting of appropriate standards will usually ensure that other objectives can be met simultaneously. Where drinking water abstraction is not important, objectives in the UK have tended to be less precise and explicit, although it is quite usual for non-potable water to be managed against an objective of protecting or even creating fisheries, both game and coarse. Again, the achievement of that objective will often ensure that other aims relating to the protection of the aquatic environment are simultaneously met, as well as the protection of aquatic ecosystems. This third objective, taken by itself, has not traditionally figured very much in UK practice; where neither potable abstractions nor fisheries have been important, the implied aim has sometimes been little more than to prevent a river becoming an unsightly or smelly public nuisance.

Until recently, there has been no concerted attempt in the UK to set out explicitly any objectives or hierarchy of objectives for inland waters as a whole. The various control authorities have tended, understandably, to regard objectives as self-evident in important instances (e.g. potable abstraction), and perhaps unnecessary where quality management was a question of spending whatever money could be afforded in order to achieve some improvement in quality, or at least to prevent unacceptable deterioration. But this picture is now changing. A minor influence, which I mention first only because it came first in point of time, is the approach of the EEC. As I have mentioned, the Community Environment Action Programme foresees a number of Directives relating water quality to particular use-based objectives, and some of these have already been adopted; this process has begun to concentrate the minds of water managers rather more explicitly on the underlying aims of their management policies.

But by far the most significant move has been the decision of the Water Authorities, announced in the National Water Council's document 'River Water Quality, the Next Stage' (National Water Council, 1978) to specify river quality objectives for surface waters. This is a massive exercise, involving, in essence, the identification of the present and potential future uses of all significant stretches of inland waters in England and Wales, and the specification of short- and long-term objectives for each stretch. The intention is that the setting of explicit and detailed objectives will form the basis for the setting of individual discharge consent conditions, both now and in future, designed to achieve the stated objectives. This important undertaking was still going on at the time of writing, but most Water Authorities had already produced draft objectives for public consultation, and in some cases the process had been completed with the endorsement of the proposed objectives by the Water Authority members. The indications were that most Water Authorities had decided to adopt, as a convenient method of describing their objectives, the suggested classification of river quality in the National Water Council's 1978 document. This classification (which is shown in Table 7.5) allocates rivers to one of five main categories, each category corresponding to a set of present or potential water uses and each having attached to it a statement of the more important quality criteria appropriate to that class (e.g. dissolved oxygen and biochemical oxygen demand levels). By the end of 1979, it should be possible, for the first time in this country, for anyone to see a complete, detailed, and explicit picture of the quality objectives, present and proposed, for each stretch of water.

It is interesting to note that the attitude of the EEC to quality objectives fits well with this new domestic approach, since each Member State is expected to identify and designate stretches of water appropriate to particular use-based objectives.

WATER QUALITY STANDARDS AND THE CONTROL OF POLLUTION

If the setting of objectives, in the sense of statements of aims, is a relatively straightforward business about which there is broad agreement, the question of how to achieve those objectives, whilst ensuring proper control of pollutants, is a very different matter. Questions arise at every stage of the process. Decisions have to be made about what quality standards are needed to ensure that a stretch of water meets the objective stated for it, how those standards are to be defined, and how controls can be set for individual emissions to ensure that the necessary standards are reached. Each stage of the process raises further questions about the scientific basis for proceeding, the practicability and cost of what is proposed, and its acceptability and enforceability. A number of different approaches are possible, but to set them in context I propose to consider first the situation in the UK.

Quality standards in Britain have traditionally depended upon a pragmatic approach related to local conditions and circumstances. Standards have been related, often implicitly, to generalized water use objectives, and have been applied by a statutory system of consents to each individual discharge. The oldest and most widely used standard has been that suggested in the eighth report of the Royal Commission on Sewage Disposal (1912). This suggested that, for sewage effluent, the biochemical oxygen demand should not exceed 20 mg l^{-1} and the suspended solids content should not exceed 30 mg l^{-1}. At first sight, this looks remarkably like a uniform standard for application everywhere, and it has often been misunderstood in this way. But the Commission was rather more subtle than that. They based their suggestion on the notion that a river or stream in which the BOD in the water did not exceed 4 mg l^{-1} would usually not show obvious signs of pollution—that is, they had both a general objective (the avoidance of noticeable pollution) and a general water quality standard (4 mg l^{-1} of BOD). Their suggested emission standard was the final stage in their argument, and was designed to produce the water quality necessary to meet their objective. For this reason, it was subject to the very important proviso (another fact sometimes forgotten) that the standard would be appropriate in streams where the available dilution was at least 8:1, and the BOD of the diluting water was less than 2 mg l^{-1}. The Commission made it clear that if these conditions were not met, a higher standard of effluent quality might be required; conversely, they felt that where a greater dilution was available it might be possible to tolerate a lower quality effluent.

Practice over the years has indeed reflected these general views of the Commission. There are numerous instances of consents for sewage effluent discharges with conditions stricter than the Royal Commission standard, either because, as the Commission suggested, the low dilution available makes this necessary or, in some cases, because the water quality required in the receiving stream in relation to its use is actually higher than that assumed as generally acceptable by the Royal Commission—often in cases where drinking water abstractions are to be made downstream of the effluent discharge. It is interesting that the Royal Commission standard appears to have proved generally satisfactory, at least for the control of sewage effluent, over a long period. But the relative sophistication of argument supporting it has undoubtedly faded into the background; in practice, this has not mattered in most cases and, even where the standard has been applied as if it were a uniform emission standard, the outcome has none the less been satisfactory in water quality terms. The current major exercise, mentioned earlier, in which Water Authorities are defining explicit quality objectives for individual stretches of river, will, however, ensure that there is no misunderstanding of the approach to water quality in the UK–that is, that local quality standards are set in relation to locally determined objectives and that conditions of consent for individual discharges are then set to ensure compliance with the quality standard.

There are several features about this 'environmental quality objective' approach which call for brief comment. First, it assumes that it is acceptable and legitimate to make proper use of the assimilative capacity of the receiving water. Britain is perhaps fortunate in that such assimilative capacity is widely available in many watercourses, which provide the necessary flow and dilution to deal with quantities of common pollutants without adverse effects on the general water quality required to meet the local objective. But it is a process which could easily get out of control, and one of the benefits of setting explicit objectives across the whole country is that control Authorities and the public will be constantly reminded of the minimum quality standards necessary to meet the local objective, and thus of the limits to the permissible use of local absorptive capacity. Secondly, there is the difficulty of setting appropriate consent conditions for each discharge. This subdivides into two distinct problems. First, the control Authority must ensure that total discharges to a particular stretch of water will not cause the quality standard to be exceeded; this may not be a particularly straightforward task, especially in view of the variety of types of discharge and local environmental conditions which may be encountered. Secondly, and much more difficult, is the question of dividing the permissible total of effluent amongst all those wishing to make a discharge—the problem of so-called 'even handedness'. The difficulties here arise from the different views which different interests may take of the meaning of even handedness. An economist would look for a least-cost solution, as described in Chapter 9, in which each discharge was adjusted to ensure that the total treatment costs of all discharges in order to meet the conditions of consent were minimized; upgrading of quality would be secured by requiring improvement to those discharges which had the lowest marginal improvement costs. But that would not be the view of industrialists, who would tend to claim that they should not be required to spend more on treatment costs than their neighbours, particularly if their neighbours happened to be direct competitors. And there is the additional difficulty that Water Authorities themselves will often be the major dischargers in a particular area. The most likely short-term solution to these difficulties lies in a continuation of practice hitherto, in which improvements to individual discharges are secured by a combination of negotiation, mutual willingness, and considerations of what is technically and economically possible for individual dischargers at any time (see also Chapter 9). The third feature of the environmental quality objective approach is its dependence upon adequate monitoring of receiving waters. Again, in the UK we are perhaps fortunate in having a long-established, elaborate and reliable network of monitoring practices inherited or set up by the Water Authorities. A great deal of information about ambient water quality is readily available (as, for instance, in the water quality reports published by the Authorities in various forms), and will become even more widely accessible when the public registers under the Control of Pollution Act are set up in the near future. Thus, the observance of quality standards and objectives is easy to check, both for the control Authorities and for the interested public.

Fourthly, environmental quality objectives do require a considerable degree of public understanding. They are not easy to grasp either in concept or in practice, and individuals and pressure groups can understandably become alarmed at an apparently low-standard discharge, even though a high-grade quality standard and objective may none the less exist in the locality.

Fifthly, environmental quality objectives call for considerable sophistication in their application to specific pollutants such as heavy metals and chemicals. But the approach is not, for that reason, invalid in such cases. The essential requirement is to analyse carefully what information is needed to apply quality objectives, so that the necessary research can be undertaken. The first requirement is careful analysis of any adverse effects which the substance in question may have in relation to the general objective for the locality; for example, the levels at which it may become a danger to human health if abstraction for drinking water is the objective. This in itself may be a difficult process, since it will probably call for proper medical consideration of a safe level of intake for individuals from all likely sources, including water, before a safe level for water itself can be set. Once that level has been decided, it is then possible to work through the process as with other pollutants in order to decide the permissible level in any particular discharge and to reflect this in the consent condition. In some cases, this task may be somewhat simplified if it is clear that only one discharge is a possible source of the substance in question. It is also worth noting that, contrary to a belief sometimes expressed, the quality objective approach is perfectly capable of dealing with situations in which the acceptable level of a particular substance turns out to be very small or even nil; it is essentially a matter of setting the appropriate water quality standard and reflecting this in the consent conditions. A useful summary of the behaviour and effects of a wide range of substances in water is given by the Water Research Centre (1976).

An alternative approach which avoids some of the difficulties of environmental quality objectives, though at the price of importing others, is the application of uniform emission standards. The concept is relatively easy to understand. The pollution control authority determines limits for specified substances and pollutants in effluent, these values being applicable to every discharge covered by the scheme of standards, regardless of its location. The system is not, of course, as crude as that brief statement might imply. There is scope within the setting of the standards to take account in a general way of the likely effect of discharges upon the environment, and indeed standards are often set by reference to the general pollution potential of particular substances. But, by definition, there is no specific linkage with local water quality standards or local objectives. The result may be that in some areas water quality turns out to be inadequate, thus leading to the need to tighten up discharge standards, whilst in others the standards may prove unnecessarily stringent for achieving the required local quality. Thus, at least in theory, there would be a considerable waste of resources by dischargers. This is likely to be a particular danger when standards are set, as they sometimes are, by reference to the best available technology for minimizing the amount of a substance in a discharge (although this approach is usually applied to the kinds of substance where the minimum possible quantities would be reckoned desirable whatever system of setting quality standards was in use). None the less, the system can have its advantages in appropriate circumstances. It avoids problems of even handedness (though, as I have indicated, probably at a price); it obviates some of the difficulties which may be encountered in working back from objectives and quality standards to discharge consent conditions; and it is relatively easily understood by the public. It may be a particularly useful approach to the control of pollution in international watercourses, where the difficulties of sharing out the absorptive capacity of the water may otherwise be insuperable and it may indeed be a simpler method to apply in areas where the river system has little absorptive capacity in the first place. And in areas where pollution problems are particularly severe and widespread, it may be a useful rough and ready way of securing quick initial improvement where it is reasonably obvious that all dischargers must take some action.

For various reasons, many Member States in the EEC favour the uniform emission standard approach. There was considerable debate in the Council of Ministers, particularly during the negotiations leading to the adoption of the Directive on dangerous substances, about the relative merits of uniform emission standards as against environmental quality objectives. Although this was seen by some as a battle for supremacy between the two systems, the decision of the Council on that Directive (a decision likely to be reflected in later Directives where the same issues are at stake) was that Member States should be free to choose whichever system of control they wished to adopt; this would seem to reflect the commonsense view that local needs and circumstances must be taken into account in deciding the appropriate method of control.

A third approach is based on the rather loosely named 'zero emission' requirements. The idea here is that some substances are so potentially harmful that no discharge at all should be allowed. This counsel of perfection can rarely be achieved in practice (unless the control authority is prepared to require the complete abandonment of any use of the substance, with all the implications of this policy for industrial processes and costs). More often, the practical outcome is the use of 'best available technology' to reduce discharges as far as possible.

The zero emission/best available technology approach is used in some circumstances by some of our EEC partners, and is reflected occasionally in EEC Directives. But, in the latter case, the requirement is usually confined to new plant built and commissioned after the date of imposition of the requirement, thus avoiding the most severe implications which the method could have. And a wide interpretation can be put on 'available', so that it can, for example, include a realistic assessment of cost implications, with an impossibly or disproportionately expensive technology being ruled out as not 'available'. (A good example of this approach, in practice, is to be found in the EEC Directive on titanium dioxide waste in relation to new plants or major extensions of existing plants which are in many ways equivalent to new plants.)

CONCLUSIONS

What general conclusions can be drawn from this somewhat variegated picture? In the UK, at least, the most important is that environmental quality objectives offer the best way through to the achievement of desired environmental standards at the least cost. This is evidently so when one is dealing with degradable pollutants which can be absorbed or diluted in the local aquatic environment and with a relatively uncomplicated pattern of discharges.

It is an approach less easy to apply in other cases. The setting of acceptable quality standards for the more toxic, persistent, and bio-accumulable substances is often far from easy, and the expression of such standards in particularly useful ways can be just as difficult. Sometimes, indeed, the end result may be indistinguishable to the casual observer from the application of uniform or even 'zero' emission standards—e.g. the actual requirement on a discharger may in effect be to keep the amounts of a substance in his effluent to the lowest practicable figure in order to meet the quality standard, just as he would have to do to meet a uniform or 'zero' standard.

Further difficulties can arise where there are multiple discharges to the same stretch of water, especially if their sources are a mixture from industry and public authority. If the focus of interest is on improving the quality of the receiving water to a desired standard, it may well be that several different routes to the same goal can be theoretically devised, each involving different rates and amounts of improvement for the various dischargers. Furthermore, it should be possible to select the least-cost pattern of improvement from amongst these alternatives, given a knowledge of the marginal unit cost of improvement to each discharger. But practical problems can set in when an attempt is made to put all this into effect. Individual dischargers often do not see why they should invest more than their neighbours, particularly their Water Authority neighbours, in further cleanup. Water Authorities can have their own spending constraints placing limits on what they can do.

A different kind of complication arises over the application of quality objectives to rivers flowing through different countries. The river as 'received' at the border of one country may already be at, or even below, the minimum quality which the receiving country is prepared to accept through the activities of upstream dischargers in other countries (who may in theory at least be maintaining standards acceptable in their own country). At worst, therefore, a country could find itself with no means of reaching its own quality objectives; even at best the ability of its own dischargers legitimately to use the absorptive capacity of the river may be limited by the discharges already taking place upstream.

In the face of these difficulties, it is sometimes argued that the quality objective approach is useful as a theory, but not much help in some circumstances in actually reducing and controlling pollution. Uniform and 'zero' emission standards, it could be said, with all their imperfections at least give the best practical hope of quick and effective action.

Experience in the UK is that quality objectives are a rational and effective control method which will always tend (even if imperfectly at times) to minimize the resources cost of achieving a given environmental standard, as well as offering a real hope of actually being able to reach a desired standard. Other systems may be more appropriate in other circumstances. Perhaps the 'live and let live' approach now being exemplified in the EEC is the only workable answer.

CHAPTER 9

The Economics of Water Quality

D. J. Storey

INTRODUCTION

Surprise is often expressed by both natural scientists and engineers that economists should be interested in water pollution. Even where there is some acknowledgement of economic involvement, this is normally assumed to be with the cost and/or performance characteristics of individual treatment plants, rather than with major issues of policy.

Such misunderstandings are the fault of economists. They have failed to inform natural scientists of the inherent economic dimension of the water pollution problem and of the unique contribution that economics can make to its analysis. In the United States, it is, arguably, the failure of economists to assert any leadership in environmental issues, despite their active interest, that has resulted in vast expenditure on treatment facilities failing to yield any recognizable improvements in water quality. By contrast, there has been a significant improvement in the quality of the waterways of England and Wales since 1951, despite the modest expenditure by both industry and public authorities.

Nevertheless the relevant economic question to be asked of all expenditure on treatment is: has it been spent so as to achieve a maximum improvement in water quality? Conversely, could present water qualities have been achieved at lower cost? In a nut-shell, are we getting value for money in expenditure on pollution control? These questions are applicable to catchments where improvement in quality has taken place, as well as to those where there has been no measurable improvement, or even a decline.

It will be the purpose of this chapter to demonstrate that economists have a comprehensive body of theory which can be applied to environmental problems, since these are, subject to minor amendments, inherently economic. In the next section, the concept of an optimal level of pollution is outlined, together with its limitations. The data on pollution costs, both in the UK and elsewhere, are then examined and an assessment of the policies for 'clean-up' is undertaken. The conclusion reached is that both effluent charges, much favoured by economists, and policies of physical regulation, which are described later in this chapter, have disadvantages, and that only the combination of both policies, certainly in the UK, can ensure water quality is achieved in an efficient, effective, and equitable manner.

AN OPTIMAL LEVEL OF POLLUTION?

There are two categories of pollution costs incurred by society. The first is the cost of pollution damage: namely, the loss of recreational benefits, or fish killed through river pollution. As the level of pollution rises, such costs are also likely to rise. The second cost of pollution is that of installing treatment or abatement facilities to improve or reduce discharges. Empirical studies indicate that most pollutants are relatively cheap to reduce at first, but that costs rise sharply with successive reductions. These costs are expressed algebraically in equation (1)

$$\mathrm{TC(P)} = \mathrm{CC(P)} + \mathrm{DC(P)} \tag{1}$$

where TC is the total cost of pollution, CC are control costs, DC are damage costs, and P represents pollution. Total control costs and total damage costs vary with (are a function of) the level of pollution. To minimize the total costs of pollution, the first-order conditions are shown in equation (2).

$$\frac{\mathrm{dTC}}{\mathrm{dP}} = \frac{\mathrm{dCC}}{\mathrm{dP}} + \frac{\mathrm{dDC}}{\mathrm{dP}} = 0$$

$$-\frac{\mathrm{dCC}}{\mathrm{dP}} = \frac{\mathrm{dDC}}{\mathrm{dP}}. \tag{2}$$

These are very familiar conditions in economics. In words, they state that the optimal level of pollution obtains when the marginal costs of control (MCC) equal the marginal damage costs (MDC), where marginal may be thought of as 'incremental'. This same concept is demonstrated in Figure 9.1. Here, it is postulated that the marginal cost of control curve rises because it becomes *progressively* more expensive to remove each unit of pollution. It is also postulated that the damage caused by each unit of pollution also rises *progressively*. Hence, there is an optimal level of pollution q^*.

This concept of optimality has been challenged by those who argue that it is not possible, in practice, to estimate the MDC curve. Of those studies which have been attempted, the most frequently quoted is that of the Delaware River, in the USA. This argued that water pollution damage could be divided into three categories; first reduction of the recreation/amenity value of the river, secondly the damage to commercial fisheries, and, thirdly, the costs imposed

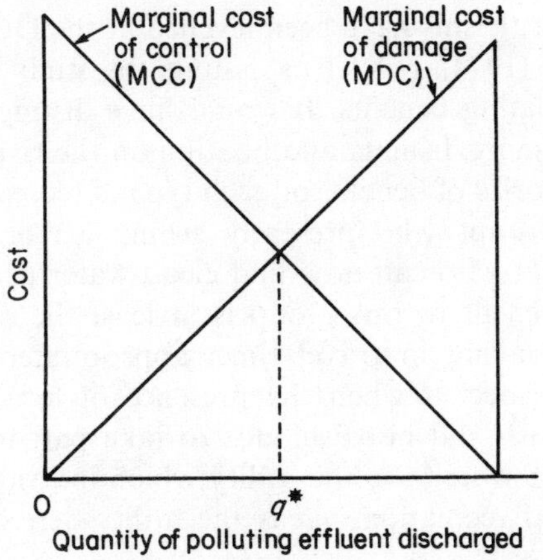

Figure 9.1. The 'optimal' level of pollution

upon industry which had to install plant to treat abstracted water. The estimated monetary values of this damage for the Delaware River are shown in Table 9.1, with water quality I being the highest standard and water quality IV being the lowest. Two points are immediately apparent. First, total damage increases (benefits are forgone) as water quality falls. Secondly, the main decrease in the benefits of clean water (i.e. the increase in costs of dirty water) is due to loss of recreation opportunities, yet it is the method of calculating such losses that has been the subject of intense criticism.

Table 9.1. Present value of all net quantifiable benefits: study of Delaware River

Objective	Type of estimate	Value ($ million—1964 prices)			
		Recreation	Commercial fishing	Industrial*	Total
I	Max.	355	12	−15	350
	Min.	125	9		160
II	Max.	320	12	−13	320
	Min.	135	9		140
III	Max.	310	7	−10	310
	Min.	125	5		130
IV	Max.	280	5	−7	280
	Min.	115	3		120

* The negative numbers are due to cleaner water imposing extra costs upon industrial users due to *increased* corrosion rates on pipework. It is presumed that rounding errors account for the sum of the component parts not equalling the total.

Many powerful criticisms have been levelled at the Delaware Study, notably by Ackerman *et al.* (1974), but let us assume the study had used the 'correct' procedure for estimating benefits. It would have distinguished between those wishing to go swimming, fishing, and boating on the river. It would then have identified two categories of benefit for each type of recreation; namely, first, the benefits to those groups who presently swim, fish or boat in clean water further upstream, but who can now find clean water nearer their homes. It is assumed that the benefit to this group is at least the savings in travel costs, plus an estimated saving in travel time, appropriately costed. The second category of benefits occurs when the presence of local clean water induces groups who previously did not fish, etc., to take part in these activities. The benefits from such a switch are the utility which individuals now derive from the new water-based recreation, *minus* the utility derived from the next best (i.e. previous) activity.

Although it is possible, in theory, to place monetary values upon the benefits of cleaner water, the results are subject to considerable uncertainty. For example, it is very difficult to know the value individuals place on travelling to and from the river. It is also difficult to estimate the *extra* utility derived by those prepared to switch to water-based recreations, if such facilities were provided locally. Yet these are crucial questions.

These difficulties are so considerable that they have caused many economists to abandon a rigorous notion of an optimal level of pollution. The comments by Baumol and Oates (1975) summarize the matter well:

> '... It is hard to be sanguine about the availability in the foreseeable future of a comprehensive body of statistics reporting the marginal net damage of the various externality-generating activities in the economy. The number of activities involved and the number of persons affected by them are so great that, on this score alone, the task assumes Herculean proportions. Add to this the intangible nature of many of the most important consequences—the damage to health, the aesthetic costs—and the difficulty of determining a money equivalent for marginal net damage becomes even more apparent'.

(An externality is a direct, but unpriced effect imposed by one individual or firm upon another individual or firm. Pollution is often regarded as the classic externality since the damage which waste discharges from a factory cause to the community is not reflected in the price of the final product produced. For a comprehensive discussion of externality theory, see Baumol and Oates (1975).)

The solution proposed by Baumol and Oates, and accepted by the present writer, is that debate should take place, possibly assisted by some economic estimates of damage, on acceptability standards, e.g. dissolved oxygen in a river should always exceed 5 mg l^{-1}. To ensure meaningful debate, it is necessary to obtain an estimate of the costs of reaching these standards, and it is to this matter which we now turn.

THE COSTS OF WATER POLLUTION CONTROL

Data on the costs incurred by both public and private industry in Britain in controlling water pollution are patchy. The most comprehensive source is 'Report of a River Pollution Survey of England and Wales 1970' (Department of the Environment, 1973) from which Tables 9.2 and 9.3 are derived. These tables illustrate a number of important aspects of policy on water pollution control in Britain.

Table 9.2 shows that between 1960 and 1970 approximately £100 million was spent by industry on capital equipment for treating discharges of aqueous effluent. Secondly, it shows capital expenditure per unit of process water varies according to the ultimate location of the discharge, since expenditure is substantially higher for discharges to the public sewer than for discharges to rivers or canals. It should be noted that the higher capital expenditure by firms discharging to the public sewer is *not* due to the direct payment of charges to the Authority, but rather to the stimulation which such payments have in inducing the introduction of capital abatement equipment. The charges are, however, included in the operating costs incurred by dischargers. The table also shows that capital expenditure, per unit of process water, is higher for discharges to non-tidal waters than to tidal waters. This reflects the policy of encouraging major polluting dischargers to locate on the banks of estuaries rather than alongside fresh water, since the latter are often used for abstraction purposes. All the above comments can be seen to apply to both capital and to operating costs.

Table 9.3 shows that capital expenditure per unit of process water is, however, not solely dependent upon the location of the discharge, and some industries contradict the general conclusions of Table 9.2. For example, for three out of the five comparable industries, expenditure per unit of process water is higher for discharges to tidal than to non-tidal waters. Also, in three of the industries, expenditure per unit of process water is higher for discharges to non-tidal waters than discharges to the public sewer. Table 9.3 also shows that there is considerable variety in levels of expenditure by various industries. Coal mining, for example, spends ten times that of chemical and allied industries on controlling discharges to non-tidal rivers, whereas matters are reversed for discharges to the public sewers. Hence, it would be dangerous to assume that expenditure by a particular industry can be fully 'explained' in terms of the ultimate location of the discharge.

Several comments need to be made in qualifying the validity of such data. First, it was actually collected in 1968 and distinguished between actual

Table 9.2. Costs of disposal of industrial effluent

Discharges of effluent to	Capital expenditure on treatment plant 1960–70 (£ million)	Capital expenditure 1960–70 on treatment plant per gallon per day of process water (£ million)	Operating costs 1968 prices (£ million)	Operating costs per gallon per day of process water 1968 prices (£ million)
Non-tidal rivers	34.297	0.08475	5.999	0.0148
Tidal rivers	13.752	0.03606	1.503	0.0039
Canals	2.428	0.0542	0.315	0.0070
Public sewers	48.793	0.1447	8.481	0.0252
Total	99.270		16.298	

Table 9.3. Costs of effluent disposal to specific industries (using CBI code)

Industry	Discharges to non-tidal rivers		Discharges to tidal rivers		Discharges to public sewers	
	Capital expenditure on treatment plant 1960–70 (£ million)	Capital expenditure on treatment plant per unit of process water (£ million)	Capital expenditure on treatment plant 1960–70 (£ million)	Capital expenditure on treatment plant per unit of process water (£ million)	Capital expenditure on treatment plant 1960–70 (£ million)	Capital expenditure on treatment plant per unit of process water (£ million)
Chemical and allied	3.948	0.04442	5.702	0.057730	15.405	0.52631
Coal mining	13.327	0.42595	—	—	0.005	0.04807
Engineering	1.174	0.23197	0.073	0.57031	6.315	0.12316
Food processing	1.950	0.14177	0.931	0.04612	7.270	0.13521
Iron and steel	2.578	0.04767	3.598	0.02447	0.718	0.07919
Paper and board making	2.886	0.03398	2.842	0.05019	9.977	0.29416

expenditure 1960–68 and projected or planned expenditure 1969–70. Since nearly half of the expenditure is in the latter category, and since no further surveys have been undertaken, there can be no certainty that these expenditures were actually incurred. Secondly, the survey was carried out by the Confederation of British Industry (CBI), who make no claim to a complete coverage of all firms in all industries. Information from the survey was also classified according to a CBI code, which is not strictly compatible with the Standard Industrial Classification (SIC) used in most Government surveys. Finally, it must be recalled that the data refer to a period in which concern for the environment was only beginning to be manifest. It would be interesting to see whether the trends of the 1960s were continued into the present decade.

A more recent study has been undertaken by Atkins and Lowe (1977), but this covers only the small geographical area of the East Midlands, UK (Nottingham and Leicester areas), thus limiting the generality of their findings. In addition, the East Midlands contains few direct discharges, and hence the authors concentrate on the charges incurred by firms discharging to the public sewers. Only five firms in their sample of 58 pre-treated waste before disposal.

Data for other countries, especially the United States, are both more recent and more comprehensive. Table 9.4 (taken from Leone, 1976) shows the average impact on selling prices in four US industries of the 1972 Federal Water Pollution Control Amendment Act. This Act required all firms in an industry to reach a standard by 1977 whereby 'best practicable' abatement technology was introduced, i.e. taking account of the financial position of firms in the industry. By 1983, the Act required the installation of 'best available technology'—i.e. irrespective of the consequences for the firm, it had to install sufficient treatment capacity to bring it into line with the best firms (in terms of effluent treatment) in the industry. The ultimate aim of the Act was that, by 1985, there would be zero discharges. Table 9.4 shows that the cost consequences could be considerable if such a comprehensive policy of effluent reduction were to take place.

Table 9.4. Cost (imposed on four industries as a percentage of selling costs) of the United States 1972 Federal Water Pollution Control Amendment Act

	1977 standard	1983 standard	1985 standard
Oil refining	2.4	6.3	8.0
Paper and pulp	1.8	2.5	3.0
Steel	1.8	2.7	na
Aluminium	na	1.0	na

na = not available.

Given the level of costs, both in absolute terms and in their impact on selling prices, it is important for policy instruments to be concerned with the cost to society as a whole, and with the distribution of these costs, in achieving any specified quality.

A REVIEW OF POLICY INSTRUMENTS: LEGAL AND ADMINISTRATIVE OPTIONS

Given that a specific water quality has been chosen, a number of policies have been proposed to ensure achievement. This section briefly reviews those policies proposed by lawyers and administrators, in order to highlight later the approach adopted by economists.

Where an individual B (the polluted) has no power to prevent discharges from A, the latter will discharge more than where account of B's preferences has to be taken. In practice, the interests of B are often protected by his being allocated rights, enforceable through law. The method of allocating varies between societies, and here only the briefest description may be given.

Common Law

The common law is the product of disputes resolved in the courts. It has evolved through historical precedent, and the custom of certain parties being allocated rights and enforcing them through the courts. In both the United Kingdom and the United States, the most frequent forms of common law action are for nuisance, where some physical/financial damage has to be shown, and for trespass, which only requires evidence of entry onto private land. The common law is, however, unsatisfactory in dealing with pollution in a modern industrial state, for a number of reasons. First, litigants normally require substantial resources to go to court. Secondly, the law can only function once pollution has occurred, yet retrospective compensation may be less desirable than prevention. Finally, only certain groups are granted common law rights. In England and Wales, only a riparian landlord can bring a suit against a polluter, under common law, whilst walkers on a river bank have no such privilege, unless the Attorney General institutes proceedings on their behalf. The best recent study of the role of common law rights in water pollution is presented in Ogus and Richardson (1977).

Citizen Suits

Recognizing that nuisance and trespass were rather narrow grounds on which to bring cases against large polluters, lawyers (especially in the United States) began to search for alternative legal instruments. One favoured policy is to assert that US citizens, under their constitution, have a right to a clean environment, and hence a right to bring citizen suits against any infringer of those rights. Rights have also been incorporated into many State Environmental Acts in order to encourage more vigorous law enforcement: for example, the Illinois Environmental Protection Act 1970, Section 1030(b). Klein (1973) in reviewing this Act states, that:

> 'Under this section, any citizen may file, with the Board, a complaint against any person allegedly violating the Act, or any rule or any regulation adopted thereunder, and unless the Board determines that the complaint is either frivolous or duplicative, a hearing on the charge must be held. The clear value of such a system is that it serves as a safeguard against governmental inaction attributable either to budget limitations or the political considerations of an elected prosecutor'.

Nevertheless, citizen suits still encounter the difficulty of large numbers of people having to club together to bring retrospective action.

Prohibition

In the United States, the Federal Water Pollution Control Act Amendment 1972 has, as its objective, the total elimination of water pollution by 1985. It has already been shown that economic theory predicts that a zero level of discharges, for all firms, is likely to impose high costs upon dischargers, in comparison to the benefits obtained. This seems now to have been accepted by US administrators, and the implementation of the programme has certainly 'slipped' a number of years.

Regulation

Assuming that the objective of a control policy is to achieve a low but positive level of pollution, most countries have used physical controls over effluent discharged from point sources. Different countries have pursued these aims in different ways. As noted above, since 1972 the United States has directed its policy towards requiring specific cutbacks from various industries. Similar logic appears to lie behind the policies employed by the EEC in its proposed Directives against various industries. In England and Wales, the discharge of polluting material is prohibited, except with the specific permission of the Regional Water Authority. This permission is normally granted subject to restrictions imposed on the quality and quantity of effluent. These restrictions are embodied in a document called a consent, the specific terms of which are generally negotiated between Authority and discharger, so that controls vary both between industries and between rivers. The consents procedure is also described and discussed in Chapters 7 and 8.

All these legal remedies are criticized by economists, since there is no provision made for pollution control to be cost effective, i.e. under such systems, the chosen water quality could be achieved at lower cost or, for a given sum of money, a higher quality of water could be obtained. Only by taking into account that dischargers are likely to have different costs of abatement will any policy be cost effective.

THE CONCEPT OF COST EFFECTIVENESS IN WATER QUALITY

A system of control is defined to be cost effective if it achieves a given water quality objective at minimum resource cost to the *whole community*. Resource costs may be approximately thought of as market prices, less any Government taxes and subsidies.

As a simple numerical example of the economist's logic, let us assume there are two firms, A and B, discharging a single pollutant into a body of water, such that the effect on water quality of a unit of A's effluent is equal to that of a unit of B's. In this example, it is assumed that the desired water quality can be achieved by the total load discharged being cut by half, and that the costs to both firms of reducing discharges are as shown in Table 9.5.

Assume, at present, that 12 units are being discharged to the river and that the objective of policy is to reduce this to 6. Examination of Table 9.5 shows that the total costs to firm A of removing the units of polluting effluent are much higher than for firm B. The incremental cost column shows the costs of removing one extra unit, and may be thought of as an approximation to the marginal abatement cost of the firm.

Two policies might be introduced to ensure that firms A and B reduce their combined discharges to 6 units. The first is a type of administrative/legal policy, whereby both dischargers have to reduce by the same amount, i.e. by 3

Table 9.5. Costs of abatement to firms A and B

	Firm A		Firm B	
Units of polluting effluent discharged	Total cost units	Incremental cost units	Total cost units	Incremental cost units
6	0		0	
5	4	4	1	1
4	14	10	4	3
3	31	17	10	6
2	59	28	20	10
1	104	45	42	22
0	183	79	92	50

units. The total cost of this would be $31+10=41$ cost units. This is *not*, however, a least cost solution, since if B were to cut back by 4 units and A by only 2 units, the same total discharge could have been achieved at a total cost of $14+20=34$ cost units.

This is obviously a contrived example, but the principle of cost effectiveness should be apparent; that when decreases in load are required, those for whom it is cheapest carry out the abatement. Cost effectiveness is not concerned with who bears the costs, merely that they are, in aggregate, to be minimized. The numerical example also demonstrates the crucial condition for cost effectiveness in this context, namely, that the incremental (marginal) costs of both firms are identical when the required total cutback is achieved. In the example, the incremental cost to both firms is 10. A moment's reflection suggests the explanation: recall that with a cost effective policy the firm which finds it cheapest to undertake abatement will be required to install equipment. From the numerical example, if a step-by-step procedure were adopted, firm B would remove the first and second units, the third would be removed by A, the fourth by B and units 5 and 6 by each firm. At this point, there can be no redistribution of abatement costs which will achieve a total reduction of 6 units at a lower cost. For a more mathematical exposition of these conditions see Elliott and Storey (1979).

Given that cost effectiveness is an important (but not the only) criterion for comparing policies (for other criteria, see Storey and Walker (1978)), economists generally regard policies which take no account of the difference in firms' costs as unsatisfactory. In the United States, the 1972 Act has been fiercely criticized, and economists have, almost unanimously, proposed the introduction of effluent charges. Hence, instead of industry being directed to install abatement equipment to meet specified effluent standards, it has a tax imposed upon each unit of effluent which it discharges. The individual firm can then determine the effluent which it can afford to discharge. If the total polluting load discharged by all firms to a river exceeds that compatible with a specified water quality standard, the charge/tax will be raised.

Figure 9.2 plots the data from Table 9.5, and assumes the total and marginal cost curves are continuous. Again, assume the objective of policy is to reduce the aggregate discharges of both firms by 6 units and that the authority imposes a tax of 10 cost units per unit of polluting effluent discharged. If firm A is a cost minimizer, it will reduce its discharges from 6 units to 4 units since this minimizes its total (taxes plus abatement) costs. Again, this is easy to demonstrate. Table 9.6 shows that the least cost position to firm A is the removal, through abatement, of 2 units.

Close scrutiny of Table 9.6, however, shows that total costs are 54 units at a discharge of both 4 and 5 units, but we have argued that the firm discharges only 4 units. This difference occurs because of the use, for ease of exposition, of an incremental rather than continuous marginal cost function. To show that

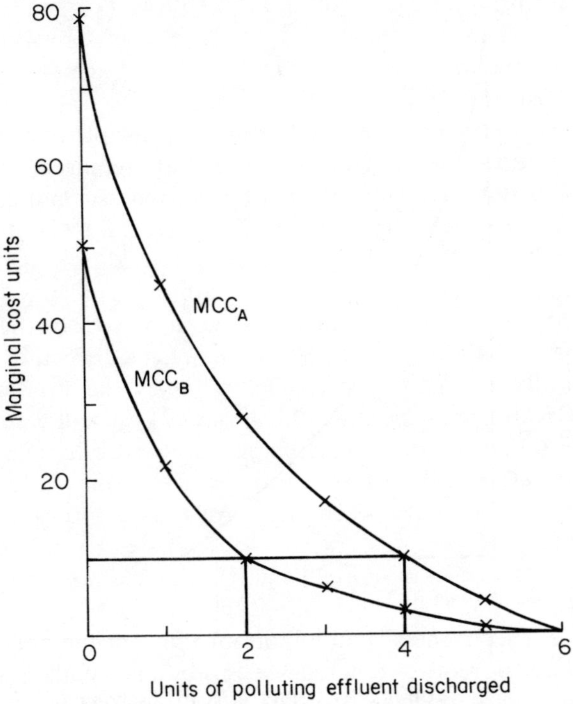

Figure 9.2. Marginal costs of control for firms A and B

the cost minimizing firm will, in fact, discharge only 4 units, we can examine, in more detail, Figure 9.3. This shows the (continuous) marginal abatement curve for firm A. Assume the firm discharges 4.5 units. Here it will pay in total the extra amount shown by the shaded triangle ABC. If, however, it reduces

Table 9.6. Costs to firm A facing an effluent tax of 10 units

Units of polluting effluent removed	Units of polluting effluent discharged	Tax units paid	Abatement cost units	Total cost units
0	6	60	0	60
1	5	50	4	54
2	4	40	14	54
3	3	30	31	61
4	2	20	59	79
5	1	10	104	114
6	0	0	183	183

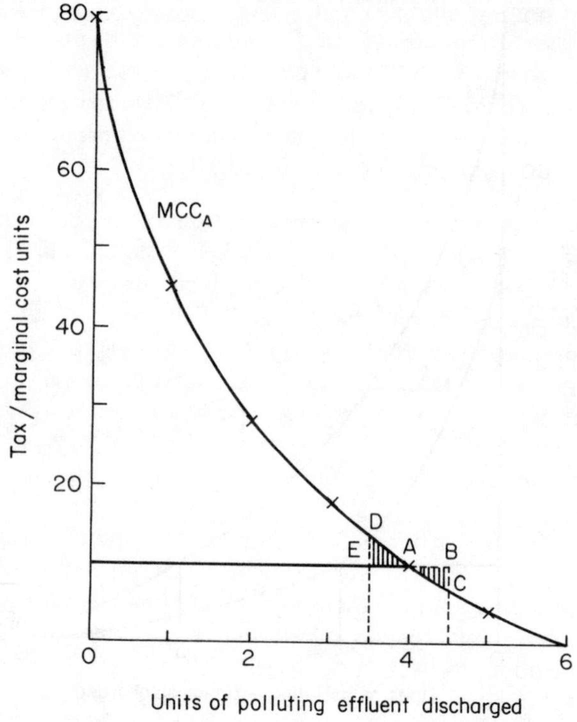

Figure 9.3. Firm A, subject to an effluent tax

discharges to less than 4 units, it pays the extra costs shown by the shaded area ADE.

From Table 9.6 and Figure 9.3, it should be clear that total costs to the firm are minimized where it equates its marginal (incremental) abatement cost to the tax rate. Since the tax rate is the same for both firms, their marginal abatement costs will be equal. This was shown above to be the condition guaranteeing cost effectiveness.

It is this property which makes effluent taxes attractive to economists, since, if firms are cost minimizers, and if the correct tax rate is levied, this will yield any chosen water quality in a cost effective manner.

Any policy with these advantages has to be seriously considered, but, in addition, charges are argued to offer a number of other advantages over administered solutions.

FURTHER SUPPORT FOR EFFLUENT CHARGES

Baumol and Oates (1975) make a persuasive case for charges on the basis that less information is required by the enforcing authority to ensure a cost effective solution than with direct controls.

'... the information needed for iterative adjustments to the tax rate would be easy to obtain; if the initial taxes did not reduce the pollution of the river sufficiently to satisfy the pre-set acceptability standard, one would simply raise the tax rate. Experience would soon permit authorities to estimate the tax levels appropriate for the achievement of a target reduction in pollution.'

Their argument is based on the presumption of market economists that allocation of most goods and services through a decentralized price system requires less information than the use of physical controls.

Most advocates of effluent charges also believe that one of its major advantages is that it reduces the opportunity for bureaucratic intervention, the objections to which are manifold. Some economists argue that bureaucrats do not concern themselves with maximizing societal welfare. Rather, the objective of individual bureaucrats is to maximize the size of their bureau, since it is this which ultimately determines their salary; the most comprehensive exposition of this hypothesis is found in Niskanen (1971). The introduction of charges will reduce the opportunity of bureaucrats to exercise 'discretion' in negotiations over the quality and composition of effluents, since the charge will be uniform for all dischargers for each pollutant. Majone (1975) has even argued that it is only this self-interest of the bureaucrat that has prevented effluent charges being widely introduced.

The effect of bureaucratic involvement is argued to introduce substantial delays into the implementation of policy. Kneese and Schultze (1975) state:

'... Studies of how the regulatory process has worked in general ... reveal that it is ... cumbersome, corruptible, arbitrary and capricious in its impact.

The regulatory approach suffers from an inescapable dilemma. If the system is simple enough to be handled by a central bureaucracy ... it is bound to be inefficient. But if it seeks to accommodate the tremendous diversity of the economy, and tries to devise standards that minimise costs, the regulatory task becomes insurmountable. The resulting—and unavoidable—arbitrariness and inequity will, in turn, make the enforcement of regulatory limits in the courts, difficult, time consuming and ineffective.'

There is little doubt that significant delays in implementing environmental policy in the United States have occurred. Freeman (1971) reports:

'In 1956 a rendering plant in Maryland was permanently enjoined by court order from releasing 'noxious and offensive gases and odours'. But not until 1970, when the US Supreme Court denied the company's last appeal, was the plant finally closed.'

Advocates of the effluent charge argue that the present discretionary system operated by the bureaucracy only encourages firms to delay, by appeals and prevarication, the introduction of abatement facilities. In essence, firms find it cheaper to employ lawyers to oppose the bureaucracy than to employ scientists and technologists to devise new methods of reducing discharges. If, however, the bureau is unable to discriminate, or if courts of law cannot allow more favourable conditions for individual dischargers, then it is not worthwhile for the firm to oppose the bureau. Since such powers would not exist with effluent charges, which would apply uniformly, there would be, within firms, a shift in expenditure from lawyers to technologists. The reduced expenditure by firms on lawyers would also be complemented by a reduced demand for legal advice from within the bureau. It is the opinion of the author that a major factor determining expenditure by a firm is the expected profitability, as perceived by the industrialist, of entering into a conflict with the enforcing authority. These views are expressed at length in Storey (1978).

Increased expenditure by firms on abatement technology would be supplemented by another advantage claimed for effluent charges—that they stimulate innovation. Beckerman (1975) states:

> 'Firms will have a continuing incentive to experiment and seek new and more economic methods of reducing pollution, for the more they do so the more they save on pollution charges.... By contrast if they are given directions to reduce pollution by a certain amount ... they have little further incentive to reduce pollution even more than the regulations require.... Technical progress in reducing pollution will therefore be much faster with pollution charges than with direct regulations. Consequently the amount of pollution will tend to be lower.'

Indeed many economists would now agree with Orr (1976), who says:

> 'It seems to me that the greatest advantage of effluent charges, relative to alternative control mechanisms, is their provision of a decentralised incentive for technical change.'

Finally, effluent charges are claimed to yield a benefit to the authority which can use the revenue raised either for further pollution abatement or for reducing taxes elsewhere in the economy. This latter alternative is particularly attractive, since most taxes are raised on 'goods', whereas an effluent charge would be raised on 'bads'.

SOME CRITICISMS OF AN EFFLUENT CHARGE

An appropriate policy for the control of water pollution in one country may be wholly inappropriate elsewhere. In the United States, where it has been traditional for industry and government to adopt adversary postures, effluent charges may be a low cost and politically feasible instrument for improving the quality of waterways. In Britain, such posturing is less apparent. Indeed, some commentators have referred to those organs of government charged with protecting the environment as 'toothless watchdogs' (Bugler, 1971). Whilst such descriptions indicate a misconception of the role of such organizations, they demonstrate a difference in approach between the two countries, since, in the UK, industry and authority discuss major items of abatement with the objective of finding a mutually satisfactory solution.

Given these differences, this section will examine the relevance of effluent charges as a policy tool for improving the quality of British waters. The main justification for a charging policy is its cost effectiveness, subject to the proviso that firms are cost minimizers. If firms operate in imperfect markets, however, it may be difficult to predict their reactions to given tax rates. Some firms may, for example, have difficulty in raising sufficient capital to install abatement equipment. Hence, the payment of a tax from revenue may be attractive, and thus the firm may discharge more effluent, and pay more tax than if it reacted in a cost minimizing manner.

The major criticism of the effluent tax scheme, especially of the iterative variety, is that the information required to set a tax which will exactly achieve the desired water quality is considerable, yet the consequences of not collecting that information are potentially disastrous. It will be recalled that, in the numerical example in the section on cost effectiveness in water quality, the authority chose a tax rate of 10 cost units. It will also be recalled that Baumol and Oates argued that it made little difference whether the 'correct' rate was chosen initially, since the rate could always be adjusted to achieve the desired water quality.

To demonstrate the naivety of this argument, assume an authority selects initially a tax rate of 5 cost units, and that firm A reacts to that tax rate, in what it perceives to be a cost minimizing manner, by installing a by-product recovery procedure. After a given period, the authority will find that the chosen water quality has not been achieved by that level of tax. Assume now the tax is raised to 10 units, but at this rate firm A may find that it would have been better, originally, to have installed a totally new treatment technology but, since the by-product recovery process is already installed, this precludes, at least in the short-run, the new treatment technology. Hence, it is vital for the authority not to present dischargers with a misleading tax rate, since this can result in costly errors in investment decisions and a delay in the achievement of a given water quality. This argument is explained in detail in Walker and Storey (1977).

The effluent tax becomes more complicated once spatial considerations are included, since the impact of a unit of effluent upon water quality will differ according to the location of the discharge. These matters are normally handled by a water quality model (see Chapter 10), from which it is possible to derive a matrix of coefficients which describe the impact of a unit discharge at point i on water quality in stretch j of the river. Economists have recognized the importance of these spatial considerations, and Tietenberg (1974) has shown it is theoretically possible to weight a uniform effluent tax by the coefficients derived from such a model. The problem which arises is that, unless detailed information is collected on a firm's costs of abatement, it is not apparent which stretch of water is to be used as a base for setting the tax. It is inconceivable for the authority to experiment in selecting the appropriate stretch.

Discussion has thus far been in terms of a unit of pollution, but, in practice, there are likely to be a number of pollutants discharged to a stretch of water. In principle, there is no problem in taxing each pollutant, but difficulties do arise where the authority has very little information about firms' costs, and where procedures may be installed which remove more than a single pollutant. This is called the joint cost problem, and means the authority has to select, not only the 'correct' tax rate for each pollutant, but also ensure these are the 'correct' relative rates. Otherwise, in achieving the desired level in one parameter, this may lead to substantial overachievement in others. The likelihood of this being achieved by an iterative procedure as outlined by Baumol and Oates (1975) would seem to be remote.

THE ROLE OF INFORMATION IN WATER POLLUTION IN THE UNITED KINGDOM

To ensure that a pricing system achieves a specified water quality standard at minimum resource cost to the community, the following information has to be collected from all dischargers:

(1) Present discharge levels of all pollutants.
(2) Equipment/treatment procedures already installed to achieve present discharges. Data on purchase price, replacement costs, and length of life of equipment are necessary together with operating costs.
(3) Expected changes in output of primary product together with data on expected untreated waste load.
(4) Data on treatment technologies that would have to be installed to meet specified reductions in effluent load.
(5) For each treatment technology, data would have to be provided on installation and operating costs of equipment together with expected length of life.
(6) A water quality model would be required to determine the impact of reductions in effluent load on water quality.

If Water Authorities in the UK collected such information, if such information was correct, and if firms reacted as if they were cost minimizers, then the Authority could calculate a level of tax which, if imposed, would achieve any specified water quality at minimum resource cost to the community. The number of provisos in the above statement makes it clear that, in practice, the introduction of an effluent tax is not as simple as Baumol and Oates suggest. Indeed, the whole procedure begs the question of why, if an effluent tax requires the collection of large quantities of information, cannot that information be used to set consents in an efficient manner. In other words, given that information has to be collected, would it not be preferable to use that information to set consents rather than levy taxes? There are a number of advantages in such a procedure. First, imposing physical controls upon dischargers would reduce uncertainty on firms' reaction to tax rates. Secondly, it would enable the authority to set consents, a process with which it is familiar, rather than levy taxes which raise a number of political issues such as the ultimate purpose for which the revenue raised would be used. Thirdly, the consent allocation procedure, whereby conditions are varied according to the nature and character of the discharge and of the receiving waters, is the traditional way in which the British have dealt with pollution problems. Both the regulators and the dischargers find the system 'reasonable'; although charges have been imposed for discharges to the domestic sewer system, even these are reinforced with physical controls.

AN ALTERNATIVE POLICY

It would be incorrect to assume that there is a complete consensus that the *present* British consent procedures are the best available. A number of significant criticisms have been made. First, since little information is collected on firms' costs of abatement, there can be no guarantee that any given water quality is achieved cost effectively. Studies which have been undertaken in the United States indicate there are considerable opportunities for saving by the introduction of cost effective policies (Dorcey, 1973; Herzog, 1976). For example, the cost of achieving given standards of quality can be halved by ensuring that those dischargers who find it relatively cheap carry out most of the abatement.

A second criticism of present British policy is that there is insufficient opportunity for public involvement and debate (Gunningham, 1974). Interested parties were not, until recently, allowed access to information on the actual quantities of effluent discharged—i.e. the monitoring of effluent. Hence, it was not possible to evaluate the performance of the Water Authority in enforcing consents.

It is to overcome the dual problem of inefficiency in the consent allocation procedure, together with a need for greater political involvement by interested parties, that the case for an informed consents policy is now briefly reiterated (for a comprehensive exposition of informed consents see Storey (1979)). The purpose is to formulate a least cost set of consents through a cost abatement model constructed from the type of information necessary to introduce an effluent taxation policy. These consents would be the basis of discussions between the authority, the dischargers, and other interested parties. Debate would take place on the ultimate quality of water required and the expected date of achievement. Dischargers would have the opportunity to highlight the implications of consent conditions on the viability of the firm, the reductions in profit margins, the possible redundancies or any other factor which it felt that the Water Authority should consider. Having collected both cost information, and obtained an impression of the social consequences of cutback, the Water Authority is in a position to build in such equity considerations as it considers appropriate. For example, it may feel the least cost method of achieving the specified water quality imposes excessive burdens upon a single firm, and that it is appropriate for that burden to be distributed more equitably.

When making such 'equity' judgements, the Water Authority has to bear in mind two crucial points. First, that any movement away from the least cost solution, by definition, adds to the total cost to society of achieving the specified water quality. Hence these decisions must be taken overtly. Secondly, if one discharger is allowed a less stringent consent than under the least cost solution, decision-takers in the Water Authority must realize that a compensatory increase in the severity of the consent of another discharger is necessary. The allocation of consents is therefore a zero-sum game.

There may be a tendency for decision-takers to try to escape making unpopular decisions, so it is important that all consents, so far as possible, are negotiated simultaneously to ensure that trade-offs are made explicitly. For example, if consents were allocated on an individual basis, some dischargers might be given more favourable treatment, whilst others might be allocated the least cost solution consent. The effect would be, of course, that the water quality objective would not be met. Hence, environmental interest groups have to be involved in the consent allocation procedure.

A final word is perhaps relevant on the procedures for enforcement. In 1970, over half of all consented industrial discharges in the UK were classified as unsatisfactory, but Water Authorities have always been reluctant to prosecute dischargers who flagrantly breached their consent limits. There have been a number of explanations offered, but most involve a recognition that prosecution is likely to endanger the good relations which have been established between authorities and polluters over a number of years. It is argued that prosecutions are likely to be less effective than what has been called 'arm-twisting', defined as the non-court pressures which the Water Authority can exert upon the discharger. Hence, it is not necessary to prosecute the discharger in order to encourage him to install pollution control equipment. If, however, the Water Authority had to prosecute for each consent breach, this would reduce its opportunity for discretion. Thus, at present, the Water Authority exercises discretion over both the allocation of the consent conditions and the decision to enforce any breaches of those consent conditions which it detects. It therefore combines, uniquely, the functions of a legislature and of an enforcement agency. This means that any manifestations of inefficiency in its enforcement activities may be attributed to its exercise of discretion.

The only discretion which the Water Authority would exercise under a system of informed consents would be the setting of the consent. Once these conditions have been established, the Authority would be expected to enforce the consent rigorously. Whilst this may initially increase the number of prosecutions undertaken, ultimately the number would probably not exceed present totals, since all dischargers would recognize that consent conditions cannot be lightly breached.

CONCLUSIONS

Throughout this work, differences between the policies pursued in the United States and in the United Kingdom have been articulated. The alternative to effluent charges in the United States is the manifestly utopian 1972 Federal Water Pollution Control Amendment Act, and it seems that a switch to the use of charges might be beneficial. In the UK, the conditions imposed upon major discharges are varied according to local conditions. Hence, it would seem a logical step, in keeping with British traditions, to require Water Authorities to discriminate between firms, not simply on physical grounds, but also on costs of meeting specified standards. If that discrimination could be carried out in a politically acceptable manner, it would cement the traditional close relationship between industry and Water Authorities, whilst ensuring that any chosen water quality is achieved at as low a cost to the community as possible.

ACKNOWLEDGEMENTS

In preparing this chapter, I would like to acknowledge the assistance of the editor of this volume. Thanks also, to my former colleague, David Elliott, Lecturer in Environmental Control Engineering, University of Newcastle Upon Tyne, for commenting valuably on one of the many earlier drafts.

CHAPTER 10

Water Quality Modelling

A. James

INTRODUCTION

Mathematical modelling has been used extensively in the management of water resources for hydrological purposes. It is now being used for management of water quality, and has proved to be a useful tool for forecasting the effect of different management policies, determining the least cost solution or testing the sensitivity of a situation to different strategies.

Models of water quality can be at all levels of complexity. An extremely simple model of change in water quality downstream from a discharge would be to assume an exponential decay process, and may be formulated immediately, i.e.

$$C_x = C_0 e^{-kt}$$

where C_x is the concentration at any point x, C_0 is the concentration at the point of discharge, k is the decay rate, and t is the time taken for flow from the discharge point to point x. More complicated models need to be developed in a logical manner, as shown in Figure 10.1.

An example of this process is illustrated in Figure 10.2, which shows a proposed discharge of coke-oven waste some distance upstream of a water supply abstraction. The questions posed are whether the proposed discharge should be permitted and, if so, what is the critical parameter that will limit the discharge? The steps in producing the model may be summarized as follows:

Formulation of problem	Phenol is the critical parameter. At levels above 0.001 mg l^{-1} it causes an unpleasant taste. Therefore, $C_t \not> 0.001$.
Review of theory	Phenol is miscible with water. Concentration decreases due to initial mixing and due to subsequent decay. The decay rate may be represented as proportional to the remaining concentration. Longitudinal dispersion may be ignored. Travel time is proportional to flow rate.

Formulation of equations

(a) Initial dilution

$$C_{\mathrm{m}}=\frac{Q_{\mathrm{EF}}\times C_{\mathrm{EF}}}{Q_{\mathrm{EF}}+Q_{\mathrm{FW}}}$$

(b) Decay

$$C_t=C_m\mathrm{e}^{-kt}$$

(c) Travel time

$$t=f/Q_{\mathrm{EF}}+Q_{\mathrm{FW}}$$

which relates travel time to some function (f) of combined flows.

Information flow

Set Q_{EF}, C_{EF}, k, and f.
Calculate C_{m}.
Calculate t.
Calculate C_t.

Method of solution

Analytical.

Preparation of programme

```
   10 INPUT K, F
   20 Q1=0.005
*  30 FOR Q2=5 TO 1 STEP -1
†  40 FOR C1=1 TO 100
   50 C2=(C1 * Q1)/(Q1+Q2)
   60 T=F/(Q1+Q2)
   70 C3=C2 * EXP ↑ (-K * T)
   80 PRINT C2, C3
   90 IF C3>0.001 THEN 120
  100 NEXT C1
  110 NEXT Q2
  120 END
```

Testing of model

Comparison of observed and predicted values of C_{m} and C_t.

* Line 30 explores a range of fresh water flows in decreasing order.
† Line 40 explores a range of effluent phenol concentrations in increasing order.

The programming of the model allows for steadily increasing values of phenol in the effluent to be tested at a range of flows, with the model automatically stopping at a level in excess of 0.001 mg l^{-1}. It is arranged so that k and f can be easily varied in the model. This would be done to test the sensitivity of the output to variations in their values. For example, if a small change in k caused a significant change in C_t, it would be extremely important to have a correct value for k.

The above example can also be used to illustrate some of the terminology connected with modelling. The above model is:

(1) *Steady state* because input values and coefficients do not change with time. (The alternative is a dynamic model.)
(2) *Deterministic* because there is a fixed relation between the inputs and

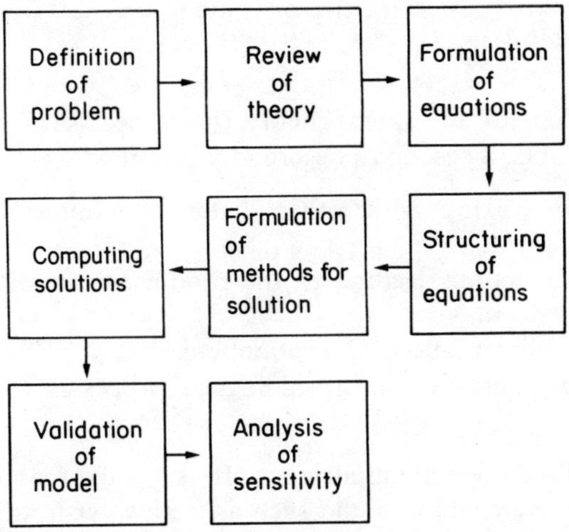

Figure 10.1. Steps in creating a mathematical model

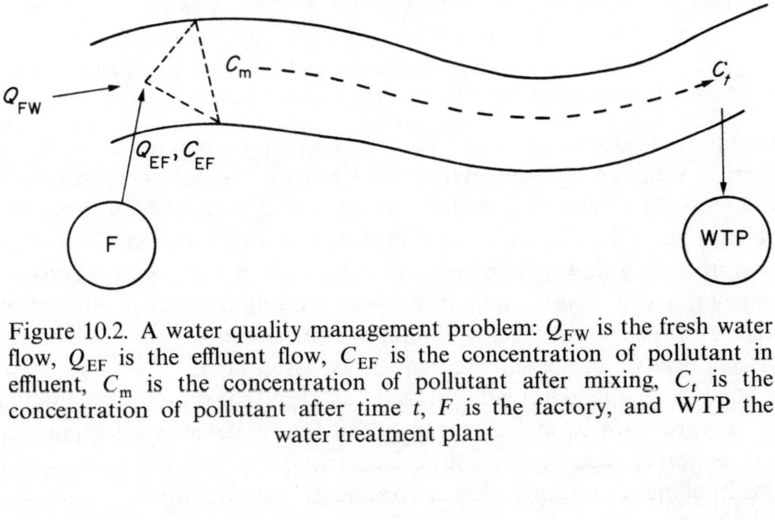

Figure 10.2. A water quality management problem: Q_{FW} is the fresh water flow, Q_{EF} is the effluent flow, C_{EF} is the concentration of pollutant in effluent, C_m is the concentration of pollutant after mixing, C_t is the concentration of pollutant after time t, F is the factory, and WTP the water treatment plant

outputs. (The alternative is a stochastic model where random elements are introduced into inputs and relationships.)

(3) *Internally descriptive* because the processes causing the change in concentration are explicitly described. (The alternative is a black-box model which uses statistical analysis to connect output with input.)

The following notes describe the basic ideas used in water quality modelling and give examples of models of natural waters and treatment processes.

BASIC THEORY

Simulation

In simulating changes in water quality, it is necessary to represent the processes responsible. These may be broadly classified as:

(1) *Hydraulic*, i.e. mixing with other water containing higher or lower concentrations.
(2) *Chemical*, i.e. reactions leading to the production or disappearance of a substance.
(3) *Physical*, i.e. sedimentation, adsorption, etc.
(4) *Biological*, i.e. micro- or macro-biological processes causing uptake or release.

Mixing is of fundamental importance. It is brought about by advective processes (bulk movements of liquid such as freshwater flow, tidal oscillation, etc.) or non-advective processes (transfer due to molecular or eddy diffusion). The degree of mixing varies, but may often be represented by the two extremes, i.e.

(a) *plug flow reactors* — where there is no internal mixing, and any quality changes are due to physical, chemical or biological processes, and
(b) *completely mixed reactors* — where the contents of the reactor are completely and instantaneously mixed.

Intermediate situations between these two types of reactors may be represented by a number of completely mixed reactors in series. Occasionally, in environments like estuaries, where mixing patterns are both complex and important, it may be necessary to represent this explicitly. Hydraulically, the situation in a plug flow reactor is straightforward. Changes in a conservative substance influence quality and cause corresponding changes in effluent quality after the appropriate time delay. Changes in non-conservative substances are due to chemical or biological change as discussed later.

Quality changes in completely mixed reactors may be studied analytically where the variations in input concentration follow some regular pattern. For example, a step change in the input concentration will produce an exponential increase in effluent concentration according to the equation:

$$C_E = C_I + (C_0 - C_I)\,e^{-Qt/v}$$

where C_E is the concentration in effluent (at time t), C_1 is the concentration in influent, C_0 is the concentration in effluent (at start), Q is the flow rate in and out, and V is the volume. But the analytical approach rapidly becomes cumbersome, if not impossible, where two or more factors, like flow rate or influent concentration, are varying with time.

A more general approach is to draw up a mass balance on the substance concerned, and to solve the resulting equations by numerical methods. A completely mixed reactor may then be represented as shown in Figure 10.3.

The mass balance for the liquid in the reactor is given by:

(Rate of change of volume) = (Flow in) − (Flow out)

i.e.

$$\frac{dV}{dt} = Q_1 - Q_2$$

and the mass balance for a conservative substance in solution is given by:

(Rate of change of concentration in reactor) = (Flux in) − (Flux out)

i.e.

$$\frac{dC_2}{dt} = \frac{Q_1 C_1}{V} - \frac{Q_2 C_2}{V}.$$

Chemical reactions may raise or lower the concentration. Some substances are so inert in solution that their concentration does not sensibly change due to chemical reaction (so-called conservative substances). The concentration of other substances may increase or decrease due to reactions, and the kinetics can be classified as shown in Table 10.1.

Third- or higher-order reactions are generally too slow to be of interest. Much more interesting are reactions in series, e.g. consecutive first-order:

$$A \rightarrow B \rightarrow C.$$

This has been used extensively for representing substances being produced and removed as in the case of Streeter–Phelps (Velz, 1974) where the dissolved oxygen deficit is being increased by oxidation of organic matter (first-order)

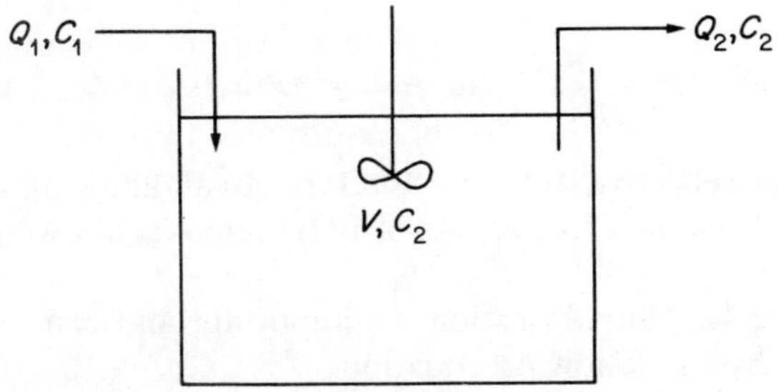

Figure 10.3. Mass balance on a completely mixed reactor: Q_1, Q_2 are inflow and outflow, respectively, C_1, C_2 are concentrations in inflow and outflow, respectively, and V is the volume of the reactor

Table 10.1 Kinetics of chemical reactions

Order of reaction	Type of reaction	Method of representation
Zero	Fixed rate regardless of concentration	$\frac{dA}{dt}=K$
First	Rate proportional to concentration of one reactant	$\frac{dA}{dt}=K(A)$
Second	Rate proportional to concentrations of two reactants	$\frac{dA}{dt}=K(A)K(B)$

and reduced by re-aeration (first-order with respect to deficit), so

$$\frac{dD}{dt}=K_1L-K_2(C_s-C)$$

where D is the dissolved oxygen deficit, L is the BOD, C_s is the saturation concentration, and C is the existing concentration of dissolved oxygen. Also, K_1 and K_2 are rate coefficients for de-oxygenation and re-aeration, respectively.

Physical processes such as sedimentation, adsorption, etc., may also play important roles in reducing the concentration of substances in suspension or solution. Modelling of sedimentation in relatively static waters, like lakes and reservoirs, is straightforward, provided the settlement characteristics of the suspended matter can be defined. Modelling of the behaviour of suspended matter in flowing waters is generally not possible on a theoretical basis, and must depend on arbitrary or experimentally determined coefficients of settlement, resuspension, e.g., the integrated form of the Streeter–Phelps equation becomes:

$$D_t=\frac{K_dL_o}{K_2-K_r}(e^{-K_rt}-e^{-K_2t})+D_0e^{-K_2t}$$

where K_d is the effective rate of BOD removal allowing for decay and deposition, and K_r is the effective rate of BOD removal allowing for decay and resuspension.

Physical processes like adsorption are important and can be approximated mathematically by the following equation:

$$Y=KC^{1/n}$$

where Y is the amount of material adsorbed per unit weight of adsorbent, C is the concentration of solute after adsorption, and K and n are constants.

Adsorption has chiefly been used in models of water and waste treatment units, but the phenomenon is widespread, affecting, for example, the concentration of heavy metals in solution in inshore waters.

Diffusion helps to govern rates of transfer of substances in solution wherever concentration gradients exist. At the microscopic level, transport of solutions into bacterial films is due to molecular diffusion. At the macroscopic level, transfer of salt, BOD, etc., in estuaries is in part brought about by turbulent diffusion. In both cases, the processes may be described by:

$$\frac{\partial c}{\partial t}=D_{\mathrm{m}}\left(\frac{\partial^2 c}{\partial x^2}+\frac{\partial^2 c}{\partial y^2}+\frac{\partial^2 c}{\partial z^2}\right)$$

$$N_x=-D_{\mathrm{m}}\frac{\partial c}{\partial x}$$

where N_x is the rate of mass transport in the x direction, $\partial c/\partial x$ is the concentration gradient in the x direction, and D_{m} is the molecular diffusion coefficient.

Biological processes, especially micro-biological ones, can bring about considerable changes in the concentration of nutrients, organic matter, etc. Also, water quality is often concerned directly with the concentration of micro-organisms, e.g., algal concentrations in lakes. It is therefore useful to understand the kinetics of microbial growth and to be able to represent it mathematically. The distinction between plug flow and completely mixed reactors is again useful, so microbial kinetics of these are discussed separately below.

The general features of microbial growth in a plug flow reactor are shown in Figure 10.4, which relates mainly to bacterial growth, although most of the features, apart from the stationary phase, are also found in algal growth. The major difference between the heterotrophic and autotrophic organisms is that growth of the former is always related to availability of food, but, with the latter, inorganic nutrients may be present in excess, whereupon growth becomes limited by other external factors such as light (due to self-shading)

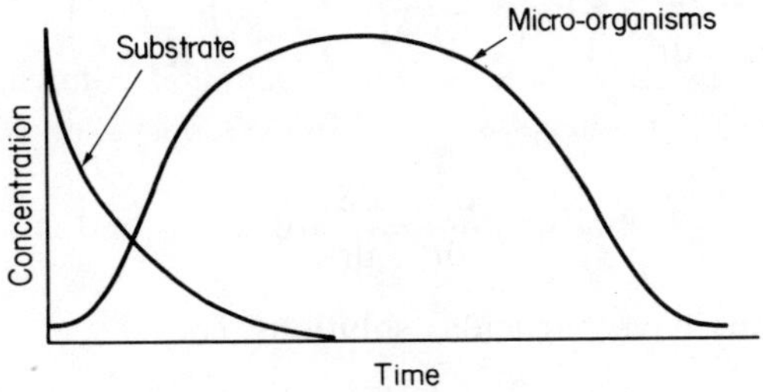

Figure 10.4. Microbial growth curve in a batch culture

and carbon dioxide. The rate of growth may be represented by:

$$\frac{dN}{dt}=RN\frac{(N_m-N)}{N_m}-KN$$

where dN/dt is the growth rate, N is the concentration of organisms, N_m is the maximum number, K is the death rate, and R is the intrinsic growth rate.

The rate of substrate removal may be given as:

$$\frac{dS}{dt}=-KS$$

where K is the reaction coefficient, and S is the substrate concentration. Here K is related in some way to the uptake per organism per unit of time, although this relationship may not be a fixed one.

In completely mixed reactors, the input and output play an important part in determining the concentration of both substrate and micro-organisms. The mass balance for micro-organisms may be written as:

$$\frac{dN}{dt}=\frac{Q_1}{V}N_1-\frac{Q_2}{V}N_2+RN-KN.$$

In order to write the mass balance for the substrate, we need to relate uptake to micro-organisms, which is usually done by using the concept of yields, Y:

$$Y=\frac{\text{weight of micro-organisms produced}}{\text{weight of substrate used}}.$$

Thus, we have

$$\frac{dN}{dt}=-Y\frac{dS}{dt}.$$

Also, the growth rate of micro-organisms is often related to the concentration of substrate by Monod kinetics as shown in Figure 10.5 (Rich, 1973), so the substrate mass balance becomes:

$$\frac{dS}{dt}=\frac{Q_1}{V}S_1-\frac{Q_2}{V}S_1-\frac{1}{Y}R_mN\left(\frac{S}{K_s+S}\right).$$

At steady state

$$\frac{dN}{dt}=\frac{dS}{dt}=0$$

and these equations have particular solutions, i.e.

$$\bar{N}=\frac{Y(S_1-S_2)}{1+\frac{Q}{V}K_d}$$

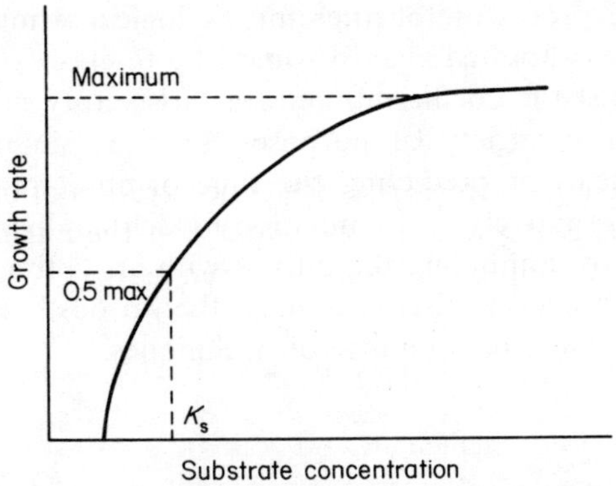

Figure 10.5. Growth rate as a function of substrate concentration

$$\overline{S}=K_s\frac{1+\frac{Q}{V}K_d}{\frac{Q}{V}(R_m-K_d)}-1$$

where S_1 and S_2 are substrate concentrations in the inflow and outflow, Q is the flow rate, V is the reactor volume, R_m is the maximum growth rate, K_d is the death rate, K_s is the Michaelis–Menten coefficient, and $\overline{N}$ and $\overline{S}$ are steady-state concentrations of bacteria and substrate, respectively.

The above functions apply chiefly to bacteria and algae, which are the principal group of organisms involved in water quality. The same basic ideas may be applied to other groups, e.g., zooplankton or fish, provided that appropriate modifications are made. These modifications are needed to take account of the more complex life histories with variation in survival rates and fecundities. For example, the kinetics of fish populations are best represented in matrix form, as follows:

$$\begin{pmatrix} S_1 & S_2 & S_3 \\ 0 & f_1 & 0 \\ 0 & 0 & f_2 \end{pmatrix}\begin{pmatrix} P_1 \\ P_2 \\ P_3 \end{pmatrix}=\begin{pmatrix} P'_1 \\ P'_2 \\ P'_3 \end{pmatrix}$$

matrix of survival and breeding success; population vector at time 0; population vector at time t_1

where S_1, S_2, ..., S_n are probabilities of survival and f_1, f_2, ..., f_n are fecundities.

Water quality models for simulation involve studying the real situation and abstracting the relevant processes—chemical, physical, and biological—which are responsible for the changes, finding the best way of representing them, and then fitting these processes together into a logical whole—the model. This model needs to be calibrated against some data to check on the output, and, if necessary, to adjust the coefficients until a satisfactory fit is obtained. It can then be used for a variety of purposes, such as simulating the effect of management policies or predicting the time of onset of damaging levels of pollution, etc. Alternatively, the simulation model may be turned into an even more useful tool by combining the output with an optimization routine. The use of simulation models is discussed under the various headings below, after a description of the basis of optimization techniques.

Optimization

Optimization is a group of mathematical techniques for selecting the most advantageous combination from a multivariate situation. There are three main groups of techniques: (1) linear mathematical programming, (2) non-linear mathematical programming, and (3) non-linear iterative methods. Although these techniques are too complex to describe in this brief review, an example from river pollution may suffice to illustrate the principles. Figure 10.6 shows a section of a river divided into two reaches, each of which receives a discharge. The effect of the discharges on the water quality of each stretch can be determined using a simulation model. This can be represented as a matrix of steady-state transfer coefficients:

$$\begin{pmatrix} \phi_{11} & \phi_{12} \\ \phi_{21} & \phi_{22} \end{pmatrix}$$

when ϕ_{12} represents, for example, the change in dissolved oxygen in mg l^{-1} in reach 2 for a unit change in load from waste 1 in kg day^{-1} BOD. The cost (C) of a unit change in load (ΔW) at each treatment plant can be given by:

$$\Delta W_1 = C_1$$

$$\Delta W_2 = C_2$$

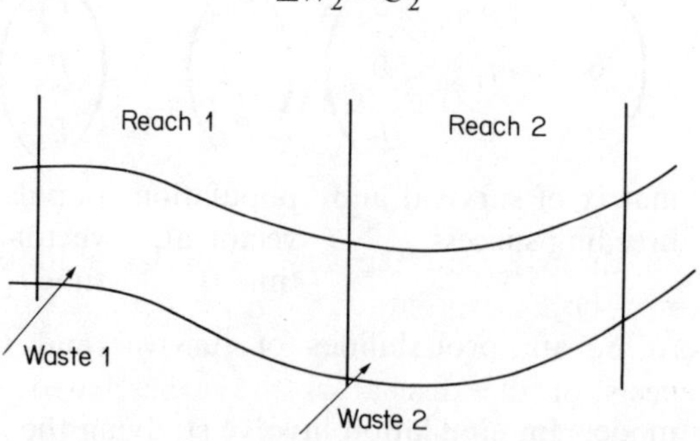

Figure 10.6. Optimization problem

and the increase in dissolved oxygen (DO) desired in each reach can be written as:

$$\Delta(\text{DO in reach 1}) = D_1$$

$$\Delta(\text{DO in reach 2}) = D_2.$$

The objective function expresses the overall aim, which is to minimize the cost of achieving the improvement, i.e.

$$\text{minimize } C = \sum \Delta W_1 C_1 + \Delta W_2 C_2$$

subject to the constraints that

$$\phi_{11}\Delta W_1 + \phi_{21}\Delta W_2 \geqq D_1$$

$$\phi_{12}\Delta W_1 + \phi_{22}\Delta W_2 \geqq D_2.$$

This can be set out in a Simplex tableau as follows:

	R_1	R_2	
W_1	ϕ_{11}	ϕ_{12}	D_1
W_2	ϕ_{21}	ϕ_{22}	D_2
$-C$	C_1	C_2	

Manipulation of this tableau by Simplex rules, or in some cases graphical solution, will give the minimum cost solution.

The different methods of optimization are used for solving linear or non-linear relationships. Alternatively, the choice of method may depend upon achieving a rapid optimization especially where the cost response surface is highly convoluted.

APPLICATIONS

Estuary Models

Many estuaries in the UK are the centres of large industrial complexes. Because of this, and because pollution control in tidal waters has been less stringent, estuaries are generally the most polluted stretches of British rivers. The polluting effects of the discharges are exacerbated by the hydraulic conditions giving long retention times, upstream transport, and poor mixing. The main problem is generally de-oxygenation, although toxic effects are often present. The modelling of water quality in estuaries has concentrated on dissolved oxygen, since standards for estuarine waters are chiefly based upon this criterion. The three main types of model are statistical, segmental, and dispersion.

Statistical Model

The statistical model is based upon regression analysis relating dissolved oxygen concentration at various sections in the estuary to important environmental variables. These are generally temperature, freshwater flow, and tidal amplitude, so that a typical regression equation has the form:

$$DO_z = A \pm B_1X_1 \pm B_2X_2 \pm B_3X_3$$

where DO_z is the concentration of dissolved oxygen at station z, $B_1, B_2, \ldots, B_n$ are regression coefficients, $X_1, X_2, \ldots, X_n$ are levels of independent variables, and A is a constant.

The relationships need to be established for each section in the estuary using multiple regression analysis of DO survey data. It has the advantages of mathematical simplicity, and very minimal data requirements. The main drawback is that the equation is only for interpolation, and can therefore only be used to explore combinations of temperature, river flow, and tidal amplitude within the observed ranges. Extrapolation outside the range, or to other circumstances like reduced discharges, are not valid. Nevertheless, as on the Firth of Clyde, this model has been used successfully for management decisions like the siting of a new power station discharge (Mackay and Fleming, 1969).

Segmental Model

The segmental model regards an estuary as a series of continuously stirred reactors. It is therefore limited in application to those estuaries that are well mixed vertically. Such estuaries may be visualized as shown in Figure 10.7 as a series of tanks whose length is approximately half the tidal excursion.

The fresh water is used as a tracer to predict the behaviour of the effluent. The latter will come to an equilibrium concentration such that in the outfall segment:

$$(\text{Concentration of effluent}) = \left(\frac{S_{\text{sea water}} - S_{\text{segment}}}{S_{\text{sea water}}}\right) \times \left(\frac{\text{Flow}_{\text{effluent}}}{\text{Flow}_{\text{fresh water}}}\right).$$

For a conservative substance, the calculation of concentration in downstream segments is the same as above, but in an upstream direction, the concentration

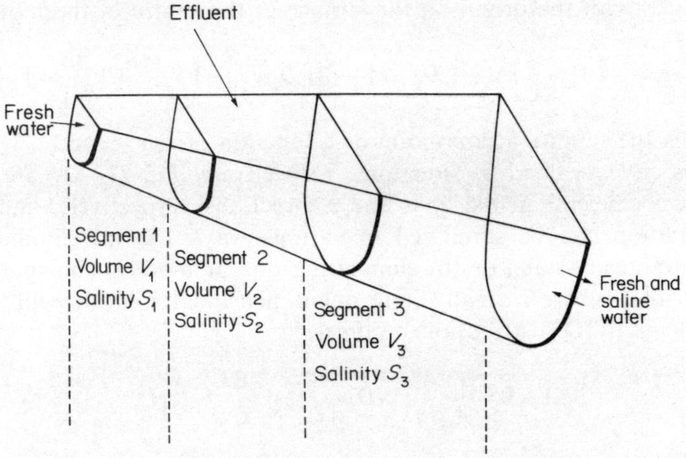

Figure 10.7. Estuarine segmental model

will fall off at a rate similar to the fall in salinity, so that:

$$(\text{Concentration in segment } n) = (\text{Concentration in segment } n+1) \times \left(\frac{S_n}{S_{n+1}}\right).$$

In the case of non-conservative substances, these equations need to be modified to take account of the rate of decomposition or the rate of die-off. For example, the concentration of BOD in the outfall is given by:

$$C_n = \frac{E_n}{2Q} \, e^{-k(R_n + R_{n-1} + \ \cdots \ + R_{n-1})} \, (1 + e^{-kR_n}) \left(\frac{S_0 - S_n}{S_0}\right)$$

Where E_n is the rate of discharge of BOD into outfall segment.

This model was derived by Pearson and Pearson (1965) for the estuary of the Great Ouse, and gives a useful technique for first approximation of estuarine concentration patterns. It has the great advantage of requiring only minimal survey data, but does not give precise concentration profiles, and nor can it be used for stratified estuaries.

Dispersion model

In the dispersion model, mixing in estuaries can be regarded as due to both advective and non-advective processes. It is possible to write the mass balance equation for salt (or other conservative substances) using a system of Cartesian coordinates with the origin on the surface at the centre of the mouth:

$$\frac{\partial s}{\partial t}=\frac{\partial}{\partial x}\left(D_x\frac{\partial s}{\partial x}\right)+\frac{\partial}{\partial y}\left(D_y\frac{\partial s}{\partial y}\right)+\frac{\partial}{\partial z}\left(D_z\frac{\partial s}{\partial z}\right)-\left(V_x\frac{\partial s}{\partial x}+V_y\frac{\partial s}{\partial y}+V_z\frac{\partial s}{\partial z}\right)$$

where s is the salinity at any point as a function of x, y, z, and t, V_x, V_y, V_z are velocities in x, y, and z directions, respectively, D_x, D_y, D_z are turbulent diffusion coefficients in the x, y, and z directions, respectively, and t is time. This equation can be simplified in various ways, e.g. (a) eliminating time, reduces to steady-state, or (b) eliminating one or more of the spatial dimensions, so that where a steady-state one-dimensional solution will suffice, the equation for BOD distribution becomes:

$$0=\frac{1}{A}\frac{\mathrm{d}}{\mathrm{d}x}\left(AD_x\frac{\mathrm{d}L}{\mathrm{d}x}\right)-V_x\frac{\mathrm{d}L}{\mathrm{d}x}-K_1L$$

where A is the cross-sectional area, L is the concentration of organic matter (BOD), and K_1 is the decomposition coefficient. This may be solved numerically by finite-difference techniques given suitable boundary conditions. The details of numerical methods such as finite difference and finite element are beyond the scope of this note. Further details can be obtained from books on numerical methods, e.g. Khabaza (1969).

River Models

There are many aspects of river water quality which may be of interest, depending on whether the river is used for fishing or recreation, or for abstraction of water supply. Models may be used in most cases to simulate the water quality changes and to assist in the complex decision-making process such as location, number and scale of treatment plants, location of intakes and reservoirs, and flow regulation. In large-scale schemes with numerous alternative strategies, both combined optimization and simulation models are used.

Despite the wide variety of water quality criteria that can be employed, by far the commonest type of model is for dissolved oxygen. This is a field where modelling of water quality first began with the classic work of Streeter and Phelps as previously described.

Although this approach to DO modelling has proved extremely popular, there are some serious short-comings, namely (a) it is a steady-state approach with fixed values for the parameters, and (b) only two processes are taken into account, and in polluted rivers these are often not the most important processes affecting DO.

For these reasons, various other approaches have been used. These may be classified into modifications of oxygen sag, stirred tank models, and segmental models. The first group use ideas of deposition and resuspension to modify the BOD removal term, K_1—see earlier section in this chapter.

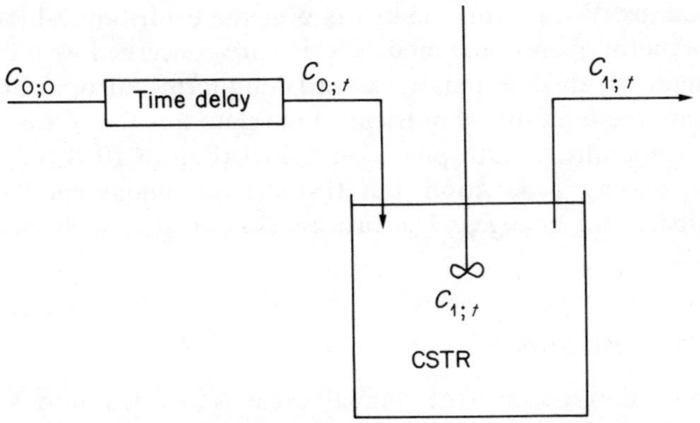

Figure 10.8. Diagrammatic representation of water quality changes in a river reach: $C_{0,0}$ is the quality of influent at time 0, $C_{0,t}$ is the quality of influent after time delay t, $C_{1,t}$ is the quality of effluent after mixing and time delay, and CSTR is a continuously stirred tank reactor

Whilst improving the model, this does not overcome all objections. Stirred tank models and segmental models permit conditions to vary with time and also with distance along the river, so that all the processes contributing to the oxygen regime can be taken into account.

An example of a stirred tank model is illustrated in Figure 10.8. The river is divided into reaches such that each may be regarded as a stirred tank with the appropriate time delay. This type of model caters for longitudinal dispersion, which can be important is some polluted situations. Segmental models can also allow for dispersion together with all other processes affecting DO. The model represents the river flow as discontinuous blocks of water. A mass balance on DO is carried out on each block each hour. The blocks are then advanced downstream with coefficients being adjusted appropriately for varying conditions such as re-aeration and photosynthesis (see Beck, 1978).

River models of all three types have been used for a variety of management purposes. The oxygen sag type of model gives a very useful first approximation for projected discharge situations where data are minimal. Segmental model gives a rather more accurate answer, provided that information is available to calculate coefficients of respiration, photosynthesis, etc. The stirred tank model has been developed for operational purposes where advanced warning of polluted water is required to close downstream intakes.

Lake Models

Although lake models can be concerned with almost any aspect of water quality, such as turbidity changes or bacterial reduction, in practice, most interest has been shown in predicting algal growth and relating this to changes in nutrient status. Within this field, it is possible to distinguish two types of model—short-term operational models which are concerned with detailed day-to-day changes in algal populations, and long-term eutrophication models which simulate the build-up of nutrients and relate this to increases in general algal levels. Generally, a eutrophication model (Figure 10.9) only covers the period of increasing production, but there is no fundamental difficulty in following subsequent changes. The processes associated with eutrophication are as follows.

Mixing and Thermal Stratification

Where stratification occurs, the annual cycle is divided into summer (i.e. stratified) and winter (i.e. non-stratified) periods. Mixing of the influent is assumed to occur with the eplimnion and the effluent is also assumed to come from this layer. Water in the epilimnion is subject to nutrient uptake by algae (which may also store nutrients), whereas nutrient concentrations in the hypolimnion are affected by bacterial release from the dead algae which have sedimented from the epilimnion. The bacterial release of nutrients varies in aerobic and anaerobic conditions, and therefore dissolved oxygen must be modelled in the hypolimnion. The equations describing the processes are as follows:

(1) nutrient balance for whole lake (winter)

$$\frac{\mathrm{d}C_t}{\mathrm{d}t}=\frac{Q_1}{V}C_1-\frac{Q_2}{V}C_t+KC_t$$

(2) nutrient balance for epilimnion (summer)

$$\frac{\mathrm{d}C_\mathrm{E}}{\mathrm{d}t}=\frac{Q_1}{V}C_1-\frac{Q_2}{V}C_\mathrm{E}-K_1\left(\frac{\mathrm{d}U}{\mathrm{d}t}\right)$$

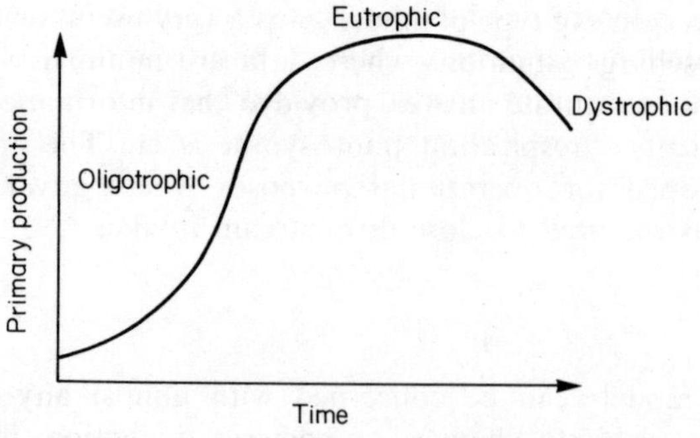

Figure 10.9. Eutrophication in lakes and reservoirs

(3) nutrient balance for hypolimnion

$$\frac{dC_H}{dt} = K_2\left(\frac{dR}{dt}\right) \times RF$$

(4) oxygen balance for hypolimnion

$$\frac{dD}{dt} = K_3(A)$$

where A is the DO in hypolimnion, C_t is the concentration of nutrient in lake, C_E is the concentration of nutrient in epilimnion, dU/dt is the rate of uptake, C_H is the concentration of nutrient in hypolimnion, dR/dt is the rate of release, F is the release coefficient, dD/dt is the rate of change of oxygen deficit, and K, K_1, K_2, K_3 are reaction coefficients.

Between winter and summer, there is no break in the continuity of lake water quality as stratification begins, the lake is simply divided into two layers of equal quality. During the summer, water quality in the layers changes and, at the end of the summer, a significant change occurs when the two layers once more mix. Water quality in the whole lake at the beginning of winter is the resultant of the qualities in the two layers and their respective volumes.

Algal Growth

This is a function of light, temperature, sedimentation, and nutrients. The activities of grazers are generally not included in this type of model, although they would be extremely important in short-term operational models. The calculation of primary production is the most complex aspect of the model, since it is so closely controlled by light intensity which is a function of the angle of incidence, cloudiness, depth, and turbidity, and the length of day which varies seasonally. Some simplifying assumptions are required in a long-term model to avoid using an integration step shorter than one day. Daily light patterns are often represented by sine or cosine functions.

Primary production is a balance between photosynthesis and losses due to respiration and sedimentation; growth of algal biomass is also constrained by nutrient limitation. Changes in algal numbers may be represented by the following equations:

$$\frac{dn}{dt} = GP - R - S - U$$

(1) daily gross production

$$GP = \frac{nP_m\Delta}{k} 1.2\, F(\phi)\left(\frac{c}{K_s + c}\right)$$

(2) respiration

$$R = 24nP_{\mathrm{m}}r$$

(3) sedimentation

$$S = K_i n$$

(4) outflow

$$U = \frac{Q}{V}n$$

where

$$F(\phi) = 1.33 \tan^{-1}(0.5\phi) \log_e[1 + (0.5\phi)^2]$$

and P_{m} is the maximum rate of photosynthesis, n is the algal concentration, Δ is the day length, K_{s} is the Michaelis–Menten coefficient, c is the concentration of limiting nutrient, r is the relative rate of respiration, and K_i is the coefficient of sedimentation dependent on type of algae.

This simple approach is usually modified in practice to take account of other important processes such as:

(1) luxury uptake and storage of nutrients enabling algae to continue growing for some time during periods of nutrient depletion, and
(2) grazing by zooplankton.

This eutrophication model may be used in a number of ways. It can be used to predict the onset of algal concentrations which may cause difficulty in treatment, or problems with amenity and fishing. It may also be used for investigating the possibilities of eutrophication due to storage augmentation by pumping from other river sources.

Models of Treatment Processes

The wide variety of processes used for altering water quality are dealt with in Chapter 7. In waste treatment reduction in parameters such as BOD and suspended solids are sought to make the water acceptable for discharge to sewers, streams or coastal waters. In water treatment, it is necessary to remove pathogens, colour, and turbidity. Modelling of these processes has been actively studied during the last ten years to help in the design of individual treatment units and as a means of optimizing the choice of units and the cost. Developments in modelling have been heavily biased towards waste treatment, and so far very few water treatment processes have adequate simulation models. An example of a simulation and optimization model in waste treatment is described below.

In a simulation model, the activated sludge process is usually represented as a completely mixed reactor followed by a gravity separation stage with sludge recycle as shown in Figure 10.10. Activated sludge is controlled in terms of sludge age, because the most critical phase of the process is the gravity separation. The sludge age affects the settleability of the bacterial flocs in the mixed liquor, thus influencing the BOD and suspended solids levels in the effluent.

Sludge age is given by:

$$\text{Sludge age} = \frac{\text{bacteria in aeration tank}}{\text{rate of removal in effluent and sludge}}$$

i.e.

$$\theta_s = \frac{VX_1}{(Q_1 - Q_W)X_e + Q_W X_R}$$

and the bacterial mean balance by incorporating the reactor kinetics, as outlined:

$$\begin{pmatrix}\text{rate of change of mixed}\\ \text{liquor suspended solids}\end{pmatrix} = \begin{pmatrix}\text{rate of}\\ \text{growth}\end{pmatrix} - \begin{pmatrix}\text{rate of}\\ \text{death}\end{pmatrix} - \begin{pmatrix}\text{rate of loss}\\ \text{in effluent}\end{pmatrix} - \begin{pmatrix}\text{rate of loss in}\\ \text{waste sludge}\end{pmatrix}$$

i.e.

$$V\frac{dx_1}{dt} = \mu_m\left(\frac{C_1}{K_s + C_1}\right)X_1V - K_dX_1V - (Q_1 - Q_W)X_e - Q_WX_R.$$

(See Figure 10.10 for explanation of symbols.) At steady state

$$V\frac{dx_1}{dt} = 0$$

$$\frac{1}{\theta_s} = \mu_m\frac{C_1}{K_s + C_1} - K_d$$

$$C_1 = \frac{K_s(1 + K_d\theta_s)}{\theta_s(\mu_m - K_d) - 1}.$$

These and similar equations may be used to explore the effect of design variables such as the volume of aeration tank and operational variables like sludge recycle ratio.

The capital cost of waste treatment is largely dictated by the choice of treatment unit and the sizing of each unit. A complex optimization model for both has been attempted by CIRIA (Construction Industries Research and Information Association) but has not yet proved successful, due to the lack of understanding of many of the unit processes. For example, the relationship between BOD removal and variables like retention time in sedimentation are generally understood, but the details are not sufficiently known to obtain precise predictions. However, until models like the CIRIA one become fully developed, it is possible to have the designer choose the type of treatment units and then to optimize using some simple relationships. An example of this approach can be found in Knapton (1978).

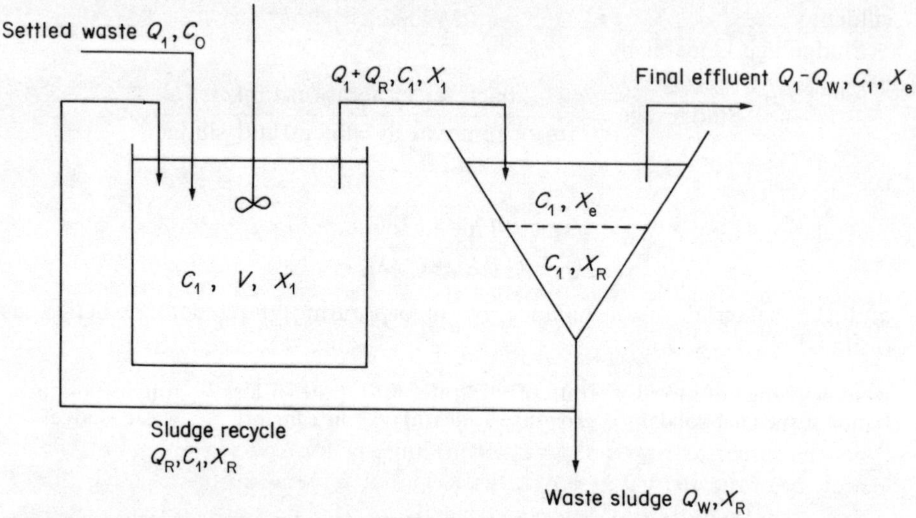

Figure 10.10. Diagrammatic representation of an activated sludge model: Q_1 is the flow rate of settled sewage, Q_W is the flow rate of waste sludge, Q_R is the flow rate of recycled sludge, V is the volume of aeration tank, C_1 is the concentration of substrate in areation tank, C_0 is the concentration of substrate in settled sewage, X_1 is the concentration of bacteria in aeration tank, X_c is the concentration of bacteria in final effluent, and X_R is the concentration of bacteria in return sludge

FUTURE TRENDS

The use of modelling techniques has become established in water quality management, mainly for simulation of quality in natural waters and optimization of abstraction/discharge networks. Acceptance of the value of modelling has stemmed from the absence of any other techniques for tackling these problems. Provided that modelling excesses are avoided, future developments in this field seem assured. Improvements in detailed knowledge and understanding of the chemical, physical, and biological transformations that take place in natural waters can be incorporated into models to improve precision and reliability.

Acceptance of modelling as an aid to design of water and wastewater treatment plants has not been so universal, mainly because the basic processes in many treatment units are insufficiently understood. Further improvements in the simulation of unit processes are required if the developments in optimization are to be of benefit.

CHAPTER 11

Water Quality and the Planning Process

Timothy O'Riordan

The relationship between water quality and planning should be a topic on which much research has been done. However, when I approached the key organizations in the UK interested in this matter (the Countryside Commission, the Water Space Amenity Commission, the Water Research Centre, the British Waterways Board), I found that very few studies have in fact been undertaken and that no-one knew of anybody who was specifically working on this topic.

Yet the subject is ripe for management analysis for four reasons. First, under Section 20 of the 1973 Water Act, the Regional Water Authorities have a duty to put their waters and associated lands to the best use for recreational purposes as far as is reasonably practicable. Secondly, the National Water Council (1978) has recently issued guidelines to assist Regional Water Authorities to determine river quality objectives for all waters under their jurisdiction. (Although this matter is discussed in Chapter 7, the policy implications for planning will be discussed later in this chapter.) Part of their evaluation must include a review of existing or potential recreational uses of water space, so the Regional Water Authorities ought to accommodate the needs of recreation when assessing the quality characteristics of direct discharges in the future.

Thirdly, the European Commission in Brussels is anxious that Britain conforms with her continental member states in establishing regional water quality standards for, amongst other things, bathing and the protection of fresh water fish. (The distinction between ambient quality standards and objectives is dealt with in Chapter 8. Basically it is a matter of flexibility. When standards are set, there is a legal obligation to respect them, and interested parties can take legal action against pollution control authorities and/or polluters if such standards are persistently violated. Objectives, on the other hand, indicate targets or desirable levels of quality toward which pollution control authorities strive to attain. So long as they can show a record of steady improvement, even if the objectives are not attained, neither the courts nor other responsible authorities are likely to harass them.) The British water industry is currently resisting the principle of clearly defined water quality standards due to the legal and economic implications involved, and because the principle of fixed standards does not conform to the practice of pollution control in this country. Nevertheless, in principle at any rate, the EEC Directives and the new National Water Council (NWC) guidelines should result in Water Authorities having to relate current and future polluting discharges to expected uses of their waters. In many areas, one of these uses will certainly be recreation, whether for swimming, boating or fishing.

Fourthly and finally, the link between water quality management and planning is steadily being forged in the growing use of upland catchment reservoirs for public enjoyment and amenity. This is a Water Authority responsibility under Section 20 of the 1973 Water Act, but it often involves a close liason with local planning departments, many of which now employ recreation officers whose duties encompass recreational access to water areas and forecasts of potential recreational demand.

In my view, the policy environment is favourable for a much closer working relationship between authorities responsible for water quality and those interested in planning especially for amenity and recreational interests. Some Water Authorities are very much alive to this fact, while others may find that, in the formulation of their river quality objectives, certain recreational interests (especially angling groups) may use the opportunity for public consultation to encourage their officers to take more seriously the relationship between water quality, the siting and composition of direct effluent discharges, and the planning and provision of recreational facilities.

RECREATION AND WATER QUALITY

It is generally accepted that when people seek a recreational experience in summer time, they like to do so in or adjacent to water. Why this is the case, nobody is exactly sure. It may be that the land–water interface has a peculiar psychological attraction, it may be that water surfaces provide reflections and colour patterns that give people pleasure, it is possible that water spaces permit a greater variety of activities to watch or to participate in than are provided by purely land areas. The fact remains that, in terms of outdoor recreational activities, water-related uses such as swimming, fishing, and visits to the seaside are enormously popular. Statistics are hard to come by, but the latest available figures are presented in Table 11.1. These are drawn from the 1973 general household survey which analyses the attitudes, activities, and expenditure patterns of a carefully selected cross section of the British population (see also Veal, 1977; Countryside Review Committee, 1977).

From these surveys, it appears that about 20% of the public participate in outdoor recreation on a regular basis. The most popular pastime is the visit to the countryside. In fact, it is estimated that, in any year, up to 75% of the population will take one trip to the countryside. No-one knows precisely how

Table 11.1. Leisure patterns in Britain (from Birch, 1976)

	Participation in the four weeks prior to interview (adjusted for season)				Extrapolated frequency of participation for whole population	
	Males	Females	All	Overall Rank*	Days per 1000 people	Overall rank*
Passive pastimes						
Visits to the country	15.4	15.8	15.6	4	5740	10
Visits to the seaside	17.1	18.6	17.9	2	3650	13
Visits to parks	4.6	5.7	5.2	16	2260	16
Active pastimes						
Watching football	13.6	1.9	7.4	1	1970	1
Swimming outdoors	5.2	3.9	4.5	2	1220	5
Swimming indoors	4.1	3.2	3.7	3	1190	6
Fishing	5.8	0.8	3.2	4	990	9

*The overall ranking is for the whole survey. The most popular uses of leisure time (excluding watching television) were gardening (men), needlework (women), games of skill, and do-it-yourself handiwork.

much of this is water-related in one way or another because systematic surveys to ascertain this have not been undertaken. What is known, however, is that active participation in water sports (e.g., fishing and boating) is undertaken by a relatively modest proportion of visitors to country parks (around 5–6%; see Tanner, 1977b). But this does not tell us anything about the passive, aesthetic enjoyment of water space in such areas. This is probably very important if a recent survey by Stabler and Ash (1978) is anything to go by. They found that 80% of visitors to a sample stretch of canal were there for informal purposes, such as walking, picnicking or simply relaxing.

In terms of general sport, however, the general household survey found that swimming outdoors was the second most popular pastime after watching football. Because the survey concentrated on the recreational behaviour of adults over 18, this may be a serious underestimate, given the popularity among children of river and pond swimming. Coppock and Duffield (1975) review a number of studies that have attempted to predict recreational participation rates in Britain over the next 25 years. Decreased estimation is a notoriously difficult task at the best of times, but in the recreational sector it is compounded by the problem that the provision and quality of supply (the availability of recreational facilities) considerably influences demand. Thus, one has to distinguish between existing demand (based on visitor counts) and latent demand. The latter consists of both deferred demand (demand not realized because of lack of facilities) and potential demand (demand not realized due to the personal circumstances—age, mobility, income, social habits—of likely participants). Even with these forecasting difficulties, such studies all point to a continued growth in the popularity of swimming and sailing in addition to golf, horse riding, and winter sports, so there is little doubt that the degree of attractiveness of water sites for recreation should continue to influence local demand for recreation in the future.

Take the example of fishing, one of the most widely enjoyed sports in the country. The National Angling Council (1971) estimate that there are over three million anglers in Britain, some of whom fish regularly during the summer in rivers and canals. Their national angling survey indicated that as many as one million people fish on a typical weekend in the summer (more than watch football matches on a winter Saturday). The popularity of this sport continues to grow despite the increasing cost of tackle and the rise in numbers of pleasure boats frequenting the same waterways. This persistent popularity is partly because fishing clubs provide a useful social function in addition to securing rights of access, and partly because angling is a great family activity which is handed down from father to son and from uncle to nephew. Yet, survey evidence (O'Riordan and Paget, 1978) indicates that the act of catching fish is not necessarily the primary motive for angling. A pleasant countryside setting beside a peaceful river bank away from the noise and stress of everyday home and work life are equally important, and serve to emphasize the point that it is the quality of the whole recreational experience that is so important in understanding motivation and in predicting demand.

Despite this fairly obvious conclusion, there have been surprisingly few studies of how recreational use of water is affected by changes in water quality. Turner (1976, 1977) has surveyed the literature and has found little of consequence but much which is methodologically suspect. Certainly in Britain, it seems, there have been no properly conducted empirical studies of sufficient magnitude conclusively to indicate the response by the public to improvements in water quality for recreation. There are three problems here. First, water quality is only one factor influencing recreational choice: availability of suitable alternatives (and knowledge that these exist), habit, and sensitivity to symptoms of deteriorated water quality are also very important variables. Secondly, public understanding of water quality is distorted by what information is made public, how it is reinterpreted by word of mouth, and by the extent to which poor water quality is actually perceived. Research by Barker

(1967) and David (1971) found that most people interpreted water quality by sight and smell. Algae were a prime indicator since their decaying detritus on the shoreline satisfied both sensory criteria. Turbidity, debris, and the presence of dead fish were also important indicators. Needless to say, water chemists are quick to point out that some characteristics of polluted water are not recognizable to the senses, especially the presence of pathogenic bacteria and toxic substances. The fact remains that Water Authority officials can do little to stop children from bathing in bacteria-ridden river water on hot summer days, except by posting notices and trusting that both the children and their parents read them and take appropriate action. This is one of many reasons why the EEC want to see uniform standards for various parameters, including bacteria and toxic substances, in waters used for bathing and fishing.

The third difficulty in researching the relationship between recreation and water quality is the lack of any suitable measures to value the gains accruing to society when water quality is improved, or, vice versa, the losses resulting from water quality deterioration. In economic terms, these gains and losses are simply illustrated (see Figure 11.1) by a shift in the demand curve consequent upon a change in water quality. But in practical terms, it is all but impossible to measure or predict the shift in participation rate (on the horizontal axis) for the reasons already cited, to which can be added the matter of cost, for any investigation of this kind would have to be undertaken over a period of years and on a reasonably large scale. Also troublesome is the methodological difficulty of isolating the difference in willingness to pay associated with changes in water quality. American studies encouraged by the recreation industry, have imputed large values for the economic benefit of a day on, or by, the water (see Davidson *et al.*, 1966; US Water Resources Council, 1972; Turner, 1978). These range from $6.00 to as high as $20.00 per person, depending on the particular activity pursued. Given these figures, it is not difficult to boost the benefits from improved water quality simply by predicting increased use. In fact, much of the benefits attributed to the costly programme of sewage works improvement encouraged by the Water Pollution Control Act of 1972 are being attributed to increased recreational use and greater user satisfaction (Turner, 1978). Coy *et al.* (1973) feel that this is an example of cost–benefit analysis going mad. 'Recreation is far more important to the citizens of the United States as a social process than as an economic development process', they argue, 'and should be so valued.' They conclude that research should be devoted more to the satisfaction people gain from a day in pleasant surroundings by, or in, clean water, and that planners should strive to improve levels of enjoyment as a desirable social good in its own right. It is true that any research evidence as to a clear relationship between social judgement about water quality and recreational choice is inconclusive, and attempts to produce a 'water quality index' against which to evaluate recreational potential have generally failed (Turner, 1978).

These difficulties outlined above are not easy to resolve. No-one wants to waste public money, so some kind of cost accounting in relation to anticipated gains is desirable. On the other hand, survey methodologies are ill developed and costly, and little is known about the motivations underlying recreational choice. It would seem desirable to embark on a number of different pilot schemes linking water quality improvement to recreation use and monitor these with care. Meanwhile, I tend to side with Coy and his colleagues (1973). Water quality management and recreational provision are more politically than economically determined activities, because they involve matters of social value and subjective judgement about what is good for society at large.

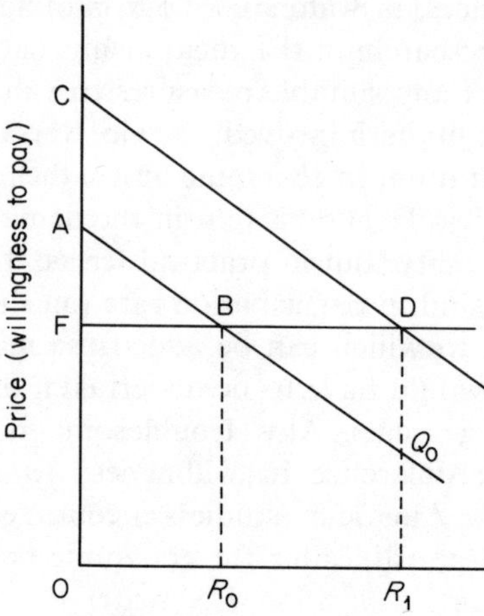

Figure 11.1. The effects of water quality change on the demand for recreation. AB is the demand curve (adjusted for income) assumed on the basis of what recreationists would be willing to pay for a day's use of a water space of a certain quality Q_0:R_0 people would turn up. If this water space was to be cleaned up to quality Q_1, then R_1 people would use it and would be willing to pay more for their enjoyment. So the income-compensated demand curve shifts to the right. The value of this benefit is the area ABDC. The text discusses how difficult it is to identify this benefit. The horizontal line FBD represents the various kinds of costs incurred by recreation users in travelling to the site, costs which are spent irrespective of whether or not there is an entrance fee

The past decade has witnessed a surge in public interest in environmental amenity generally brought in by the vociferous yet persuasive activities of conservationist groups plus important changes in legislation and administration. Even during the period of economic difficulties experienced between 1973 and 1976, the River and Water Authorities maintained their good record in improving water quality throughout the UK as a result of strong public support (Department of the Environment, 1978). People want better water quality and they would appreciate more recreational facilities. This powerful combination of sentiment is not yet ostensibly on the wane, so the Regional Water Authorities and planning authorities still have plenty of opportunity to capitalize upon it and publicize their success stories. These will do more to foster public support than a lot of dressing up in pseudo-economic accounting that often confuses more than it illuminates.

An illustration of the kind of pilot scheme I have in mind is that recently promoted by the Inland Waterways Association (IWA) in cooperation with Shell Oil. The IWA believes that the pleasures of being beside many of the British waterways could be enhanced at modest cost if the general amenity of the water space and the surrounding lands was improved. Stretches of urban canals are particularly suitable targets, since they offer excellent opportunities for casual strolling, dog walking, and jogging—all enormously popular activities in built-up residential areas. In many cases, all that is needed is a limited amount of funding and volunteers to clear debris, and clean up or remove eyesores. Under the Shell–IWA scheme, £20000 is now available (1979) to enable local groups to restore and maintain selected stretches of waterway. Four categories of scheme are eligible for grants:

(1) restoration and improvement of navigation and navigational work on rivers and canals;
(2) restoration and improvement of other stretches of water;
(3) restoration and conservation work on buildings, such as water mills and warehouses; and
(4) improvement and enhancement of the water environment, for example, the clearing of towpaths, the provision of waterside facilities, and the creation of amenities.

This is an important initiative, and though the funds are small, they represent a considerable improvement in the £6000 that were available in 1977, sufficient only to finance 50 relatively modest projects. The IWA is anxious that local organizations take advantage of this new scheme, though it is well aware that much depends on the goodwill of local volunteers. It is to be hoped that local planning authorities will seize this opportunity to grant aid to some of these developments in their own areas and possibly to seek additional funds from the Countryside Commission which can provide financial assistance under certain conditions (Countryside Commission, 1976).

The one agency that has seized the initiative here is the British Waterways Board (BWB) which has encouraged local authorities to clean up stretches of derelict canals. Many London boroughs and the West Midlands County Council have responded well to this suggestion by developing canalside facilities for recreational and educational purposes, and by removing debris so that the waters become more safely navigable. The active role of both the BWB and the Inland Waterways Amenity Advisory Council in encouraging these developments is to be welcomed.

It is intuitively obvious, but not well researched, that people will come to a water body of dubious quality if the surrounding land is reasonably safe and attractive to look at. The relationship between shoreline amenity, water quality, and recreational use is one that requires more investigation. After all, of some 2500 km (1560 miles) of canal towpath, 600 km (376 miles) have been established as public right of way (Tanner, 1977a). But there are difficulties associated with the improvement of urban canal sites other than those of cost. There is the legal position relating to public access which can frustrate many of local authority, and there is the matter of liability over public safety, especially for children in residential areas. However, some urban authorities such as Manchester and Birmingham are developing linked networks of canalways which combine casual recreational access with urban amenity improvement (Tanner, 1977a). In most cases, the water quality of the canal is not specifically upgraded, though debris removal and bankside restoration can have a remarkable influence on people's recreational enjoyment.

WATER QUALITY MANAGEMENT AND PLANNING

The obvious link between water quality management and planning lies in the planning controls over the siting of activities that would produce undesirable effluent discharges, given the existing quality of the water space and the uses to which it can be put. The normal practice in Britain is for planning authorities to pass to the appropriate division of the Water Authority all planning applications involving a sewer-link or a direct outfall. If the Water Authority believes that any proposed development would result in an effluent that could seriously lower existing water quality, it can advise the local planning authority to impose a development restriction order—in effect, to refuse planning consent. (The pathway for a discharge consent application is charted in Figure 11.2, and the whole process is being studied by the Pollution Research Unit at the University of Manchester.) In the vast majority of cases, the planning authority is guided by such advice though it has the statutory powers to override it. In some cases, the planning authority will try to force a Water Authority to improve a municipal sewage works by proposing new housing schemes in areas where existing sewage works are unable to cope with current demand. Generally, however, the Water Authorities and planning authorities

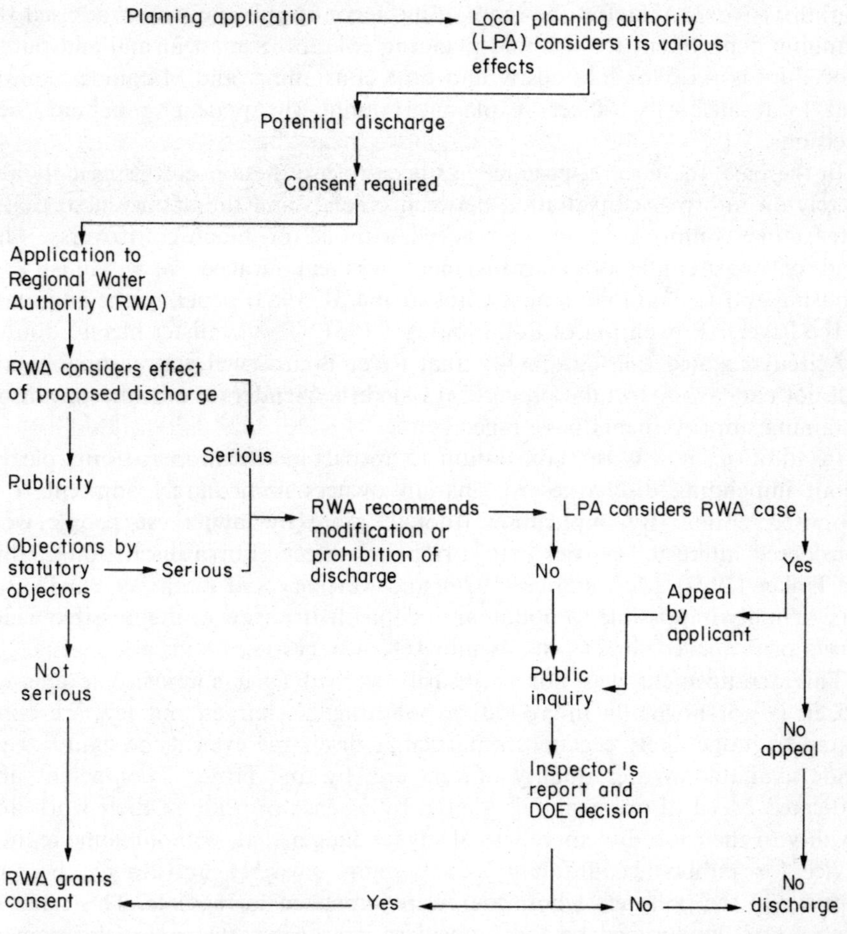

Figure 11.2. The various connections between local planning authorities and Regional Water Authorities when consents for proposed direct effluent discharges are considered

have developed reasonable working arrangements (though the effectiveness of these vary regionally) coupled with advice for local recreational interests. Where important fisheries are involved, the general policy is to connect industrial wastes to sewage works, so that as few as possible direct outfalls have to be controlled. In addition, sewage wastes in general receive considerable dilution.

If planning consent is refused or if a Regional Water Authority and planning authorities disagree, there can be an appeal which would result in a public inquiry at which the Water Authority would have to give evidence. The Water

Authority would therefore be required to account publicly why it advised the planning department to withhold planning consent. Such a formal and public procedure is rare, for it is costly and time consuming, and of course requires that local authority officers explain in public the reasoning behind their decisions.

In the past, the general practice has been to use these powers sparingly and to rely on informal consultation between officials and the applicant so that a satisfactory compromise can be reached without too much controversy. This kind of quasi-confidential arrangement was encouraged in a Ministry of Housing and Local Government Circular (64/51, 1951) issued in the aftermath of the Rivers (Prevention of Pollution) Act 1951. 'The Minister has no doubt,' the circular stated (paragraph 14), 'that River Boards will act reasonably and will not endeavour to take proceedings against offenders until other means of obtaining improvements have failed.'

In addition, it was the convention to restrict notification to third parties about impending discharges to riparian owners immediately adjacent to a proposed outfall. By implication (though more by law) these people were considered interested parties with a right to object. (For a discussion of this, see Tinker (1972), McLoughlin (1976), and Macrory and Zaba (1978).) Thus it was virtually impossible to obtain an independent review of the possible wider effects of proposed direct outfalls into UK waterways.

This arrangement was not necessarily as bad as it sounds. The average citizen is not normally interested in polluting discharges and few are competent to judge likely effects of a particular discharge even if the figures were made available. Water quality officers are, by and large, a competent and dedicated breed of professionals, united by a sense of pride in their work and loyality to their job. But there was always a danger that, without some regular device for public accountability, they might, possibly unwillingly, become captives of the polluters whom they were supposed to regulate. This sense of unease was encouraged by four considerations. First, there was the tremendous degree of confidentiality that surrounded negotiations between dischargers and polluters. Secondly, there was the custom of using the process of persuasion through cooperation rather than legal enforcement. Thirdly, there was the growing resentment felt by many industrialists of the common law powers of riparian owners to prosecute when suffering from pollution. And fourthly, there was the fact that, after reorganization of the water industry in 1975, the new Water Authorities were responsible for managing and regulating sewage works and their outfalls, widely regarded as the most serious contributors to water pollution in the country.

All of this led to pressures from interest groups and prominent individuals to open up the link between water quality and planning. The first salvo was fired by the Royal Commission on Environmental Pollution, which, in its second report (1972), noted that some industries discharge products which are potentially dangerous to human health, and certainly detrimental to aquatic flora and fauna. The Commission observed:

> 'The waters into which the pollutants are discharged are public property; interest in the nature and quality of these pollutants is entirely justified. Moreover, since many industries are going to great trouble and expense to abate pollution, they do themselves a disservice by needless secrecy. We believe that public confidence in the concern these industries have for the environment would be strengthened if this needless cloak of secrecy were withdrawn.'

The chairman of the Commission, Lord Ashby, along with others, fought hard during the Committee stage of the Control of Pollution Bill (introduced in the Lords) to promote the principle of public disclosure of effluent discharge consents (permits), and details of the actual quality of these discharges as sampled from time to time. Their efforts were rewarded by Section 41 of the 1974 Control of Pollution Act which requires Water Authorities to keep a public register of all discharges, except where confidentiality has been specifically granted (with reasons publicly stated) by the Secretary of State for the Environment in England, and the Welsh and Scottish Secretaries in Wales and Scotland, respectively.

One outcome of this important change in water quality management practice is that interested individuals and groups now have access to information about the actual quality of a discharge and its legally defined level of purity. Where these two diverge, there could be a case in law for prosecution against the offending discharger (or the Water Authority for failure to undertake its statutory duty). Macrory and Zaba (1978), who have provided a useful service in translating the Act into readable English along with annotated remarks for those interested in their environmental rights, also caution against the free use of prosecution. 'Interpretation of pollution statistics is notoriously difficult, and it is not sensible to make play of facts discovered from the register unless you are completely sure of your case.'

In my own view, the use of civil action is likely to be restrained partly because of cost, partly because of the difficulties in ensuring evidence of proof, and partly because the tailoring of discharge consents to clearly defined and politically sanctioned river quality objectives should provide the water quality authorities with a good defence. In any case, in many areas, the most serious offender is the sewage treatment works (see the annual report of any Regional Water Authority) and it is politically unlikely that the owning Water Authorities will be able to raise sufficient funds from local ratepayers to improve all the troublesome works in the foreseeable future, court cases or not. River quality improvement will have to take its place alongside other demands for capital expenditure by local authorities.

Partly as a response to this increased facility for civil prosecution, but also because of pressure from the European Commission to force its member states to accept uniform air and water quality standards, the National Water Council (1978), the parent body of the English and Welsh Water Authorities, has accepted the principle that all Water Authorities should specify river quality objectives for all their waters. The idea is revolutionary in British water management terms, though not particularly new in international circles. Basically the aim is to upgrade the overall quality of the receiving waters and to accommodate discharge consents to defined river quality objectives, rather than to treat discharge consent applications on a case-by-case basis. This means that Water Authorities will be expected to sample extensively the water quality parameters of their rivers and lakes, derive implicit, if not explicit, water quality models of these watercourses, and thus be able to compute the effect of any new (or existing) discharge on water quality. How well they will be able to do this depends upon the manpower and budgeting resources at their disposal and the degree of pollution already in existence. In the first instance, much will be based on experience and rule of thumb. But a more systematic approach may be required for the law is quite explicit. To quote Macrory and Zaba (1978).

> 'If you have evidence of a discharge causing harm to the flora and fauna of a particular location, but discover from the register that no samples from that stretch of water have been taken by the Authority, this could be the basis of a campaign. You should be able to pressurize the Authority into sampling, and when the results appear on the register this should at least provide some hard evidence...'

Though the matter of water quality standards and objectives is dealt with in Chapter 8, it is developed here because the establishment of river quality objectives should have profound implications for the relationship between water quality management and planning. In the short term, the Water Authorities are placed in a bit of a dilemna since many effluents, particularly those from sewage treatment works, do not conform to existing consents. (Again, for evidence of this see the annual report of any Regional Water Authority.) In some instances, this situation is not as bad as it appears, since many consents were too strict given the degree of dilution available. Nevertheless, faced with the possible (but rather unpredictable) threat of civil prosecution, Water Authorities have little option but to alter consents or conditions attached to consents (such as sampling variability) to conform more closely to existing conditions. Though some critics (e.g. Bugler, 1978) regard this as a major concession to the polluter, it should be noted that in most cases this arrangement merely alters figures on a sheet of paper and does not change existing water quality, and that this is an interim measure giving the Water Authorities a period of grace to negotiate new consent conditions in the context of river quality objectives.

In the longer term, the Water Authorities will have to ensure that all direct discharges in combination do not lead to a deterioration of water quality below a level for which a prescribed use is established. The procedures followed are outlined in Figure 11.3 which also indicates where the relationship between water quality and planning becomes pertinent. The most difficult matter is how to determine the 'needs' (i.e. uses) for which water

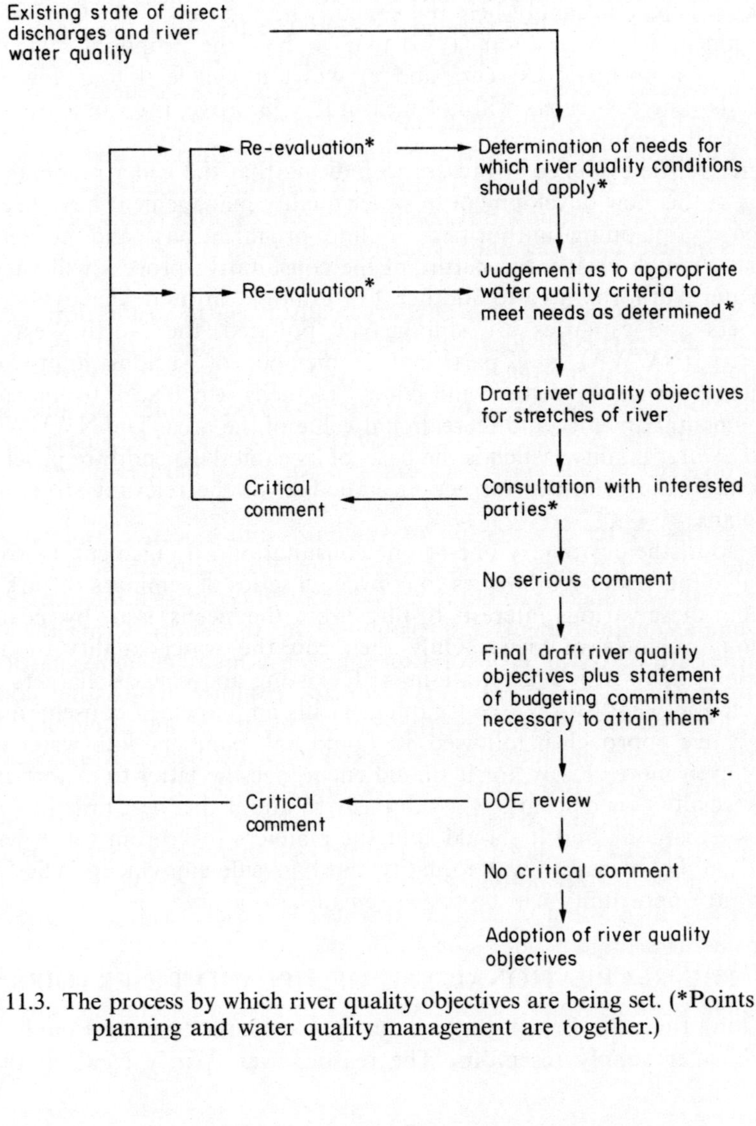

Figure 11.3. The process by which river quality objectives are being set. (*Points where planning and water quality management are together.)

quality criteria should be established. As far as one can see, this will be undertaken rather unimaginatively and will be based largely upon present uses of water bodies combined with scientifically determined water quality parameters as accepted by water chemists and the international scientific community. In short, river quality objectives will be established on the cautious basis of experience and knowledge. To be sure, draft objectives are now circulating among interested parties who have been invited to comment, but unless they have a degree of expert advice it may be difficult for them to negotiate substantial alterations. Unless, too, there is a significant change of procedure here, one cannot help but concur with the view of the Committee on Environmental Conservation (1977) that 'at best the proposed review is no more than a cosmetic exercise, and at worst it will lead to a lowering of standards. The best rivers will not measurably improve; the worst will remain as chemical conduits.'

There is little evidence at present to indicate that the wider planning implications of this new development in water quality management have been fully grasped by planning authorities, angling organizations, and conservation interests, though clearly the nature of the consultative process will vary from one Water Authority area to another. For example, around Merseyside, where the rivers and estuaries are still heavily polluted, the North-West Water Authority (NWWA) is a participating member of various joint working committees set up by district and county councils which seek to improve the environmental amenity and recreational value of the area. The NWWA policy for river water quality, which is the basis of its capital expenditure programme, is linked to the priorities for action embodied in the relevant structure and local plans.

Instead of the customary one-to-one consultation arrangement, it would be preferable for Water Authorities to convene a series of seminars or task forces involving these various interests to thrash out the 'needs' issue by means of a common exchange of views. Only then can the water quality models be established upon which negotiations with existing and new dischargers can be conducted in what Storey (see Chapter 9) calls an 'informed consent' arrangement. If this approach is followed, it should link planners with water quality officers even more closely, for it should encourage the latter to exhort existing and especially new dischargers assiduously to adopt the 'best practical means' for waste disposal, and it should spur the planners to keep in touch with the general public interest in water quality and bankside enjoyment. Whether this important opportunity will be seized remains to be seen.

THE RECREATIONAL USE OF UPLAND RESERVOIRS

For a long time, the British water industry was loathe to permit public use of upland water supply reservoirs. The reasons were partly legal, in that the

Authorities controlled the water surface and the adjacent lands and so had powers to prohibit trespass, and partly psychological, since many officials felt uncomfortable about people bathing and boating in waters that eventually were to be consumed by the general public. The main reason, however, was probably a practical one, in that long-term storage in remote catchments killed off bacteria so that, in the early days before chemical treatment was developed, only mechanical sand filtration was required before the water was considered potable. By the end of the nineteenth century, urban populations were drinking virtually pure water while improved sewage treatment works were being devevoped, so there was a laudable reduction in water-borne diseases amongst the British population. A tradition grew up among water officials that prohibiting public access from upland gathering grounds and reservoirs was the first line of defence in the battle to protect public health, and a desirable thing in its own right.

However, the combination of improved treatment facilities, public willingness to consume chlorinated waters, and increasing pressures for access to upland water catchments for various recreational needs, has led to a change of thinking amongst many water management officials. This has taken its time because many officials still see control of contamination as a key factor in their obligation to protect the 'wholesomeness' of public water supplies (as required in Section 31 of the 3rd Schedule of the 1945 Water Act). In 1948 and in 1972, respectively, major investigations of the whole question of public use of upland catchments (Central Advisory Water Committee, Gathering Grounds Sub-Committee 1948; Institution of Water Engineers, 1972) concluded on the cautious note that recreational enjoyment of such areas had to be tailored to the size and purpose of the reservoir and the cost and quality of its treatment facilities. Where reservoirs were large and/or used for regulation or compensation purposes, recreational use including bathing could be permitted. Where water supply reservoirs were associated with treatment facilities, the 1972 report concluded that the increased risk due to recreational use would normally be quite insignificant. But if poor or no treatment works were available, pollution stemming from recreational activities could be unacceptable.

The position changed somewhat with the passage of the 1973 Water Act which, as already noted, required Water Authorities to put their waters and associated lands to the best recreational use as far as is reasonably practicable (Section 20). Since then, the water industry has cooperated with various local and national bodies to develop suitable plans that would permit recreation without endangering the wholesomeness of water supply. (For a recent report on the recreational use of water supply reservoirs in England and Wales, see Tanner (1977c).) Probably the most progressive Authority in this respect is the North-West Water Authority which has endeavoured to establish a consistent policy between its water supply duties and its amenity recreation responsibilities. The NWWA scheme is based on tailoring recreational use to the level of treatment. The principle is to establish adequate 'lines of defence' based on the provision of a comprehensive disinfection process in combination with classification. To quote the Authority's own words (North-West Water Authority, 1977):

> 'The procedure for applying (this principle) to actual cases in a realistic, reliable and consistent manner in the face of the many practical difficulties, has to rely heavily on subjective judgement by experienced experts. Nevertheless the whole procedure for applying the policy to actual cases must earn credibility in the eyes of inexpert members of the public...'

The procedures referred to consist of (1) a codification of the requisite standards for water treatment, (2) a method for rating the effectiveness of water treatment plants according to their technical attributes, (3) a specification of the minimum ratings of treatment to permit recreational access, and (4) a judgement as to what recreational uses would be permitted in the context of the treatment available. This last matter includes the importance question of whether, and to what extent, existing treatment facilities should be *upgraded* to accommodate specific recreational needs. The NWWA has grasped this nettle by asking its Water Management Committee and Recreational Subcommittee to coordinate their activities by recommending what forms of recreation could become permissible following low (less than £5000) and moderate (up to £15000) expenditure to improve the treatment rating. This would clearly depend upon a whole series of other matters, such as reservoir size, location *vis à vis* roads, settlements and other water space, the extent to which the local planning authority and land owners would be willing to cooperate, and some indication of recreational need.

Although the recommended expenditures are modest, the NWWA seems to be leading the field in planning water quality to suit recreational demands. Its chairman, Mr P. Liddell (Liddell, 1978) even hints that where such expenditures are required then 'there might be a sound argument for charging the whole of these additional costs to those (people) benefitting from the recreation'. Presumably any expenditures associated with management and supervision would also be levied against the users in the form of access fees or car park charges.

The determination of the NWWA remains to be tested. Upon inquiry elsewhere, I found that water quality officers tend to take the view that enhancement of treatment facilities purely for recreational and/or amenity purposes would be regarded as a low priority given the facts (1) that there is much suspicion among local ratepayers about rising Water Authority expenditures, (2) that the Department of the Environment, which vets all Authority expenditure plans, has made it clear that major capital investments must be justified in terms of need, and (3) that most Water Authorities are still proceeding cautiously with respect to Section 20, so would not at present wish to embark on expensive commitments on facilities where the benefits are not readily defined and may not always accrue to the local ratepayers alone. Liddell (1978) stated that due to these constraints the much needed improvements to the water treatment facilities at Thirlmere, in the Lake District, have to be delayed, thereby affecting the scale and quality of recreational use on this important lake.

Given these sentiments, it is unlikely that much will happen in the short term with respect to public access to water supply reservoirs except where existing treatment facilities are adequate. Certainly there is scope here for an imaginative relationship to be made between local planning authorities, the water industry, and many local clubs and organizations interested in the recreational use of such areas. The most promising developments tend to have taken place in National Park areas where planning departments are usually eager to encourage public use of upland areas. For example, on the Burrator Reservoir in the Dartmoor National Park, the Dartmoor National Park Authority has developed a sensitive recreational plan for bankside strolling and picnicking (coupled with properly designed toilets). The Burrator plan was prepared on the basis of a report by a working party of Water Authority, National Park and County Engineering Department officers. This group accepted as a basic policy 'that the beauty and peace and quiet of Burrator are its prime assets and should not in any way be prejudiced by too much recreational use or by the wrong type of recreational activity'. Thus, the slightly narrow and winding access roads were deliberately left in a condition that discouraged their use by causal holidaymakers, though restrictions in vehicular access would not be ruled out if surveys indicate that these would be warranted. Only canoeing and possibly sub-aqua water sports are to be permitted on the reservoir, even when adequate treatment facilities are prepared. Sailing and rowing are not to be permitted because (1) other opportunities exist in adjacent waters, and (2) activities associated with clubhouses and the use of rescue boats would disturb the peaceful character of the area. Motor boating is to be strictly prohibited, a ban that can be enforced very readily by those clubs permitted to use the reservoir.

Much of the management of recreational activities in water supply reservoirs is being left to responsible clubs in collaboration with regional sports authorities and planning departments. It is evident that, where there is a prescriptive right of navigation (as, for example, is the case on many rivers), separation of incompatible uses by zoning procedures cannot be conducted by statutory powers, so only voluntary means are acceptable. On new reservoirs, however, Water Authorities and planning authorities have no such legal restriction and should be able to cooperate in the development of recreational schemes that meet local needs yet do not jeopardize water quality. The Burrator example illustrates how recreational management must take into account wider considerations than those of water quality. It is probably fair to say that this principle is being applied more generally in water quality policy making.

SOME CONCLUDING COMMENTS

This chapter began by stating the reasons why I thought the relationship between water quality and planning was an important one, and it developed the argument that much remains to be done at this interface in the light of new legislation, new management philosophies, and the sustained public interest in maintaining or enhancing an environment that is healthy and enjoyable. By way of a conclusion, it is perhaps worth observing that the British custom of widespread but selective consultation with statutory bodies and other interested parties, combined with a gradual and incremental approach toward change, is not all bad. The hallmarks of British management practice are reliance on the best scientific evidence available and the careful use of expert judgement made by people with experience tailored to the particular circumstances. This combination is frustrating to those who want more openness in administrative decision making and a greater say in the resolution of issues. But one virtue of this custom, for all its faults, is that the wider implications of possible decisions are aired, admittedly not always publicly, and sincere attempts are made to take these into consideration under the umbrella of discretion that is usually granted administrative authorities in this country.

Compare this with the experience in the United States where strict water quality standards are laid down nationally for all rivers and where dischargers must prove (in the courts, if necessary) that they have used the best practicable technology currently in existence, and, by 1985, the best available technology economically available even though that technology does not currently exist. The Administrator of the US Environmental Protection Agency has established national standards of performance for each of 27 categories of polluting sources. Only for 1977 standards were individual plants allowed to vary their actual discharges from these regulations, but by 1983 new dischargers will have to prove publicly that they cannot meet the best available technology then in operation without becoming bankrupt. Note that this ruling, backed by strict legal interpretation (Wayland, 1977), applies regardless of the assimilative capacity of the receiving water, or to the uses to which it is put. If this principle holds, then it may well be that water quality officials in the US will become *de facto* regional planners for they would have the powers to refuse planning permission and possibly drive an established industry out. Whether this draconian state of affairs will actually come about remains in doubt, for there is ambiguous evidence about how strictly these rules will be applied in economically depressed areas during a time of recession. One should also consider that this is the kind of approach which is in the minds of European Commission officials in Brussels. Already, with some justification, they have established this principle with respect to the so-called 'black list' of toxic substances (*Official Journal*, **129**, 18 May 1976); the danger is that they will try to apply this to a much wider array of pollutants, some of which are degradable in the right conditions.

These illustrations should cause us to reflect that, whatever the inadequacies of the British approach, the water quality management tail is not yet wagging the development control dog. But there is scope for considerable improvement in the relationship between these two functions, and imaginative authorities should at least put the water quality function on an equal footing with planning. If, in addition, the special merits of each particular situation can be considered and any preliminary decisions are subjected to political accountability before final judgements are made, then the link between water quality and planning should be properly forged. In my view, the major challenge is to hoist up the water quality function to something like equivalent status with planning, and to grant interested members of the public the opportunity to comment, question, and prod when practice fails to match stated objectives.

References

CHAPTER 1

Ackermann, W. C. (1969). Scientific hydrology in the United States. In *The Progress of Hydrology, Proceedings of the First International Seminar for Hydrology Professors, Urbana, Ill*, vol. I, pp. 50–60, US National Science Foundation, Washington DC.

Barnes, I (1965). Geochemistry of Birch Creek, Inyo County, California, a travertine depositing creek in an arid climate. *Geochim. Cosmochim. Acta*, **29**, 85–112.

Best, G. R., and Monk C. D. (1975). Cation flux in hardwood and white pine watersheds. In Howell, F. G., Gentry, J. B., and Smith, M. H. (eds), *Mineral Cycling in Southeastern Ecosystems*, pp. 847–61, US Energy Research and Development Administration, Washington DC.

Biesecker, J. E., and Leifeste, D. K. (1975), *Water Quality of Hydrologic Bench Marks—an Indicator of Water Quality in the Natural Environment*, Circular no. 460-E, US Geological Survey, Washington DC.

Blakey, J. F. (1966). *Temperature of Surface Waters in the Conterminous United States*, Hydrologic investigations atlas HA-235, US Geological Survey, Washington DC.

Bolt, G. H., Bruggenwert, M. G. M., and Kamphorst, A. (1976). Adsorption of cations by soil. In Bolt, G. H., and Bruggenwert, M. G. M. (eds), *Soil Chemistry, A. Basic Elements*, Elsevier, Amsterdam.

Bricker, O. P., and Garrels, R. M. (1967). Mineralogic factors in natural water equilibria. In Faust, S. D., and Hunter, J. V. (eds), *Principles and Application of Water Chemistry*, pp. 449–69, Wiley, New York.

Brown, G. W. (1971). Water temperature in small streams as influenced by environmental factors and logging. In *Proceedings of a Symposium on Forest Land Uses and Stream Environment*, pp. 175–81, Oregon State University, Corvallis, Oregon.

Brown, G. W., and Krygier, J. T. (1970). Effects of clearcutting on stream temperature. *Water Resour. Res.*, **6**, 1133–40.

Brune, G. M. (1951). Sediment records in midwestern United States, *Int. Assoc. Hydrol. Sci. Publ.*, **33**, 29–38.

Carlisle, A., Brown, A. H. F., and White, E. J. (1967). The nutrient content of tree stem flow and ground flora litter and leachate in a sessile oak woodland. *J. Ecol.*, **55**, 615–27.

Cawse, P. (1977). Deposition of trace elements from the atmosphere in the UK. *Conference on Inorganic Pollution and Agriculture*, Agricultural Development and Advisory Service, Paper SS/OC/77/3.

Chebotarev, I. (1955). Metamorphism of natural waters in the crust of weathering. *Geochim. Cosmochim. Acta*, **8**, 22–48, 137–70, 198–212.

Cole, D. W., and Johnson, D. W. (1977). Atmospheric sulfate additions and cation leaching in a Douglas Fir ecosystem. *Water Resour. Res.*, **13**, 313–7.

Curtis, W. R. (1973). Effects of strip mining on the hydrology of small mountain watersheds in Appalachia. In Hutnik, R. J., and Davis, G. (eds), *Ecology and Reclamation of Devastated Land*, pp. 145–57, Gordon and Breach, New York.

Dickinson, W. T., and Whiteley, H. (1970). Watershed areas contribution to runoff. In Results of research on representative and experimental basins, Proceedings of the Wellington symposium 1970, *Int. Assoc. Hydrol. Sci. Publ.*, **96**, 12–26.

Dochinger, L. S., and Seliga, T. A. (eds) (1976). *Proceedings of the First International Symposium on Acid Precipitation and the Forest Ecosystem*, US Dept of Agriculture, Northeastern Forest Experiment Station, Upper Darby, Pennsylvania.

Douglas, I. (1968). The effects of precipitation chemistry and catchment area lithology on the quality of river water in selected catchments in Eastern Australia. *Earth Sci. J.*, **2**, 126–44.

Downing, R. A., and Williams, B. P. J. (1969). *The Groundwater Hydrology of the Lincolnshire Limestone*, Publication no. 9, Water Resources Board, Reading.

Dutt, G. R. (1962). Prediction of the concentration of solutes in soil solutions for soil systems containing gypsum and exchangeable Ca and Mg. *Soil Sci. Soc. Am. Proc.*, **27**, 627–30.

Fritz, P., Cherry, J. A., Weyer, K. U., and Sklash, M. (1976). Storm runoff analysis using environmental isotopes and major ions. In *Interpretation of Environmental Isotope and Hydrochemical Data in Groundwater Hydrology*, pp. 111–42, International Atomic Energy Agency, Vienna.

Gambell, A. W., and Fisher, D. W. (1966). *Chemical Composition of Rainfall, eastern North Carolina and southeastern Virginia*, Water supply paper no. 1535-K, US Geological Survey, Washington DC.

Georgii, H. W., and Weber, E. (1960). *Chemical Composition of Individual Rainfalls*, Report no. AD266 285, Armed Services Technical Information Agency, Arlington, Virginia.

Gibbs. R. (1970). Mechanisms controlling world water chemistry. *Science*, **170**, 1088–90.

Glover, B. J. and Johnson, P. (1974). Variations in the natural chemical concentrations of river water during flood flows and the lag effect. *J. Hydrol.*, **22**, 303–16.

Gregory, K. J., and Walling, D. E. (1973). *Drainage Basin Form and Process—a geomorphological approach*, Edward Arnold, London.

Hall, F. R. (1970). Dissolved solids–discharge relationships. 1 Mixing models. *Water Resour. Res.*, **6**, 845–50.

Hall, F. R. (1971). Dissolved solids–discharge relationships: 2 Application to field data. *Water Resour. Res.*, **7**, 591–601.

Harmeson, R. H., and Larson, T. E. (1969). *Quality of Surface Water in Illinois* 1956–66, Bulletin 54, Illinois State Water Survey. Urbana, Illinois.

Heidel, S. G. (1956). The progressive lag of sediment concentration with flood waves. *Trans. Am. Geophys. Union*, **37**, 56–66.

Hem, J. D. (1971). Preface to *Nonequilibrium Systems in Natural Water Chemistry*, Advances in Chemistry Series no. 106, American Chemical Society, Washington DC.

Hewlett, J. D., and Hibbert A. R. (1967). Factors affecting the response of small watersheds to precipitation in humid areas. In Sopper, W. E., and Lull, H. W. (eds), *International Symposium on Forest Hydrology*, pp. 725–36, Pergamon Press, Oxford.

Ineson, J., and Downing R. A. (1963). Changes in the chemistry of groundwaters of the Chalk passing beneath argillaceous strata. *Bull. Geol. Surv. GB*, **20**, 176–92.

Johnson, N. M., Likens, G. E., Bormann, F. H., Fisher, D. W., and Pierce, R. S. (1969). A working model for the variation of stream water chemistry at the Hubbard Brook Experimental Forest, New Hampshire. *Water Resour. Res.*, **5**, 1353–63.

Junge, C. E., and Werby, R. T. (1958). The concentration of chloride, sodium, potassium, calcium and sulphate in rain water over the United States. *J. Meteorol.*, **15**, 417–25.

Kennedy, V. C. (1977). Hypothesis to explain the rapid release of solutes from soil during storm runoff. *Eos*, **58**, 386.

Langbein, W. B., and Schumm, S. A. (1958). Yield of sediment in relation to mean annual precipitation. *Trans. Am. Geophys. Union*, **39**, 1076–84.

Lavis, M. E. and Smith, K. (1972). Reservoir storage and the thermal regime of rivers with special reference to the river Lune, Yorkshire. *Sci. Total Environ.*, **1**, 81–90.

Leak, W. B., and Martin, C. W. (1975). *Relationship of Stand Age to Streamwater Nitrate in New Hampshire.* Note NE-211, US Dept of Agriculture, Forest Service Research.

Likens, G. E., Bormann, F. H., Pierce, R. S., Eaton, J. S., and Johnson, N. M. (1977). *Biochemistry of a Forested Ecosystem*, Springer-Verlag, New York.

Miller, M. (1977). Anthropogenic influences of sediment quality at a source, Nutrients: Carbon, Nitrogen and Phosphorous. In Shear, H., and Watson, A. E. P. (eds), *Proceedings of a Workshop on the Fluvial Transport of Sediment-Associated Nutrients and Contaminents*, pp. 81–92, International Joint Commission on the Great Lakes, Windsor, Ontario.

Morgan, C. O., and Winner, M. D. (1962). *Hydrochemical Facies in the* '400-*foot*' *and* '600-*foot*' *Sands of the Baton Rouge area, Louisiana.* Professional Paper no. 450-B, pp. 120–6, US Geological Survey, Washington DC.

Nihlgard, B. (1970). Precipitation, its chemical composition and effect on soil water in a beech and spruce forest in south Sweden. *Oikos*, **21**, 208–17.

Nutter, W. L., and Hewlett, J. D. (1971). Stormflow production from permeable upland basins. In Monke, E. J. (ed.), *Biological Effects in the Hydrological Cycle, Proceedings of the Third International Seminar for Hydrology Professors*, pp. 248–58, Purdue University, West Lafayette, Indiana.

Oden, S. (1976). The acidity problem—an outline of concepts. In Dochinger, L. S., and Seliga, T. A. (eds), *Proceedings of the First International Symposium on Acid Precipitation and the Forest Ecosystem*, pp. 1–36, US Dept of Agriculture, Northeastern Forest Experimental Station, Upper Darby, Pennsylvania.

Pierce, R. S., Hornbeck, J. W., Likens, G. E., and Bormann, F. H. (1970). Effects of vegetation elimination on stream water quantity and quality. In Results of research on representative and experimental basins, Proceedings of the Wellington symposium 1970, *Int. Assoc. Hydrol. Sci. Publ.*, **96**, 311–28.

Pierce, R. S., Martin, C. W., Reeves, C. C., Likens, G. E., and Bormann, F. H. (1972). Nutrient loss from clearcuttings in New Hampshire. In *Watersheds in Transition*, Proceedings Series no. 14, pp. 285–95, American Water Resources Association, Minneapolis, Minnesota.

Piper, A. M. (1944). A graphic procedure in the geochemical interpretation of water analyses. *Trans. Am. Geophys. Union*, **25**, 914–23.

Posokhov, Y. V. (1968). Factors forming the chemical composition of precipitation. *Sov. Hydrol.*, **7**, 640–52.

Schumm,.S. A. (1954). The relation of drainage basin relief to sediment loss. *Int. Assoc. Hydrol. Sci. Publ.*, **36**, 216–9.

Skakalskiy, B. G. (1966). Basic geographical and hydrochemical characteristics of the local runoff of natural zones in the European territory of the USSR. *Trans. State Hydrol. Inst.* (*Trudy GGI*), **137**, 125–80.

Smith, K. (1975). Water temperature variation within a major river system. *Nord. Hydrol.*, **6**, 155–69.

Stevens, H. H., Ficke, J. F., and Smoot, G. F. (1975). Water temperature—influential factors, field measurement, and data presentation. *Techniques of Water-Resources Investigations of the U.S. Geological Survey*, TWI1—D5, US Geological Survey, Washington DC.

Stevenson, C. M. (1968). An analysis of the chemical composition of rain-water and air over the British Isles and Eire for the years 1959–64. *Q. J. R. Meteorol. Soc.*, **94**, 56–70.

Striffler, W. D. (1965). Suspended sediment concentrations in a Michigan trout stream as related to watershed characteristics. *Proceedings of the Federal Inter-agency Sedimentation Conference*, US Dept of Agriculture, Miscellaneous Publication 970, pp. 144–50, US Dept of Agriculture, Washington DC.

Sylvester, R. O., and Seabloom, R. W. (1963). Quality and significance of irrigation return flow. *J. Irrig. Drain. Div. Proc. ASCE*, **89**, 1–27.

Tollenaar, P. and Ryckborst, H. (1975). The effect of conifers on the chemistry and mass balance of two large lysimeters in Castricum (the Netherlands). *J. Hydrol.*, **24**, 77–87.

US Department of Agriculture (1976). *Control of Water Pollution from Cropland*, US Dept of Agriculture, Washington DC.

Vitousek, P. M. (1977). The regulation of element concentrations in mountain streams in the northeastern United States. *Ecol. Monogr.*, **47**, 65–87.

Vitousek, P. M., and Reiners, W. A. (1975). Ecosystem succession and nutrient retention: a hypothesis. *Bioscience*, **25**, 376–81.

Walling, D. E. (1974). Suspended sediment and solute yields from a small catchment prior to urbanization. In Gregory, K. J., and Walling D. E. (eds). *Fluvial Processes in Instrumented Watersheds*, Special Publication no. 6, pp. 169–92, Institute of British Geographers, London.

Walling, D. E. (1975). Solute variations in small catchment streams: some comments. *Trans. Inst. Br. Geogr.*, **64**, 141–7.

Walling, D. E. (1978). Suspended sediment and solute response characteristics of the River Exe, Devon, England. In Davidson-Arnott, R., and Nickling, W. (eds), *Proceedings of the Fifth Guelph Symposium on Geomorphology* 1977, pp. 169–97, Geo Abstracts, Norwich.

Walling, D. E., and Webb, B. W. (1975). Spatial variation of river water quality: a survey of the River Exe. *Trans. Inst. Br. Geogr.*, **65**, 155–69.

Webb, B. W., and Walling, D. E. (1974). Local variation in background water quality. *Sci. Total Environ.*, **3**, 141–53.

Weyman, D. R. (1970). Throughflow on hillslopes and its relation to the stream hydrograph. *Int. Assoc. Hydrol. Sci. Bull.*, **15**, 25–33.

White, D. E., Hem, J. D., and Waring, G. A. (1963). *Data of Geochemistry, Chapter F, Chemical Composition of Subsurface Waters*, Professional Paper 440-F, US Geological Survey, Washington DC.

Whitehead, H. C., and Feth, J. H. (1964). *Chemical Character of Precipitation at Menlo Park, California*, Professional Paper 424D, pp. 29–30, US Geological Survey, Washington DC.

Wischmeier, W. H., and Smith, D. D. (1965). *Predicting Rainfall-Erosion Losses from Cropland east of the Rocky Mountains*, Agriculture Handbook no. 282, US Dept of Agriculture, Washington DC.

Wolman, M. G. (1967). A cycle of sedimentation and erosion in urban river channels. *Geografiska Annaler*, **49A**, 385–95.

Wolman, M. G., and Schick, A. P. (1967). Effects of construction on fluvial sediment: urban and suburban areas of Maryland. *Water Resour. Res.*, **3**, 451–62.

CHAPTER 2

Burford, J. R., Greenland, D. J., and Pain, B. F. (1976). Effects of heavy dressings of slurry and inorganic fertilisers applied to grassland on the composition of drainage waters and the soil atmosphere. In *Agriculture and Water Quality*, Technical Bulletin no. 32, pp. 432–43, Ministry of Agriculture, Fisheries and Food, HMSO, London.

Burford, J. R., and Stefanson, R. C. (1973). Measurement of gaseous losses of nitrogen from soils. *Soil Biol. Biochem.*, **5**, 133–41.

Burns, I. G. (1977). Nitrate movement in soil and its agricultural significance. *Outlook Agric.*, **9** (3), 144–8.

Cooke, G. W. (1976). A review of the effects of agriculture on the chemical composition and quality of surface and underground waters. In *Agriculture and Water Quality*, Technical Bulletin no. 32, pp. 5–57, Ministry of Agriculture, Fisheries and Food, HMSO, London.

Cunningham, R. K., and Cooke, G. W. (1958). Changes in levels of inorganic nitrogen in a clay-loam soil caused by fertiliser additions, by leaching and uptake by grass. *J. Sci. Fd Agric.*, **9**, 317–24.

Foot, A. S. (chairman) (1976). *Studies on Farm Livestock Wastes*, Agricultural Research Council, London.

Foster, S. S. D. (1976). The vulnerability of British groundwater resources to pollution by agricultural leachates. In *Agriculture and Water Quality*, Technical Bulletin no. 32, pp. 68–91, Ministry of Agriculture, Fisheries and Food, HMSO, London.

Fryer, J. D. (1977). Recent developments in the agricultural use of herbicides in relation to ecological effects. In Perring, F. H., and Mellanby, K. (eds), *Ecological Effects of Pesticides*, pp. 27–46, Academic Press, London.

Gasser, J. K. R. (1959). Transformations and movements of fertiliser nitrogen in a light soil. *J. Sci. Fd Agric.*, **10**, 192–7.

Gasser, J. K. R. (1962). Effects of long continued treatment on the mineral nitrogen content and mineralisable nitrogen of soil from selected plots of the Broadbalk experiment on continuous wheat, Rothamsted. *Pl. Soil*, **17**, 209–20.

Gasser, J. K. R. (1968). Mineralisable-N in the soil under various leys and its effect on the yields of following wheat. *J. Agric. Sci., Camb.*, **70**, 323–9.

Graham-Bryce, I. J. (1977). Recent developments in the chemical control of agricultural pests and diseases in relation to ecological effects. In Perring, F. H., and Mellanby, K. (eds), *Ecological Effects of Pesticides*, pp. 47–60, Academic Press, London.

Holden, A. V. (1976). The relative importance of agricultural fertilisers as a source of nitrogen and phosphorus in Loch Leven. In *Agriculture and Water Quality*, Technical Bulletin no. 32, pp. 306–14, Ministry of Agriculture, Fisheries and Food, HMSO, London.

Hood, A. E. M. (1976). The leaching of nitrates from intensively managed grassland at Jealott's Hill. In *Agriculture and Water Quality*, Technical Bulletin no. 32, pp. 201–21, Ministry of Agriculture, Fisheries and Food, HMSO, London.

Kolenbrander, G. J. (1972). Eutrophication from agriculture with special reference to fertilisers and animal waste. *Soils Bull.* **16**, 305–27, FAO, Rome.

Lawes, J. B., and Gilbert, J. H. (1874). *Sixth Report of the Commissions Appointed in* 1868 *to Inquire into the Best Means of Preventing the Pollution of Rivers*, pp. 58–68, River Pollution Commission of Great Britain, London.

Lawes, J. B., Gilbert, J. H., and Warington, R. (1882). On the amount and composition of the rain and drainage water collected at Rothamsted. *J. R. Agric. Soc.*, **25**, 1–24.

Letcombe Laboratory (1978). *Annual Report for* 1977, Agricultural Research Council, London.

Low, A. J., and Armitage, E. R. (1970). The composition of the leachate through cropped and uncropped soils in lysimeters compared with that of rain. *Pl. Soil*, **33**, 393–411.

McAllister, J. S. V. (1976). Studies in Northern Ireland on problems related to the disposal of slurry. In *Agriculture and Water Quality*, Technical Bulletin no. 32, pp. 418–31, Ministry of Agriculture, Fisheries and Food, HMSO, London.

O'Callaghan, J. R., and Pollock, K. A. (1976). A land spreading trial with pig slurry. In *Agriculture and Water Quality*, Technical Bulletin no. 32, pp. 371–99, Ministry of Agriculture, Fisheries and Food, HMSO, London.

O'Callaghan, J. R., Pollock, K. A., and Dodd, V. A. (1971). Land spreading of manure from animal production units. *J. Agric. Eng. Res.*, **16**, 280–300.

Petersen, R. G., Woodhouse, W. W. (Jr), and Lucas, H. L. (1956). The distribution of excreta by freely grazing cattle and its effect on pasture fertility: II, Effect of returned excreta on the residual concentration of some fertiliser elements. *Agron. J.*, **48**, 444–9.

Richardson, S. J. (1976). Animal manures as potential pollutants. In *Agriculture and Water Quality*, Technical Bulletin no. 32, pp. 405–17, Ministry of Agriculture, Fisheries and Food, HMSO, London.

Shaw, K. (1962). Loss of mineral nitrogen from soil. *J. Agric. Sci., Camb.*, **58**, 145–51.

Sly, J. M. A. (1977). Changes in use of pesticides since 1945. In Perring, F. H., and Mellanby, K. (eds), *Ecological Effects of Pesticides*, pp. 1–6, Academic Press, London.

Smith, R. V. (1976). Nutrient budget of the River Maine, Co. Antrim. In *Agriculture and Water Quality*, Technical Bulletin no. 32, pp. 315–39, Ministry of Agriculture, Fisheries and Food, HMSO, London.

Stewart, W. D. P., May, E., and Tuckwell, S. B. (1976). Nitrogen and phosphorus from agricultural land and urbanisation and their fate in shallow freshwater lochs In *Agriculture and Water Quality*, Technical Bulletin no. 32, pp. 276–305, Ministry of Agriculture, Fisheries and Food, HMSO, London.

Troake, R. P. Troake, L. E., and Walling, D. E. (1976). Nitrate loads of South Devon streams. In *Agriculture and Water Quality*, Technical Bulletin no. 32, pp. 340–51, Ministry of Agriculture, Fisheries and Food, HMSO, London.

Webber, J., and Wadsworth, G. A. (1976). Nitrate and phosphate in borehole, well and stream waters. In *Agriculture and Water Quality*, Technical Bulletin no. 32, pp. 237–51, Ministry of Agriculture, Fisheries and Food, HMSO, London.

White, J. W. (1975). Relative significance of dietary sources of nitrate and nitrite. *J. Agric. Fd Chem.*, **23**, 886–91.

Widdowson, F. V. (1974). Results from experiments measuring the residues of nitrogen fertiliser given for sugar beet, and of ploughed-in sugar beet tops, on the yield of following barley. *J. Agric. Sci., Camb.*, **83**, 415–21.

Widdowson, F. V., Penny, A., and Williams, R. J. B. (1963). Experiments comparing yield, and residual effects on winter wheat, of 1-year clover, ryegrass and clover-ryegrass leys. *J. Agric. Sci., Camb.*, **61**, 397–408.

Widdowson, F. V., Penny, A., and Williams, R. J. B. (1967). Experiments comparing the effects on yields of potatoes of three methods of applying three amounts of NPK fertiliser and the residual effects on following winter wheat. *J. Agric. Sci., Camb.*, **69**, 247–53.

Wild, A., and Babiker, I. A. (1976). Winter leaching of nitrate at sites in Southern England. In *Agriculture and Water Quality*, Technical Bulletin no. 32, pp. 153–62, Ministry of Agriculture, Fisheries and Food, HMSO, London.

Williams, R. J. B. (1976). The chemical composition of rain, land drainage, and borehole water from Rothamsted, Broom's Barn, Saxmundham and Woburn Experimental Stations. In *Agriculture and Water Quality*, Technical Bulletin no. 32, pp. 174–200, Ministry of Agriculture, Fisheries and Food, HMSO, London.

CHAPTER 3

Andelman, J. B., and Suess, M. J. (1970). Polycyclic aromatic hydrocarbons in the water environment. *Bull. World Health Org.*, **43**, 479–508.

Andersson, T. (1969). Small scale variations of the contamination of rain caused by washout from the lower layers of the atmosphere. *Tellus*, **21** (5), 685–92.

Angino, E. E., Magnuson, L. M., and Stuart, G. F. (1972). Effects of urbanisation on storm water quality; a limited experiment, Naismith Ditch, Lawrence, Kansas. *Water Resour. Res.*, **8** (1) 135–40.

Barkdoll, M. P., Overton, D. E., and Betson, R. P. (1977). Some effects of dustfall on urban stormwater quality. *J. Water Pollut. Control Fed.*, **49** (9), 1976–84.

Barry, R. G., and Chorley, R. J. (1968). *Atmopshere, Weather and Climate*, Methuen, London.

Blumer, M., Blumer, W., and Reich, T. (1977). Polycyclic aromatic hydrocarbons in soils of a mountain valley: correlation with highway traffic and cancer incidence. *Environ. Sci. Tech.*, **11** (12), 1082–4.

Borch-Jensen, J. E. (1978). Runoff from motorways passing open country or open areas. In *OECD Symposium on Road Drainage, Berne*, pp. 364–83, OECD, Paris.

Brownlee, A. C., Austin, T. A., and Wells, D. M. (1970). *Variation of Urban Runoff Quality with Duration and Intensity of Storms*, Water Resources Center, Texas Technical University, Lubbock, Texas.

Bryan, E. H. (1974). Concentrations of lead in urban storm water. *J. Water Pollut. Control Fed.*, **46** (10), 2419–21.

Burt, T. P., and Day M. R. (1977). Spatial variations in rainfall and stream water quality around the Avonmouth industrial complex. *Int. J. Environ. Stud.*, **11** (3), 205–9.

Changnon, S.A. (1968). The La Porte weather anomaly—fact or fiction? *Bull. Am. Meteorol. Soc.*, **49** (1), 4–11.

Cordery, I. (1976). *Evaluation and Improvement of Quality Characteristics of Urban Stormwater*, Report no. 147, Water Research Laboratory, Manley Vale, NSW.

Cowen, W. F., and Lee, G. P. (1976). Phosphorus availability in particulate materials transported by urban runoff. *J. Water Pollut. Control Fed.*, **48** (3), 580–91.

Dauber, L., Novak, B., Zobrist, J., and Zürcher, F. (1978). Pollutants in motorway stormwater runoff. In *OECD Symposium on Road Drainage, Berne*, pp. 332–42, OECD, Paris.

Ellis, J. B. (1975). *Urban Stormwater Pollution*, Research Report no. 1, Middlesex Polytechnic, London.

Falk, H. L., Kotin, P., and Mehler, A. (1964). Polycyclic hydrocarbons as carcinogens for man. *Arch. Environ. Health*, **8**, 721–30.

Field, R., and Lager, J. A. (1975). Urban runoff pollution control—state of the art. *J. Environ. Eng. Div. Proc. ASCE.*, **101** (EE1), 107–25.

Field, R., Struzeski, J., Masters, H. E., and Tafuri, A. N. (1974). Water pollution and associated aspects from street salting. *J. Environ. Eng. Div. Proc. ASCE*, **100** (EE2), 459–77.

Gleason, G. B. (1952). Consumptive use of water, municipal and industrial areas. *Trans. Am. Soc. Agric. Eng.*, **117**, 1004–9.

Goettle, A. (1978). Atmospheric contaminants, fallout and their effect on storm water quality. *Prog. Water Tech.*, **10** (5/6), 455–67.

Graf, W. L. (1977). Network characteristics in suburbanizing streams. *Water Resour. Res.*, **13** (2), 459–63.

Grossling, B. F. (1976). An estimate of the amounts of oil entering the oceans. In *Sources, Effects and Sinks of Hydrocarbons in the Aquatic Environment; Proc. Symp.*

Washington DC, 9–11 August, pp. 6–34, American Institute of Biological Sciences, Washington DC.

Harms, L. L. (1977). *Release and Sorption of Phosphorus by Sediments in a Moving Watercourse*, OWRT Paper no. B-046-SDAK, Water Resources Research Institute, South Dakota State University, Brookings, South Dakota.

Harrison, R. M., Perry, R., and Slater, D. H. (1974). An adsorption technique for the determination of organic lead in street air. *Atmos. Environ.*, **8**, 1187–94.

Hedley, G., and Lockley, J. C. (1975). Quality of water discharged from an urban motorway. *Water Pollut. Control*, **74** (6), 659–74.

Hedley, G., and Lockley, J. C. (1978). Use of retention tanks on sewerage systems, a five year assessment. *Water Pollut. Control*, **77** (2), 178–92.

Hites, R. A. (1976). Sources of polycyclic aromatic hydrocarbons in the aquatic environment. In *Sources, Effects and Sinks of Hydrocarbons in the Aquatic Environment; Proc. Symp. Washington DC, 9–11 August*, pp. 82–94, American Institute of Biological Sciences, Washington DC.

Hites, R. A., and Biemann, K. (1972). Water pollution; organic compounds in the Charles River. *Science*, **178**, 158.

Holbrook, R. F., Perez, A. I., Turner, B. G., and Miller H. I. (1976). Stormwater studies and alternatives in Atlanta. *J. Environ. Eng. Div. Proc. ASCE*, **102** (EE6), 1263–77.

Hopkin, G. M. (1976). Problems associated with the transport of chemicals by road. *Water Pollut. Control*, **75** (5) 509–16.

Knight, E. W., and Rimkus, R. R. (1977). Effluent standards for treatment plants and stormwater overflows. *Prog. Water Tech.*, **8** (6), 17–26.

Koplyay, T., and Buxton, G. V. (1977). Practical solutions to stormwater problems in Canada. *Prog. Water Tech.*, **8** (6), 71–6.

Kuichling, E. (1909). Report on the adequacy of the existing systems of the Black Bay district of Boston, USA. In Watkins, L. H. (1962), *The Design of Urban Sewer Systems. Research into the Relation between the Rate of Rainfall and the Rate of Flow in Sewers*, Road Research Technical Paper no. 55, HMSO, London.

Landsberg, H. E. (1970). Climates and urban planning. In *Urban Climates, Proc. WMO Symp., Brussels*, Technical Note no. 108, World Meteorological Organization, Geneva.

Lavalle, P. D. (1973). *The Effects of Household Sanitary Systems on Effluent Phosphate Levels*, Research Programme for the Abatement of Municipal Pollution within the Provisions of the Canada–Ontario Agreement on Great Lakes Water Quality, Project no. 72-5-14, Research Report no. 2, Environment Canada, Ottawa, Ontario.

Lloyd, R. D. C. (1975). Improvement of quality of stormwater discharges to rivers and lakes. In *Water in the Urban Environment*, Summer School A7, Australian Water and Wastewater Association, Canberra, ACT.

McKee, H. C., and McMahon, W. A. (1967). *Polynuclear Aromatic Content of Vehicle Emissions*, Project no. 21-2139 to the Committee for Air and Water Conservation Technical Report no. 1, American Petroleum Institute, Washington DC.

Malaney, G. W., Lutin, P. A., Cibulka, J. J., and Hickerson, L. H. (1967). Resistence of carcinogenic organic compounds to oxidation by activated sludge. *J. Water Pollut. Control Fed.*, **39** (12) 2020–9.

Matson, E. A., Hornor, S. G., and Buck, J. D. (1978). Pollution indicators and other microorganisms in river sediment. *J. Water Pollut. Control Fed.*, **50** (1), 13–9.

Mattraw, H. C., and Sherwood, C. B. (1977). Quality of stormwater runoff from a residential area, Broward County, Florida. *J. Res. US Geol. Surv.*, **5** (6) 823–34.

Milaszewski, R., and Osuch-Pajdzinska, E. L. (1972). Features of rainfall generated runoff pollution. *Gaz Woda I Technika Sanitarna*, **46** (4), 112–6.

Mueller, J. A., and Anderson, A. R. (1978). A mass balance for estimating combined sewer overflow quality from sewage treatment plant data, *Prog. Water Tech.*, **10** (5/6), 727–39.

Myers, L. H. (1976). Sources and distribution of hydrocarbons in the environment. In *Sources, Effects and Sinks of Hydrocarbons in the Aquatic Environment; Proc. Symp. Washington DC, 9–11 August*, pp. 67–79, American Institute of Biological Sciences, Washington DC.

Norton, J. L. (1974). *The Identification and Measurement of Chlorinated Hydrocarbon Pesticides Accumulated from Urban Runoff*, Water Resources Research Institute, Oklahoma State University, Stillwater, Oklahoma.

Oliver, B. C., Milne, J. B., and LaBarre, N. (1974). Chloride and lead in urban snow. *J. Water Pollut. Control Fed.*, **46** (4), 766–71.

Oliveri, V. P., Kruse, C. W., and Kawata, K. (1977). *Microorganisms in Urban Stormwater*, Report EPA-600/2-77-087, Environmental Protection Agency, Cincinnati, Ohio.

Patterson, C. C. (1965). Contaminated and natural lead environments of man. *Arch. Environ. Health*, **11**, 344–60.

Pecher, R. (1974). The annual quantity and pollution of stormwater runoff from populated areas. *Osterreichische Abwasser Rundschau*, **19** (6), 89–94.

Peukert, V. (1974). Effects of atmospheric contaminants on water quality. *Wasserwirt-Wassertech.*, **24** (5), 158–61.

Pirner, S. M., and Harms, L. L. (1978). Rapid City combats the effects of urban runoff on surface water. *Water Sew. Wks*, **125** (2), 48–53.

Pluhowski, E. J., and Spinello, A. G. (1978). Impact of sewerage systems on stream base flow and ground water recharge on Long Island, New York. *J. Res. US Geol. Surv.*, **6** (2), 263–73.

Pope, W., Graham, N. J. D., Young, R. J., and Perry, R. (1978a). Urban runoff from a road surface—a water quality study. *Prog. Water Tech.*, **10** (5/6), 533–43.

Pope, W., Young R. J., and Perry, R., (1978b) Correlations of hydrological conditions with water quality in motorway runoff. In *OECD Symposium on Road Drainage, Berne*, pp. 384–400, OECD, Paris.

Pope, W., Young, R. J., and Perry, R. (1979). Aspects of water quality. *Water Pollut. Control* (in press).

Potter, H. R., Jarvie, A. W. P., and Markall, R. N. (1977). Detection and determination of alkyl lead compounds in natural waters. *Water Pollut. Control*, **76** (1), 123–8.

Pratt, M. J., and Coler, R. A. (1976). A procedure for the routine biological evaluation of urban runoff in small rivers. *Water Res.*, **10**, 1019–25.

Randall, C. W., Garland, J. A., Grizzard, T. J., and Hoehn, R. C. (1977). The significance of stormwater runoff in an urbanizing watershed. *Prog. Water Tech.*, **9** (3), 547–62.

Reed, R. R. (1976). *Stream Community Response to Road Construction Sediments*, Bulletin no. 97, Virginia Water Resources Research Center, Virginia.

Roberts, P. V., Dauber, L., Novak, B., and Zobrist, J. (1977). Pollutant loadings in urban stormwater. *Prog. Water Tech.*, **8** (6), 93–101.

Sartor, J. D., Boyd, G. B., and Agardy, F. J. (1974). Water pollution aspects of street surface contaminants. *J. Water Pollut. Control Fed.*, **46** (3) 458–67.

Scheidt, M. E. (1967). Environmental aspects of highways. *J. Sanit. Eng. Div. Proc. ASCE*, **93** (SA5), 17–25.

Schmassmann, G. (1978). Exemples pratiques de mesures pour la protection des eaux souterraines en cas d'accidents routiers avec des produits petroliers. In *OECD Symposium on Road Drainage, Berne*, pp. 426–36, OECD, Paris.

Scott, W. S. (1976). The effect of road deicing salts on sodium concentration in an urban water course. *Environ. Pollut.*, **10**, 141–53.

Söderlund, G., and Lehtinen, H. (1973). Comparison of discharges from urban stormwater runoff, mixed storm overflow and treated sewage. In Jenkins, S. H. (ed.), *Advances in Water Pollution Research, Proceedings of the Sixth International Conference*, pp. 309–25. Pergamon Press, Oxford.

Stoerner, F., Ladewski, B. G., and Shelske, C. L. (1978). Population response of Lake Michigan phytoplankton to nitrogen and phosphorus enrichment. *Hydrobiologia*, **57** (3), 249–65.

Suess, M. J. (1976). The environmental load and cycle of polycyclic aromatic hydrocarbons. *Sci. Total Environ.*, **6**, 239–50.

Ter Haar, G. L., Lenane, D. L., Hu, J. N., and Brandt, M. (1972). Composition of size and control of automobile exhaust particulates. *J. Air Pollut. Control Assoc.*, **22** (1) 39–46.

Von Kaenel, A., and Tschumi, P. A. (1978). Die einleitung autobahnabwasser in altwassergebeit suedlich von Bern. In *OECD Symposium on Road Drainage, Berne*, pp. 437–48, OECD, Paris.

Waller, D. H., and Coulter, W. A. (1974). *Winter Runoff from an Urban Catchment*, Research Programme for the Abatement of Municipal Pollution within the Provisions of the Canada–Ontario Agreement on Great Lakes Water Quality, Project no. 74-8-3, Research Report no. 41, Environment Canada, Ottawa, Ontario.

Walling, D. E., and Gregory, K. (1970). The measurement of the effects of building construction on drainage basin dynamics. *J. Hydrol.*, **11**, 129–44.

Wanielista, M. P., Yousef, Y. A., and McLlellon, W. M. (1977). Non-point source effects on water quality. *J. Water Pollut. Control Fed.*, **49** (3), 441–51.

Weber, W. G., and Reed, L. A. (1976). Sediment runoff during highway construction. *Civ. Eng. ASCE.*, **46** (3), 76–9.

Weibel, S. R., Weinder, R. R., Cohen, J. M., and Christianson, A. G. (1966). Pesticides and other contaminants in rainfall and runoff. *J. Am. Water Wks Assoc.*, **58**, 1075–84.

Welch, E. H., Sturtevant, P., and Perkins, M. A. (1978). Dominance of phosphorus over nitrogen as the limiter to phytoplankton growth rate. *Hydrobiologia*, **57** (3), 209–15.

Whipple, W. (1975). Water quality planning for the Delaware estuary—an example. *Water Resour. Bull.*, **11** (2), 300–5.

Whipple, W. (1977). Flood management for small urban streams. *J. Water Resour. Plann. Manage. Div. Proc. ASCE*, **103** (WR2), 315–24.

Whipple, W., Berger, B. B., Gates, C. D., Ragan, R. M., and Randall, C. W. (1978). Characterization of urban runoff. *Water Resour. Res.*, **14** (2), 370–2.

Whipple, W., Hunter, J. V., and Yu, S. L. (1974). Unrecorded pollution from urban runoff. *J. Water Pollut. Control Fed.*, **46** (5), 873–85.

Wisseman, R. W., and Cook, S. F. (1977). Heavy metal accumulation in the sediments of a Washington Lake. *Bull. Environ. Contam. Toxicol.*, **18** (1), 77–82.

Wolman, M. G. (1967). Two problems involving river channel changes and background observations. In *Studies in Geography*, vol. 14, pp. 67–107, North Western University, Chicago, Illinois.

Youngblood, W. W., and Blumer, M. (1975). Polycyclic aromatic hydrocarbons in the environment: homologous series in soils and recent marine sediments. *Geochim. Cosmochim. Acta*, **39**, 1303–14.

CHAPTER 4

Alexander, G. S., and Stevens, R. J. (1976). Per capita phosphorus loading from domestic sewage. *Water Res.*, **10**, 757–64.

Buswell, A. M., and Mueller, H. F. (1952). Mechanism of methane fermentation. *Ind. Eng. Chem.*, **44**, 550–2.

Central Electricity Generating Board (1976). *Statistical Yearbook 1975–6*, Central Electricity Generating Board, London.

Central Water Planning Unit (1976). *The Significance of Synthetic Chemicals in Rivers used as a Source of Drinking Water*, Technical Note no. 14, Central Water Planning Unit, Reading.

Department of the Environment (1971/2). *Report of a River Pollution Survey of England and Wales 1970*, HMSO, London.

Department of the Environment (1976). *Local Authority Waste Disposal Statistics 1974/75*, Waste Management Paper no. 10, HMSO, London.

Department of the Environment (1978a). *River Pollution Survey of England and Wales Updated 1975*, HMSO, London.

Department of the Environment (1978b). *Eighteenth Report of the Standing Technical Committee on Synthetic Detergents*, HMSO, London.

Department of the Environment/National Water Council (1977). *Report of the Working Party on the Disposal of Sewage Sludge to Land*, National Water Council, London.

Devey, D. G., and Harkness, N. (1973). The significance of man-made sources of phosphorus: detergents and sewage. *Water Res.*, **7**, 35–54.

Doneen, L. D., and Henderson, D. W. (1957). Quality of irrigation waters. *Calif. Agric.*, **11**, 4–11.

Esvelt, L. A., and Hart, H. H. (1970). Treatment of fruit processing wastes by aeration. *J. Water Pollut. Control Fed.*, **42**, 1305–26.

Garland, J. H. N. (1978). Nitrification in the river Trent. In James, A. (ed.), *Mathematical Models in Water Pollution Control*, pp. 167–92, Wiley, Chichester.

Glover, H. G. (1975). Acidic and ferruginous mine drainage. In Chadwick, M. J., and Goodman, G. T. (eds), *The Ecology of Resource Degradation and Renewal, 15th Symposium of the British Ecological Society 1973*, pp. 173–96, Blackwell Scientific, Oxford.

Greene, L. A. (1978). Nitrate in water supply abstractions in the Anglian region: current trends and remedies under investigation. *Water Pollut. Control*, **77**, 478–91.

Holden, A. V., and Marsden, K. (1966). Examination of surface waters and sewage effluents for organo-chlorine pesticides. *J. Inst. Sew. Purif.*, **65**, 295–300.

Huang Chin Pau and Ghadiran, M. (1974). Physical-chemical treatment of paint industry wastewaster. *J. Water Pollut. Control Fed.*, **46**, 2340–6.

Hughes, B. D., and Edwards, R. W. (1977). Flows of sodium, potassium, magnesium and calcium in the river Cynon, South Wales. *Water Res.*, **11**, 563–6.

Institute of Water Pollution Control (1972). *Directory of Municipal Waste Water Treatment Plants.* Institute of Water Pollution Control, Maidstone.

Isaac, P. C. G. (1966). Treatment and disposal of brewery effluents. *Brewers Guild J.*, **52**, 441–58.

James, A. (1976). Pollution of the River Tyne estuary—the use of mathematical models. *Water Pollut. Control*, **75**, 322–40.

Jones, A. N., and Howells, W. R. (1975). The partial recovery of the metal polluted River Rheidol. In Chadwick, M. J., and Goodman, G. T. (eds), *The Ecology of Resource Degradation and Renewal, 15th Symposium of the British Ecological Society 1973*, pp. 443–59, Blackwell Scientific, Oxford.

Klein, L., Aylang, M., Nash, N., and Kirschner, S. L. (1974). Sources of metals in New York City wastewater. *J. Water Pollut. Control Fed.*, **46**, 2653–62.
Lester, W. F. (1975). Polluted river; River Trent, England. In Whitton, B. A. (ed.), *River Ecology*, pp. 489–513, Blackwell Scientific, Oxford.
Little, A. H. (1973). Sampling and samplers. *Water Pollut. Control*, **72**, 606–17.
Matthews, P. J. (1974). A survey of the boron content of certain waters of the Greater London area using a novel analytical method. *Water Res.*, **8**, 1021–8.
Ministry of Housing and Local Government (1970). *Final Report of the Technical Committee on Storm Sewage Overflows and Disposal of Storm Sewage*, HMSO, London.
National Water Council (1978a). *River Water Quality, the Next Stage. Review of Discharge Consent Conditions*, National Water Council, London.
National Water Council (1978b). *Water Industry Review 1978*, National Water Council, London.
Owens, M. (1970). Nutrient balances in rivers. *Water Treat. Exam.*, **19**, 239–52.
Prater, B. E. (1975). The metal content and dispersal characteristics of steelworks effluents discharging to the Tees estuary. *Water Pollut. Control*, **74**, 63–78.
Purves, J. B. (1975). Aspects of mining and pollution control in S.W. England. In Jones, M. J. (ed.), *Minerals and the Environment*, pp. 159–83, Institution of Mining and Metallurgy, London.
Rosenfelder W. J. (1978). The industrial uses of boron chemicals. *Chem. Ind.* **1978** (12), 413–6.
Royal Commission on Sewage Disposal (1912). *8th Report: Standards and Tests for Sewage and Sewage Effluents discharging into Rivers and Streams*, HMSO, London.
Severn–Trent Water Authority (1978). *Water Quality 1977/8*, Severn–Trent Water Authority, Birmingham.
Singleton, K. A., and Hosden A. J. (1973). Developments in effluent treatment at Ciba–Geigy (UK) Ltd, Plastics Division, Duxford. *Water Pollut. Control*, **72**, 581–9.
Stumm, W., and Morgan, J. J. (1970, *Aquatic Chemistry*, Wiley–Interscience, New York.
Thackray, J. E., Cocker, V., and Archibald, G. (1978). The Malvern and Mansfield studies of domestic water usage. *Proc. Inst. Civ. Eng.*, Part 1, **64**, 37–61.
Theis, T. L., Westrick, J. O., Hdu, C. L., and Marley, J. J. (1978). Field investigation of trace metals in groundwater from fly ash disposal. *J. Water Pollut. Control Fed.*, **50**, 2457–69.
Thom, N. S., and Agg, A. R. (1975). The breakdown of synthetic organic compounds in biological processes. *Proc. R. Soc. B*, **189**, 347–57.
Tookos, I. (1977). Treatment of high organic waste waters from the food industry. *Effl. Water Treat. J.*, **17**, 449–59.
Waggott, A. (1969). An investigation of the potential problem of increasing boron concentrations in rivers and watercourses. *Water Res.*, **3**, 749–66
Wilcox, L. V. (1958). Water quality from the standpoint of irrigation. *J. Am. Water Wks Assoc.*, **50**, 650–4.
Williams, R., Williams, P. F., Benson-Evans, K., Hunter, M. D., and Harcup, M. F. (1975). Chemical and biological studies associated with the recovery of the river Ebbw Fawr (1970–1974). Water Pollut. Control, **75**, 428–44.
Wood, L. B. (1977). River Standards in the Thames Water Authority (2), Standards for Discharge to the Thames Tideway. *Prog. Water Tech.*, **8** (6), 33–40.
World Health Organization (1975). *Health Effects relating to Direct and Indirect Reuse of Waste Water for Human Consumption*, Technical Paper no. 7, World Health Organization, Geneva.
Wuhrmann, K., and Mechsner, K. (1974). *News* no. 3, Swiss Federal Institute for Water Resources and Water Pollution Control, Dubendorf.

CHAPTER 5

Alabaster, J. S. (1972). Suspended solids and fisheries. *Proc. R. Soc. B*, **180**, 395–406.
Aston, S. R., and Thornton, I. (1977). Regional geochemical data in relation to seasonal variations in water quality. *Sci. Total Environ.*, **7**, 247–60.
Balloch, D., Davies, C. E., and Jones, F. H. (1976). Biological assessment of water quality in three British rivers: the North Esk (Scotland), the Ivel (England) and the Taf (Wales). *Water Pollut. Control*, **75**, 92–110.
Brown, B. E. (1976). Observations on the tolerance of the isopod *Asellus meridianus* Rac. to copper and lead. *Water Res.*, **10**, 555–9.
Brown, B. E. (1977). Uptake of copper and lead by a metal tolerant isopod *Asellus meridianus* Rac. *Freshwat. Biol.*, **7**, 235–44.
Brown, V. M. (1978). Fish as indicators of water quality. In *Elaboration of the Scientific Bases for Monitoring the Quality of Surface water by Hydrobiological Indicators*, Pollution Report no. 3, pp. 92–108, Department of the Environment, London.
Bryce, D., Caffoor, I. M., Dale, C. R., and Jarrett, A. F. (1978). *Macroinvertebrates and the Bioassay of Water Quality: A Report on a Survey of the River Lee*, Nelpress, London.
Butcher, R. W. (1946). The biological detection of pollution. *J. Inst. Sew. Purif.*, **2**, 92–7.
Carpenter, K. E. (1924). A study of the fauna of rivers polluted by lead mining in the Aberystwyth district of Cardiganshire. *Ann. Appl. Biol.*, **9**, 1–23.
Chandler, J. R. (1970). A biological approach to water quality management. *Water Pollut. Control*, **69**, 415–22.
Cook, S. E. K. (1976). Quest for an index of community structure sensitive to water pollution. *Environ. Pollut.*, **11**, 269–88.
Davies, D. M. (1950). A study of the blackfly population of a stream in Algonquin Park, Ontario. *Trans. R. Can. Inst.*, **28**, 121–59.
Gower, A. M., and Buckland, P. J. (1978). Water quality and the occurrence of *Chironomus riparius* Meigen (Diptera: Chironomidae) in a stream receiving sewage effluent. *Freshwat. Biol.*, **8**, 153–64.
Harding, J. P. C., and Whitton, B. A. (1978). Zinc, cadmium and lead in water, sediments and submerged plants of the Derwent reservoir, northern England. *Water Res.*, **12**, 307–16.
Hargreaves, J. W., Mason, C. F., and Pomfret, R. (1979). A simplified biotic index for the assessment of biological oxidizable pollution in flowing waters. *Water Pollut. Control*, **78**, 98–105.
Hawkes, H. A. (1963). Biological detection and assessment of river pollution. *Effl. Water Treat. J.*, **3**, 651–6.
Hellawell, J. M. (1978). *Biological Surveillance of Rivers*, Water Research Centre, Stevenage.
Horne, A. J. (1977). Nitrogen fixation—a review of this phenomenon as a polluting process. *Prog. Water Tech.*, **8**, 359–72.
Jones, J. R. E. (1940a). A study of the zinc-polluted river Ystwyth in north Cardiganshire, Wales. *Ann. Appl. Biol.*, **27**, 368–78.
Jones, J. R. E. (1940b). The fauna of the River Melindwr, a lead-polluted tributary of the River Rheidol in north Cardiganshire, Wales. *J. Anim. Ecol.*, **9**, 188–201.
Mortimer, C. H. (1941). The exchange of dissolved substances between mud and water in lakes. I and II. *J. Ecol.*, **29**, 280–329.
Mortimer, C. H. (1942). The exchange of dissolved substances between mud and water in lakes. III and IV. *J. Ecol.*, **30**, 147–201.
Moss, B. (1977). Conservation problems in the Norfolk Broads and rivers of East

Anglia, England—phytoplankton, boats and the causes of turbidity. *Biol. Conserv.*, **12**, 95–114.

Nuttall, P. M. (1973). The effects of refuse-tip liquor upon stream biology. *Environ. Pollut.*, **4**, 215–22.

Owens, M. (1978). Nutrients—their sources and effects. In *Elaboration of the Scientific Bases for Monitoring the Quality of Surface Water by Hydrobiological Indicators*, Pollution Report no. 3, pp. 19–28, Department of the Environment, London.

Patrick, F. M., and Loutit, M. W. (1976). Passage of metals in effluents, through bacteria to higher organisms. *Water Res.*, **10**, 333–5.

Patrick, F. M., and Loutit, M. W. (1978). Passage of metals to freshwater fish from their food. *Water Res.*, **12**, 395–8.

Pitcairn, C. E. R., and Hawkes, H. A. (1973). The role of phosphorus in the growth of *Cladophora. Water Res.*, **7**, 159–71.

Price, D. R. H. (1978). Fish as indicators of water quality. *Water Pollut. Control*, **77**, 285–96.

Ramamoorthy, S., Springthorpe, S., and Kushner, D. J. (1977). Competition for mercury between river sediment and bacteria. *Bull. Environ. Contam. Toxicol.*, **17**, 505–11.

Resh, V. H., and Unzicker, J. D. (1975). Water quality monitoring and aquatic organisms: the importance of species identification. *J. Water Pollut. Control Fed.*, **47**, 9–19.

Sawyer, C. N. (1947). Fertilization of lakes by agricultural and urban drainage. *J. New Engl. Water Wks Assoc.*, **61**, 109–27.

Shannon, C. E., and Weaver, W. (1963). *The Mathematical Theory of Communications*, University of Illinois Press, Urbana, Illinois.

Simonsen, J. F., and Harremoës, P. (1978). Oxygen and pH fluctuations in rivers. *Water Res.*, **12**, 477–89.

Södergren, S. (1976). Ecological effects of heavy metal discharge in a salmon river. *Rep. Inst. Freshwat. Res. Drottingholm*, **55**, 99–131.

Sullivan, J. F., Atchinson, G. J., Kolar, D. J., and McIntosh, A. W. (1978). Changes in the predator–prey behaviour of fathead minnows (*Pimephales promelas*) and largemouth bass (*Micropterus salmoides*) caused by cadmium. *J. Fish. Res. Board Can.*, **35**, 446–51.

Thomas, E. A. (1969). The process of eutrophication in central European lakes. In *Eutrophication: Causes, Consequences, Correctives, Proc. Symp.*, pp. 29–49. National Academy of Sciences, Washington DC.

Vollenweider, R. A. (1968). *Scientific Fundamentals of the Eutrophication of Lakes and Flowing Waters with Particular Reference to Nitrogen and Phosphorus as Factors in Eutrophication*, Paris, OECD.

Wentsel, R., McIntosh, A., and Anderson, V. (1977a). Sediment contamination and benthic macroinvertebrate distribution in a metal impacted lake. *Environ. Pollut.*, **14**, 187–93.

Wentsel, R., McIntosh, A., and Atchison, G. (1977b). Sublethal effects of heavy metal contaminated sediment on midge larvae (*Chironomus tentans*), *Hydrobiologia*, **56**, 153–6.

CHAPTER 6

Amavis, R., Hunter, W., and Smeets, J. (eds). (1976) *Hardness of Drinking Water and Public Health. Proceedings of the European Scientific Colloquium, Luxembourg, 1975*, Pergamon Press, Oxford.

Anon (1978). Groundwater protection in the Christina basin. *Water Sew. Wks*, **125** (1), 68–71.

Baier, J. H. and Rykbost, K. A. (1976). The contribution of fertiliser to the groundwater of Long Island. *Groundwater*, **14** (6), 439–47.

Beeton, A. M. (1965). Eutrophication of the St Lawrence Great Lakes. *Limnol. Oceanogr.*, **10**, 240–54.

Bruvold, W. H. (1972). *Public Attitudes toward Reuse of Reclaimed Water*, Report no. 137, Water Resources Center, University of California, Los Angeles, California.

Butler, R. G., Orlob, G. T., and McGaughey, P. H. (1954). Underground movement of bacterial and chemical pollutants. *J. Am. Water Wks Assoc.*, **46**, 97–111.

Cantor, K. P., and McCabe, L. J. (1978). The epidemiologic approach to the evaluation of organics in drinking water. In Jolley, R. L., Gorchev, H., and Hamilton D. H. (Jr), (eds), *Water Chlorination: Environmental Impact and Health Effects*, vol. 2, Ann Arbor Science, Ann Arbor, Michigan.

Dutka, B. J. (1973). Coliforms are an inadequate index of water quality. *J. Environ. Health*, **36** (1), 39–46.

Eckholm, E. P. (1975). Salting the Earth. *Environment*, **17** (7), 9–15.

Friend, J. P. (1973). The global sulfur cycle. In Rasool, S. I. (ed.), *Chemistry of the Lower Atmosphere*, pp. 177–201, Plenum Press, New York.

Greenberg, A. E., and Ongerth, H. J. (1966). Salmonellosis in Riverside, California. *J. Am. Water Wks Assoc.*, **58**, 1145–50.

Hodges, L. (1977). *Environmental Pollution*, Holt, Rinehart and Winston, New York.

International Reference Centre for Community Water Supply (1973). *The Potential Pollution Index as a Tool for River Water Quality Management*, Technical Paper Series no. 6, International Reference Centre for Community Water Supply, The Hague.

International Reference Centre for Community Water Supply (1975). *Health Effects Relating to Direct and Indirect Reuse of Wastewater for Human Consumption*, Technical Paper Series no. 7, International Reference Centre for Community Water Supply, The Hague.

Kenny, A. W. (1978). Water quality criteria and standards in the UK and EEC: current and foreseeable problems. *J.R. Soc. Health.*, **98** (3), 116.

Metzler, D., *et al.* (1958). Emergency use of reclaimed water for potable supply at Chanute, Kansas, *J. Am. Water Wks Assoc.*, **50**, 121.

Miller, R. J., and Smith, R. B. (1976). Nitrogen balance in the southern San Joaquin Valley. *J. Environ. Qual.*, **5** (3), 274–8.

National Academy of Sciences, National Research Council, Safe Drinking Water Committee (1977). *Drinking Water and Health*, National Academy of Sciences, Washington DC.

Nriagu, J. O., and Coker, R. D. (1976). Emission of sulfur from Lake Ontario sediments. *Limnol. Oceanogr.*, **21**, 485–9.

Olson, B. H., and Pratte, J. L. (1978). Public acceptance of expanded use of renovated wastewater. *Prog. Water Tech.*, **10** (1/2), 319–27.

Packham, R. F. (1976). Potable water from sewage effluent. In *Sewage Effluent as a Water Resource, London, November 1973, Report of Proceedings*, pp. 100–11, Institute of Public Health Engineers, London.

Pitt, W. A. J., *et al.* (1976). *Groundwater Quality in Selected Areas serviced by Septic Tanks, Dade County, Florida*, US Geological Survey, Washington DC.

Saliternik, C. (1976). Groundwater pollution by nitrogen compounds in Israel. In Jenkins, S. H. (ed.), *Progress in Water Technology, Proceedings of the Seventh Conference of the International Association on Water Pollution Research*, vol. 2, pp. 171–9, Pergamon Press, Oxford.

Sawyer, C. N. (1960). *Chemistry for Sanitary Engineers*, McGraw-Hill, New York.
Schiffman, A. (1978). Controlling impoundments which may pollute groundwater. *Water Sew. Wks*, **125** (1), 62–7.
Skeat, W. O. (ed.) (1969). *Manual of British Water Engineering Practice*, 4th edn, W. Heffer for Institution of Water Engineers, Cambridge.
Swenson, H. A. (1962). The Montebello incident. *Proc. Soc. Water Treat. Exam.*, **11**, 84.
UNESCO/WHO (1978). *Water Quality Surveys*, UNESCO, Paris. WHO, Geneva.
Van Vuuren, L. R. J., Henzen, M. R., Stander, G. J., and Clayton, A. J. (1971). The full scale reclamation of purified sewage effluent for the augmentation of the domestic supplies of the city of Windhoek. In Jenkins, S. H. (ed.), *Advances in Water Pollution Research, Proceedings of the Fifth International Conference*, vol. I, Paper no. I–32, Pergamon Press, Oxford.
Walker, R. (1978). *Water Supply, Treatment and Distribution*, Prentice-Hall, Englewood Cliffs, New Jersey.
Windle-Taylor, E. (1978). The relationship between water quality and human health: medical aspects. *J.R. Soc. Health*, **98** (3), 121.
World Health Organization (1970a). *European Standards for Drinking Water*, World Health Organization, Geneva.
World Health Organization (1970b) *Fluorides and Human Health*, Monograph Series no. 59, World Health Organization, Geneva.
World Health Organization (1971). *International Standards for Drinking Water*, World Health Organization, Geneva.
World Health Organization (1973a). *Reuse of Effluents: Methods of Wastewater Treatment and Health Safeguards*, Technical Report Series no. 517, World Health Organization, Geneva.
World Health Organization (1973b). *World Health Statistics Report*, vol. 26, no. 11, World Health Organization, Geneva.
World Health Organization (1978). *Health Hazards from Drinking Water; Report of a Working Group*, Offset Document, ICP/PPE/005, World Health Organization, Copenhagen.
World Health Organization (1979a). *Consultation on the Revision of Drinking Water Standards*, Offset Document EHE/EHC/79.12, World Health Organization, Geneva.
World Health Organization (1979b). *Sodium, Chlorides and Conductivity in Drinking Water; Report of a Working Group*, World Health Organization, Copenhagen (in press).
Young, C. P., Oakes, D. B., and Wilkinson, W. B. (1977). Prediction of future nitrate concentrations in groundwater. *Groundwater*, **14** (6), 426–38.

CHAPTER 7

Anglian Water Authority (1978). Anglian's £10 million nitrate problem. *Water*, **1978** (19), 37.
Arrhenius, E. (1977). Health effects of a multipurpose use of water. *Ambio*, **6**, 59–62.
Bolton, D. H., and Ousby, J. C. (1977). The ICI deep shaft effluent treatment process and its potential for large sewage works. *Prog. Water Tech.*, **8** (6), 265–73.
Boon, A. G., and Borne, B. J. (1976). Minimising the problem of treatment and disposal of industrial effluent. In *Industrial Pollution Control Yearbook, 1976*, pp. 137–44, Fuel and Metallurgical Journals, Redhill.
Brenner, R. C. (1974). *EPA Experiences in Oxygen—Activated Sludge*, Environmental Protection Agency, Cincinnati, Ohio.
Bruce, A. M., Brown, B. L., and Mann, H. T. (1973). Some developments in the treatment of sewage from small communities. *Public Health Eng.*, **1973** (3), 116–37.

Callely, A. G., Forster, C. F., and Stafford, D. A. (1977), *Treatment of Industrial Effluents*, Hodder and Stoughton, London.

Cooper, P. F., Drew, E. A., Bailey, D. A., and Thomas, E. V. (1977). Recent advances in sewage effluent denitrification: Part 1. *Water Pollut. Control* **76** (3), 287–300.

Department of Industry (1975). *Technology and the Environment*, Reports from Scientific Counsellors Overseas no. 8. HMSO, London.

Department of the Environment (1970). *Disposal of Solid Toxic Wastes*, HMSO, London.

Department of the Environment (1975). *Programme of Research on the Behaviour of Hazardous Wastes in Landfill Sites. Interim Report on Progress*, Dept of the Environment, London.

Department of the Environment (1976a). *Local Authority Waste Disposal Statistics 1974/75*, Waste Management Paper no. 10, HMSO, London.

Department of the Environment (1976b). *The Licensing of Waste Disposal Sites*, Waste Management Paper no. 4, Appendix 1, HMSO, London.

Department of the Environment/National Water Council (1977). *Report of the Working Party on the Disposal of Sewage Sludge to Land*, National Water Council, London.

Eckenfelder, W. W. (1966). *Industrial Water Pollution Control*, McGraw-Hill, New York.

EEC (1975a). *Council Directive Concerning the Quality Required of Surface Water Intended for the Abstraction of Drinking Water in the Member States*, 75/440/EEC-OJ no. L194/26-31, EEC, Brussels.

EEC (1975b). *Proposal for Council Directive Relating to the Quality of Water for Human Consumption*, no. C214/2b, EEC, Brussels.

Environmental Protection Agency (1974). *Process Design Manual for upgrading existing Wastewater Treatment Plants*, Environmental Protection Agency, Cincinnati, Ohio.

Environmental Protection Agency (1976). *Process Design Manual for Phosphorus Removal*, Environmental Protection Agency, Cincinnati, Ohio.

Fish, H.(1978a). Water quality of rivers and estuaries. *Prog. Water Tech.*, **9** (4), 973–1000.

Fish, H. (1978b). River and effluent standards. *Prog. Water Tech.*, **10** (3/4) 233–41.

Fish, H, and Torrance, S. (1977). New safeguards for river water. *Water*, **1978** (15), 15.

Fish, H., and Torrance, S. (1978). *Control of Specific Industrial Pollution of Water Sources*, Paper Presented at the International Water Supply Association Conference, Kyoto.

Fish, H., and West, J. T. (1978). The water cycle; meeting the quality objectives. *Water*, **1978** (20), 25–30.

Flanigen, E. M., Bennett, J. M., Grose, R. W., Cohen, J. P., Patton, R. L., Kirchner, R. M., and Smith J. V. (1978). Silicalite, a new hydrophobic crystalline silica molecular sieve. *Nature*, **271**, 512–6.

Gowan, D. (1972). The disposal of agricultural waste. In Fuller, J. R. (ed.), *Water Pollution Manual*, pp. 27–34, Thunderbird Enterprises, Harrow.

Greater London Council (1973). *Annual Report of the Scientific Adviser 1973*, Greater London Council, London.

Hanbury, M. J., Rachwal, A. J., Johnstone, D. W. M., Critchard, D., and Cox, G. C. (1978). *Evaluation of the Carousel System and its Potential for Denitrification*, Paper Presented to the Institution of Water Engineers and Scientists, London.

Hemming, M. L., Ousby, J. C., Plowright, D. R., and Walker, J. (1977). Deep shaft, the latest position. *Water Pollut. Control*, **76** (4), 441–51.

Hinge, D. C. (1977). *Experiences in Continuous Monitoring of River Water Quality*, Paper Presented at Water Research Centre Seminar.

Johnstone, D. W. M., Rachwal, A. J., and Hanbury, M. J. (1978). *Settlement Character-*

istics and Settlement Tank Performance in the Carousel System, Paper Presented to the Institute of Water Pollution Control, London.
McLoughlin, J. (1976). *The Law and Practice relating to Member States of the European Communities; a Comparative Study*, Graham and Trotman, London.
Mather, J. D., and Bromley, J. (1977). *Research into Leachate Generation and Attenuation in Landfill Sites*, Paper 25, Land Reclamation Conference, Thurrock, 1976.
Mills, E. V. (1945). The treatment of settled sewage in percolating filters with periodic change in the order of filters; results of the operation of the experimental plant at Minworth, Birmingham 1940–44. *J. Proc. Inst. Sew. Purif.*, **1945**, 35–55.
National Water Council (1978a). *Water Industry Review 1978*, National Water Council, London.
National Water Council (1978b). *River Water Quality, the Next Stage. Review of Discharge Consent Conditions*, National Water Council, London.
Nicolson, N. J., and Wood, L. B. (1974). The impartial assessment of performance of the sewage treatment works of a Regional Water Authority. *Munic. Eng. Suppl.*, October 1974.
Porter, K. S., and Boon, A. G. (1971). *Cost of Treatment of Wastewater with Particular Reference to the River System of the Trent Area*, Paper no. 3B, Symposium on the Trent Research Programme, Institute of Water Pollution Control, Maidstone.
Short, C. S. (1975). *Removal of Ammonia from River Water–2*, Technical Paper TP3, Water Research Centre, Stevenage.
Simpson, J. R. (1971). Farm wastes; disposal in the next decade. *Environ. Pollut. Manage.*, **1**, 90–3.
Southgate, B. A. (1948). *Treatment and Disposal of Industrial Wastewaters*, HMSO, London.
Swanwick, J. D., Fisher, W. J., and Foulkes, M. (1972). Some aspects of sludge technology including new data on centrifugation. In Fuller, J. R. (ed.), *Water Pollution Manual*, Thunderbird Enterprises, Harrow.
Swanwick, J. D., and O'Gorman, J. V. (1973). Some effects of metals discharged in effluents and possibilities for recovery. *Chem. Ind.*, 1973 (12), 573–5.
Tomlinson, T. G. (1976). Industrial wastewaters. Outline of treatment and processing methods. In *Industrial Pollution Control Yearbook 1976*, Fuel and Metallurgical Journals, Redhill.
Truesdale, G. A., and Taylor, M. R. G. (1975). Sewage treatment to meet tomorrow's needs. *Water Pollut. Control*, **74** (4), 455–64.
Water Research Centre (1975a). *The Use of Oxygen in the Treatment of Wastewater*, Notes on Water Research no. 1, Water Research Centre, Medmenham.
Water Research Centre (1975b). *Analytical Methods for Organic Compounds in Sewage Effluents*, Notes on Water Pollution no. 70, Water Research Centre, Stevenage.
Water Research Centre (1976). *Emission Standards in Relation to Water Quality Objectives*, Technical Report TR17, Water Research Centre, Medmenham.
Water Research Centre (1977a). *Water Purification in the EEC*, Pergamon Press, Oxford, for the Commission of the European Communities.
Water Research Centre (1977b). *Economic Aspects of Sewage-Sludge Disposal*, Technical Report TR42, Water Research Centre, Medmenham.
Water Research Centre (1978). *Annual Report 1977/78*, Water Research Centre, Henley-on-Thames.
Waters, C. J., and Cockburn A. G. (1978). EEC laws on water quality. *Surveyor*, **151**, 15–8.
Wheatland, A. B., Swanwick, J. D., and Tomlinson, E. J. (1976). Some recent research and development in the field of industrial effluent treatment. *Water Pollut. Control*, **75** (3), 299–310.

Wood, L. B., King, R. P. Durkin, M. K., Finch, H. J., and Sheldon, D. (1976). The operation of a Simplex activated-sludge pilot plant in an atmosphere of pure oxygen. *Public Health Eng.*, **4** (2), 36–43, and **4**(3), 67–75.

Wood, L. B., and Richardson, M. L. (1978). The Water Industry—safety and value for money. *Chem. Br.*, **14** (10), 491–6.

Wood, L. B., and Stanbridge, H. H. (1968). Automatic samplers. *Water Pollut. Control*, **67** (5) 495–522.

World Health Organization (1970). *European Standards for Drinking Water*, 2nd edn, World Health Organization, Geneva.

CHAPTER 8

EEC (1975). *Council Directive Concerning the Quality Required of Surface Water Intended for the Abstraction of Drinking Water in the Member States*, 75/440/EEC-OJ no. L194/26-31, EEC, Brussels.

EEC (1976a). *Council Directive on Pollution Caused by Certain Dangerous Substances Discharged into the Aquatic Environment*, 76/464/EEC-OJ no. L129/23-29, EEC, Brussels.

EEC (1976b). *Council Directive on the Quality of Bathing Water*, 76/160/EEC-OJ no. L31/1-7, EEC, Brussels.

EEC (1978a). *Council Directive on Waste from the Titanium Dioxide Industry*, 78/176/EEC-OJ no. L54/19-24, EEC, Brussels.

EEC (1978b). *Council Directive on the Quality of Freshwater Required for the Protection of Fish Life*, 78/659/EEC-OJ no. L222/1-10, EEC, Brussels.

National Water Council (1978). *River Water Quality, the Next Stage. Review of Discharge Consent Conditions*, National Water Council, London.

Royal Commission on Sewage Disposal (1912), *8th Report: Standards and Tests for Sewage and Sewage Effluents discharging into Rivers and Streams*, HMSO, London.

Water Research Centre (1976). *Emission Standards in Relation to Water Quality Objectives*, Technical Report TR17, Water Research Centre, Medmenham.

CHAPTER 9

Ackerman, B. R., Ackerman, S. R., Sawyer, J. W., and Henderson, D. W. (1974). *The Uncertain Search for Environmental Quality*, Free Press, New York.

Atkins, M. H., and Lowe, J. F. (1977). *Pollution Control Costs of Industry*, Pergamon Press, Oxford.

Baumol, W., and Oates, W. E. (1975). *The Theory of Environmental Policy*, Prentice-Hall, Englewood Cliffs, New Jersey.

Beckerman, W. (1975). *Pricing for Pollution*, Hobart Paper no. 66, Institute of Economic Affairs, London.

Bugler, J. (1971). *Polluting Britain*, Penguin, Harmondsworth.

Department of the Environment (1973). *Report of a River Pollution Survey of England and Wales 1970*, vol. 3, HMSO, London.

Dorcey, A. H. (1973). Effluent charges—Information generation and bargaining behaviour. *Nat. Resour. J.*, **13** (1), 118–33.

Elliott, D. J., and Storey, D. J. (1979). The role of economic efficiency in water pollution control. *Water Pollut. Control*, **78** (1), 44–6.

Freeman, A. M. (III) (1971). *The Economics of Pollution Control and Environmental Quality*, General Learning Press, New York.

Gunningham, N. (1974). *Pollution, Social Interest and the Law*, Martin Robertson, London.

Herzog, H. (1976). Economic efficiency and equity in water quality control: effluent taxes and informational requirements. *J. Environ. Econ. Manage.*, **2** (3), 170–84.

Klein, S. N. (1973). Pollution control in Illinois: the formative years. *De Paul Law Rev.*, **22**, 759–78.

Kneese, A. V., and Schultze, C. L. (1975). *Pollution, Prices and Public Policy*, Brookings Institution, Washington DC.

Leone, R. A. (ed.) (1976). *Environmental Controls*, Lexington Books, Lexington, Massachusetts.

Majone, G. (1975). Standard setting and the theory of institutional choice: the case of pollution control. *Policy and Politics*, **4** (2), 7–34.

Niskanen, W. A. (1971). *Bureaucracy and Representative Government*, Aldine Press, Chicago, Illinois.

Ogus, A. I., and Richardson, G. M. (1977). Economics and the environment: a study of private nuisance. *Camb. Law J.*, **36** (2), 284–325.

Orr, L. (1976). Incentive for innovation as the basis for an effluent change strategy. *Am. Econ. Rev.*, **66** (2), 441–7.

Storey, D. J. (1978). Market structure and externalities: the case of water pollution in England and Wales. *J. Ind. Econ.*, **27** (2), 149–62.

Storey, D. J. (1979). An economic appraisal of the legal and administrative aspects of water pollution control in England and Wales. In O'Riordan, T., and D'Arge, R. (eds), *Progress in Resource Management and Environmental Planning*, Wiley, Chichester.

Storey, D. J., and Walker, M. (1978). Water pollution control theory: an economic taxonomy. *J. Environ. Manage.*, **7** (3), 205–17.

Tietenberg, T. H. (1974). Derived decision rules for pollution control in a general equilibrium space economy. *J. Environ. Econ. Manage.*, **1** (1), 3–17.

Walker, M., and Storey, D. J. (1977). The standards and price approach to pollution control—some problems of iteration. *Scand. J. Econ.*, **79**, 99–109.

CHAPTER 10

Beck, M. B. (1978). Modelling of dissolved oxygen in a non-tidal stream. In James, A. (ed.), *The Use of Mathematical Models in Water Pollution Control*, Wiley, Chichester.

Biswas, A. J. (ed.) (1976). *Systems Approach to Water Management*, McGraw-Hill, New York.

Franks, R. G. E. (1972). *Modeling and Simulation in Chemical Engineering*, Wiley–Interscience, New York.

Khabaza, I. M. (1969). *Numerical Analysis*, Pergamon Press, Oxford.

Knapton, J. (1978). Optimisation and its application to a unit process design problem. In James, A. (ed.), *The Use of Mathematical Models in Water Pollution Control*, Wiley, Chichester.

Mackay, D. W. and Fleming, G. (1969). Correlation of dissolved oxygen levels, freshwater flows and temperatures in a polluted estuary. *Water Res.*, **3**, 121–8.

Pearson, C. R. and Pearson, J. R. A. (1965). *Simple Method for Predicting the Dispersion of Effluent in Estuaries*, Paper 9.9 Presented at Meeting on New Chemical Engineering Problems in Utilisation of Water (AIChE–IChemE Symposium).

Rich, L. G. (1973). *Environmental Systems Engineering*, McGraw-Hill, New York.

Thomann, R. V. (1974). *System Analysis and Water Quality Management*, McGraw-Hill, New York.

Velz, C. (1974). *Applied Stream Sanitation*, McGraw-Hill, New York.

CHAPTER 11

Barker, M. C. (1967). The perception of water quality as a factor in consumer attitudes and space preferences in outdoor recreation. *MA Thesis*, University of Toronto.

Birch, F. (1976). Leisure patterns in Britain. *Popul. Trends*, **3**, 18–23.

Bugler, J. (1978). Muddy waters—1. *Vole*, 1978 (9), 3–6; Pollution standards plus. *Vole*, 1978 (12), 3–4.

Central Advisory Water Committee, Gathering Grounds Sub-Committee (1948). *Gathering Grounds: Public Access, Afforestation and Agriculture*, HMSO, London.

Committee on Environmental Conservation (1977). *8th Annual Report*, Committee on Environmental Conservation, London.

Coppock, J. T., and Duffield, B. S. (1975). *Recreation in the Countryside. A Spatial Analysis*, Macmillan, London.

Countryside Commission (1976). *Grants for Recreation Footpaths in the Countryside*, Countryside Commission, Cheltenham.

Countryside Review Committee. (1977). *Leisure and the Countryside: A Discussion Paper*, HMSO, London.

Coy, J. G., Johnston, R. A., and Richerson, P. J. (1973). Critique of Water Resources Council's proposed principles for planning water and related land resources. In Goldman, C. R., McEvoy, J. (III), and Richerson, P. J. (eds), *Environmental Quality and Water Development*, pp. 478–94, Freeman, San Francisco.

David, E. C. (1971). Public perceptions of water quality. *Water Resour. Res.*, **7**, 457–66.

Davidson, P., Adams, J., and Seneca, J. (1966). The social value of water recreational facilities resulting from an improvement in water quality. In Kneese, A. V., and Smith, S. C. (eds), *Water Research*, pp. 175–214, John Hopkins University Press, Baltimore, Maryland.

Department of the Environment (1978). *River Quality Survey, 1975*, HMSO, London.

Institution of Water Engineers (1972). *Recreation on Reservoirs and Rivers: Final Report on the Recreational Use of Waterworks*, Institution of Water Engineers, London.

Liddell, P. (1978). Towards better recreational access to north west reservoirs, *Water Space*, **13**, 34–7.

McLoughlin, J. (1976). *The Law and Practice Relating to Pollution Control in the UK*, Graham and Trotman, London.

Macrory, R., and Zaba, B. (1978). *Polluters Pay: Control of Pollution Act Explained*, Friends of the Earth, London.

National Angling Council (1971). *National Angling Survey*, National Angling Council, Peterborough.

National Water Council (1978). *River Water Quality, the Next Stage. Review of Discharge Consent Conditions*, National Water Council, London.

North-West Water Authority (1977). *Recreational Access to Reservoirs and Gathering Grounds*, North-West Water Authority, Warrington.

O'Riordan, T., and Paget, G. (1978). *Sharing Rivers and Canals*, The Sports Council, London.

Stabler, M. J., and Ash, S. E. (1978). *The Amenity Demand for Inland Waterways—Informal Activities*, Amenity Waterways Study Unit, Department of Economics, University of Reading.

Tanner, M. F. (1977a). *The Potential of Towpaths as Waterside Footpaths*, Research Report no. 1, Water Space Amenity Commission, London.

Tanner, M. F. (1977b). *Water Recreation in Country Parks*, Research Report no. 2, Water Space Amenity Commission, London.

Tanner, M. F. (1977c). *The Recreational Use of Water Supply Reservoirs in England and Wales*, Research Report no. 3, Water Space Amenity Commission, London.

Tinker, J. (1972). Nanny knows best. *New Sci.*, **53**, 530–4.
Turner, R. K. (1976). *The Recreational Response to Water Quality Change*, Public Economics Research Centre, University of Leicester.
Turner, R. K. (1977). The recreational response to changes in water quality: a survey and critique. *Int. J. Environ. Stud.*, **11**, 91–8.
Turner, R. K. (1978). Water pollution. In Pearce, D. W. (ed.), *The Valuation of Social Cost*, pp. 96–118, Allen and Unwin, London.
US Water Resources Council (1972). *Summary Analysis of Public Responses to the Proposed Principles and Standards for Planning Water and Related Land Resources*, US Government Printing Office, Washington DC.
Veal, A. J. (1977). *Leisure and Recreation in England and Wales, 1973*, Centre for Urban and Regional Studies, University of Birmingham.
Wayland, S. (1977). Federal Water Pollution Control Act—Technological and economic feasibility. *Nat. Resour. J.*, **17**, 145–8.

Index